PITT LATIN AMERICAN SERIES

The Politics of Mexican Oil

George W. Grayson

The Politics
of Mexican Oil

University of Pittsburgh Press

To Gisella A. Brissette,
in appreciation of her loving kindness
to me and my family

Published by the University of Pittsburgh Press, Pittsburgh, Pa. 15260
Copyright © 1980, University of Pittsburgh Press
All rights reserved
Feffer and Simons, Inc., London
Manufactured in the United States of America

Library of Congress Cataloging in Publication Data

Grayson, George W   1938–

    The politics of Mexican oil.

    Bibliography: p. 267
    Includes index.
    1. Petroleum industry and trade—Mexico.
I. Title.
HD9574.M6G68   338.2'7282'0972   80-5253
ISBN 0-8229-3425-6
ISBN 0-8229-5323-4 (pbk.)

Contents

Tables

Figures and Maps

Acknowledgments

Like every author I owe a heavy debt of appreciation to the many persons who have helped in various ways in the preparation of my book. Above all I wish to thank forty-four Mexican scholars, diplomats, engineers, journalists, union leaders, union members, and officials in government agencies, most notably Petróleos Mexicanos, who unselfishly, courteously, and—often—patiently shared their time and knowledge to deepen my understanding of the Mexican petroleum industry. Their assistance in the form of lengthy discussions, the provision of books and articles, and comments on chapters revealed a keen interest in overcoming impediments to communication between Mexicans and Americans in order to increase understanding of a resource that continues to play a pivotal role in relations between our governments and peoples. The names of these individuals cannot be listed here without violating the anonymity of their responses. I sincerely appreciate the graciousness with which they received me and the insights they provided.

My debt is equally great to two dozen U.S. officials in the Export-Import Bank and the departments of State, Treasury, Energy, and Defense who provided invaluable information and judgments concerning energy questions affecting Mexico and the United States. Similarly, personnel of the Canadian Embassy in Washington proved extremely helpful. As promised in our interviews, I will not cite individuals by name; however, their generosity and helpfulness must not go unacknowledged.

Although I take full responsibility for errors of fact and soundness of interpretation, I have benefited greatly from the constructive criticism of a number of uncomplaining friends and colleagues. Judith Ewell, S. Lief Adleson, Alan J. Ward, James H. Street, John S. Evans, John J. Bailey, and Robert D. Bond carefully read portions of the manuscript and contributed fresh points of view to my research.

I also wish to thank Louise Craft, managing editor of the University of Pittsburgh Press, and Abby Levine, who edited the manuscript, for the painstaking care and discrimination they devoted to improving my roughly hewn prose while tightening the style, organization, and coherence of the study. My appreciation also goes to Frederick A. Hetzel, director of the

University of Pittsburgh Press, for his advice, confidence, encouragement, and—above all—patience.

T. J. Stewart-Gordon, international editor of *World Oil,* assisted with useful advice. Anne Marie Fedder cheerfully typed the manuscript, and Deborah Meek, a student at the College of William and Mary, diligently assisted with research.

Earlier versions of chapters 7, 8, and 10 have appeared previously in other publications. An article entitled ''Mexico's Opportunity: The Oil Boom'' is adapted with the permission of *Foreign Policy* magazine No. 29 (Winter 1977–1978), pp. 65–89, copyright 1977 by the Carnegie Endowment for International Peace. Revised portions of ''Mexico and the United States: The Natural Gas Controversy,'' originally published in *Inter-American Economic Affairs,* 32 (Winter 1978), pp. 3–25, appear in this book with the permission of that journal. Finally, ''Oil and U.S.–Mexican Relations,'' which appeared in the *Journal of InterAmerican Studies and World Affairs,* 21, No. 4 (November 1979), pp. 427–56, is adapted herewith by permission of the publisher, Sage Publications, Inc. I appreciate the willingness of the editors of these journals—Sanford J. Ungar, Simon G. Hanson, and John P. Harrison—to allow the inclusion of this material in my book. Chapter 5 is based on a presentation first made to a Symposium on Structural Factors in Migration in Mexico and the Caribbean Basin, sponsored by the Brookings Institution and El Colegio de México and held in Washington, D.C., in mid-1978.

Years consumed in preparing a manuscript also create obligations of a financial nature. The College of William and Mary through its Faculty Research Committee awarded funds permitting two summers and one semester of research, including two visits to Mexico. Another trip was made possible through an invitation tendered by the University of Colorado to participate in the Second Symposium on Mexican–United States Economic Relations, held in May 1979 at the Xochimilco campus of the Metropolitan Autonomous University in Mexico City.

I also wish to acknowledge the invaluable help of my wife, Dr. Carmen Brissette Grayson, a superb historian whose intelligence, curiosity, erudition, good humor, and sensitivity make her an exceptional companion. Our children, Gisèle and Keller, also contributed to this book in their willingness to allow their father time to write, while making his breaks from the typewriter an unalloyed joy.

Introduction

This book analyzes the role of oil and the oil sector in Mexico's domestic and foreign affairs. The importance of oil to Mexicans transcends its economic worth. For nearly four decades after commercial development began, revenues from this vital resource flowed principally into the exchequers of American, British, and Dutch corporations. The arrogance and contempt with which they treated Mexicans and Mexico revealed a sense of superiority toward the people and land yielding them fortunes. Their quest for profits clashed with the government's commitment, arising from a social revolution, to uplift the working class. President Lázaro Cárdenas resolved this protracted conflict in favor of the newly formed Oil Workers' Union and the Mexican people when, on March 18, 1938, he expropriated seventeen foreign companies. This move elicited an outpouring of public support from sweat-bathed day laborers as well as bejeweled society matrons. Five days later cheering masses paraded before the National Palace in Mexico City to honor their beloved leader whose bold action provided a common symbol—one of the most important; perhaps the last—of *the* Mexican nation, a concept often distinguished more by rhetoric than collective endeavor.

The action also breathed life into a revolution, many of whose protagonists had long since deserted the barricades of social reform for bureaucratic sinecures or lucrative business ventures. In recapturing a resource considered its birthright, the nation had successfully confronted Standard Oil, Royal Dutch Shell, and other economic behemoths who sought to regain their properties and privileges through boycotts, blacklistings, diplomatic pressures, propaganda, and the threat of armed intervention. Local newspapers reported that for the first time since the fall of Tenochtitlán, the Aztec capital, in 1521, Mexico was in the hands of its people.

Could Mexicans, long dependent on foreign expertise, run the industry they had seized? Or would they live up to their caricature as a friendly people fond of dozing against sun-warmed adobe walls, massive sombreros shading their eyes? Apparently persuaded by this image, a spokesman for Standard Oil of New Jersey had smugly predicted that "Mexico cannot operate the industry successfully alone." The Oil Workers' Union played a pivotal role in maintaining production after expropriation, although it became disenchanted with a government that refused to hand over the petroleum sector to the workers therein. More important, Cárdenas formed a state monopoly called Petróleos Mexicanos (PEMEX), managed by governmental appointees and union representatives, to lift, refine, transport, and sell the nation's oil. The firm, which surmounted enormous obstacles to mold the diverse and numerous foreign companies into a single productive entity, stands as the organizational embodiment of expropriation. It demonstrated that Mexicans, conventional wisdom notwithstanding, possess the expertise, technical skills, and managerial competence to operate a highly complex industry. "In the steady decline of the Mexican revolution towards bankruptcy, PEMEX has endured as the supreme triumph of a social reorganization, as the 'unifying factor' in a collective liberation experience."[1]

As the first integrated oil company in the Third World, Petróleos Mexicanos has served as a model for the creation of national petroleum corporations in other developing nations. It has, among other achievements, become the only state company in the Third World either to discover substantial quantities of hydrocarbons or to create a proficient research and development facility (the Mexican Petroleum Institute). It has also spurred development at home by constructing roads, ports, railroad lines, and other vitally needed facilities, providing jobs, and—above all—supplying industry, agriculture, the transportation sector, government, and the armed forces with good quality oil, gas, and petrochemicals. This nurturing function has led cartoonists and editorial writers to describe PEMEX as the *vaca petrolera*. Respondents tell pollsters of their faith in Petróleos Mexicanos even as they deprecate governors, cabinet members, and high-level bureaucrats.

A testament to past accomplishments, the oil industry also offers promise for the future. A combination of factors—rising consumption, imprudent exploration policies, and the exhaustion of old fields—again made Mexico a net importer of oil in 1968. The country managed to regain its status as an exporter five years later thanks to discoveries by PEMEX in the southeastern states of Tabasco and Chiapas. The abundance of reserves in this new province as well as the location of fields in Campeche Sound permitted the state enterprise to expand its announced proven reserves from 5.4 billion barrels in 1973 to 60.1 billion barrels by the end of the decade.

Not only could Mexico satisfy its own needs with reserves estimated to last one hundred years, but it could export increasingly large quantities. By late 1980 it had become one of the world's most important possessors of hydrocarbons and the fifth largest producer. Respected analysts predicted that Mexico, with 200 billion barrels of "potential" reserves, would overtake Saudi Arabia to become the world's number-one oil power. Although not a member of the

Organization of Petroleum-Exporting Countries (OPEC), Mexico has consistently followed the cartel's pricing lead, with the result that hydrocarbon sales abroad generated over $12 billion annually in the early 1980s. The newly found deposits and the income they generated helped to revive the flagging economy inherited by President José López Portillo, who began his six-year term on December 1, 1976. They also enabled him to launch a highly publicized industrial development plan, designed to promote sustained economic growth, stimulate industries in heretofore neglected regions of the country, promote the activities of small- and medium-sized businesses, strengthen the country's export capability, and—most important of all—create employment for the millions of Mexicans who lack jobs or work only a few weeks each year. As the head of PEMEX expressed it: "For the first time in its history Mexico enjoys sufficient wealth to make possible not only the resolution of economic problems facing the country, but also the creation of a new permanently prosperous country, a rich country where the right to work will be a reality."[2]

Will oil prove a solution to the formidable economic problems facing Mexico? Will it diversify the industrial base, spur exports, and solve the unemployment problem? This book considers the experiences of Venezuela and Ecuador, two other oil-endowed Latin American nations, in illuminating the obstacles confronting politicians who attempt to use newly acquired revenues for reforms to reduce the dependence of "have nots" on domestic elites.

The issue of dependence suffuses this volume, which begins with an overview of the role of foreign entrepreneurs and oil companies. What steps led to the expropriation? How did PEMEX evolve into a viable enterprise? Why is the Oil Workers' Union such a potent force? How realistic and rigorous are Mexico's plans to "sow" its oil and gas reserves? Will Mexico affiliate with OPEC? Will the presence of massive holdings open a new era in relations with the United States? Will they help free impoverished citizens dependent on domestic elites who dominate Mexico's highly stratified social system?

When we address these questions, several points become clear. *First,* the status of the oil industry as a sacred symbol of dignity for a highly nationalistic people strongly influences public policy. For example, any negotiations with the United States over oil and gas sales—irrespective of mutual benefits involved—fall prey to charges of "selling out to imperialism." *Second,* graft and corruption within PEMEX and, especially, the union have since 1938 retarded the efficient development of Mexico's oil more than actions by transnational firms. *Third,* while an abundance of oil elevates a country's standing within the community of nations and permits greater independence and assertiveness vis-à-vis a powerful nation such as the United States, the development policies of Ecuador and Venezuela have led to increased dependence on foreign banks and international lending agencies. *Fourth,* the rush to drill as many wells as possible in pursuit of a prized nonrenewable resource may cause irreparable damage to marine life and other renewable resources. *Fifth,* though bountiful, the oil-generated resources may permit Mexican leaders to follow the example of their Ecuadorean and Venezuelan counterparts; namely, to attempt to "buy off" dissidents with conspicuous social spending—at times complemented by political changes—instead of reforming a warped economic system. *Finally,* while Mexico would be markedly worse off, perhaps bankrupt, without its new wealth, petroleum encourages politicians to fill the heads of the masses with images of plenty. Failure to satisfy these aroused expectations may lead to profound unrest, as the situation in contemporary Iran reveals.

The Politics of Mexican Oil

1

Development of Mexico's
Oil Industry

History of Mexican Oil

With the prices of petroleum, natural gas, uranium, and coal having increased dramatically in recent years, it is obvious that the possession of energy reserves enhances a nation's power, influence, and independence. Such was not the case early in this century, when entrepreneurs first discovered commercial quantities of oil in Mexico. The vital deposits served as a magnet, attracting ruthless profit seekers, avaricious transnational firms, and the diplomats and soldiers of industrialized countries. Once these foreigners had wheedled, bribed, intimidated, and coerced local politicians and their praetorian guards, the resources fell under the sway of outsiders, offending the dignity and attenuating the sovereignty of the possessing state.

The Aztecs had discovered petroleum along Mexico's heavily forested, rain-drenched coastal plain and used it before the arrival of the Spanish in 1519. They burned what later became known as *chapopote* in tribute to their gods, smeared it on their bodies as a medicine, and used it as a dye and glue. The Spanish subsequently caulked their boats with it. From Tampico to Poza Rica, the names of many localities—El Chapopote, El Chapopotal, Chapopotilla, Cerro de la Pez, Ojo de Brea, for example—mean "tar" or "pitch" and had a special significance to oil prospectors.[1]

Over three centuries elapsed before efforts focused on deriving a profit from this black gold. Five years after a railroad conductor, "Colonel" Edwin L. Drake, drilled his famous well near Titusville, Pennsylvania, in 1859, a group of Mexicans discovered and used small quantities of asphalt and crude in the Pánuco region of Tamaulipas and Veracruz states. In 1869 the Compañía Exploradora del Golfo Mexicana drilled in oil springs, discovered the year before by a Dr. Autrey, on the "Furbero" hacienda near Papantla in Veracruz. While the drilling effort proved unsuccessful, the firm did obtain some oil—later distilled into kerosene—by driving a tunnel in the vicinity of the springs. John C. Spear reported manifestations of oil on the Isthmus of Tehuantepec in 1872, and the following year Prieto described seepages along the Tamesí River in Tamaulipas. At the end of the decade, the Mexican

Treasury Department published a study of oil indications by the American geologist J. W. Foster.[2]

In 1876 a Boston sea captain so impressed financiers with tar samples from Tuxpan that he won support for a drilling operation. He found sufficient oil at about five hundred feet to supply a small refinery, which produced kerosene available for sale to local villagers. When his Boston partners refused to provide additional monies to this high-cost venture, he committed suicide.[3]

A new mining law in 1884 granted to surface owners the right to develop oil on their land. Years after this action other attempts at exploitation occurred, including that at the turn of the century by an Englishman named Burke, who—encouraged by abundant oil seepages along the Gulf Coast—persuaded the multimillionaire Cecil Rhodes to organize an exploration company. The London Oil Trust, as it was called, went bankrupt after pouring £90,000 into drilling operations in the Tuxpan region. Its legal successor, the Mexican Oil Corporation, invested £70,000 without being able to turn a profit.[4] Several leading foreign and Mexican experts claimed that the country lacked commercial quantities of oil.

Porfirio Díaz, a cunning dictator who imposed order on the country he ruled for thirty-five years after seizing power in 1876, gave a fillip to the development of Mexico's riches. A group of lawyers, economists, and intellectuals surrounded the crass *cacique,* a Mixtec Indian with little Spanish blood who "retained to the end of his life the simple and unsophisticated outlook of an Indian warrior."[5] These "*científicos,*" imbued with Herbert Spencer's Social Darwinism and August Comte's positivism, believed in the inestimable value of technology, the advancement of the country through investment in mines, railroads, factories, and harbors, and the superiority of white men over Indians and mixed breeds.

They saw in Anglo-Saxon capitalism a means to uplift their unenlightened, uncivilized nation through the introduction of scientific efficiency. Yet peace was necessary in the age of Victoria Regina and Grover Cleveland to attract the foreign capital that could stimulate economic progress and—perhaps some day—permit the luxury of democracy. A policy of *pan o palo,* bread or the club, assured this peace. Lucrative opportunities awaited allies of Díaz, who used to say that a dog with a bone in his mouth cannot steal or kill. Enemies could expect repression or extermination. Peonage was widespread, and thousands of people perished in penal colonies or while working on sisal plantations in the Yucatán.[6] Díaz distrusted the army, which he allowed to dwindle to a "clutch of mummified generals."[7] To wield the crushing *palo,* he relied on a maleficent national constabulary known as the *rurales,* immediately recognizable in their broad felt hats, silver-buttoned uniforms, and silver-embellished saddles. These officially sanctioned bandits made Mexico one of the safest places in the world—for all except Mexicans.[8] Civil order clearly took precedence over civil liberties, whose advancement was decades away. As Díaz expressed it: "The day that we may say that the fundamental charter has given us a million colonists, then we have encountered the constitution that really suits us; it will not be a phrase on the lips—it will be a plow in the hands, a locomotive on the tracks, and money everywhere."[9]

"Few politics and much administration"—*poca política, mucha administración*—became the watchword of his regime. The purpose of prudent administration was to provide a suitable climate for the investment of dollars, francs, and pounds sterling that, alloyed with limited indigenous funds, would finance the construction of highways, ports, bridges, dams, and railroads—the sinews of an emerging national economy.

The regime's "scientificism" entailed the government's protection of, but not necessarily aid to, private enterprise. The formula succeeded in sending steel rails across the country as foreigners built flourishing railroad and mining industries during the Porfirian era. Such U.S. tycoons as E. H. Harriman, Jay Gould, and Collis P. Huntington helped extend Mexico's rail network from 417 miles in 1877 to 9,600 in 1901 to 15,325 in 1911.[10] A Yorkshire contractor named Weetman Dickinson Pearson, who later won fame for his oil discoveries, reconstructed a line across the Isthmus of Tehuantepec, joining the Pacific Ocean to the Gulf of Mexico. Americans also gained dominance over the mining sector as Mexico became the world's leading silver producer and second in copper output. Led by Guggenheim interests, U.S. firms controlled 840 of the approximately 1,000 foreign mining companies operating in Mexico in 1908.[11]

The prevailing philosophy also helped three entrepreneurs develop a petroleum sector to diminish dependence on coal for railroads, mining, and electrical generation. The first was Henry Clay Pierce, who came from St. Louis to Mexico in 1885 to head the Waters-Pierce Oil Company. This firm, a Standard Oil of New Jersey affiliate in which Pierce owned a 35-percent interest, soon obtained a concession to import oil. Substantially higher tariffs on refined products led him to purchase Pennsylvania crude from which he made kerosene in his Veracruz and Mexico City refineries. As Pierce gained a monopoly of the kerosene market, the import duties he paid yielded an attractive source of revenues to the Díaz regime.

Pierce was also active in the railroad business. Southern Pacific, itself engaged in geological investigations of Mexico's Northeast, had been awarded a contract to construct a rail link between San Luis Potosí and Tampico, to be operated under the name of Ferrocarril Central Mexicano. Pierce obtained the controlling interest in this line and became the chairman of its board. Anxious to locate fuel for his locomotives, he invited Edward L.

Doheny to investigate oil discharges along the company's right-of-way. Doheny was a wily product of the rough-and-tumble California oil industry in which he had amassed a fortune after discovering oil in Los Angeles in 1892. His influence later became so great in Washington that he purchased cabinet members and senators much in the way that a Medici collected Florentine art. The millionaire oilman gained prominence during the Harding administration for sending $100,000 cash "in a little black bag" to Albert B. Fall, secretary of the interior, after having fraudulently secured a lease for the navy petroleum reserves at Elk Hills, California. Long before the infamous Teapot Dome scandal, Doheny played a critical role in the unfolding drama of Mexico's oil development.

Doheny liked what he saw in Mexico. At one spot, thirty-five miles west of Tampico, oil literally oozed from the ground. He later said:

> We found a small conical-shaped hill—where bubbled a spring of oil, the sight of which caused us to forget all about the dreaded climate—its hot, humid atmosphere, its apparently incessant rains, those jungle pests the *pinolillas* and *garrapatas* (wood ticks), the dense forest jungle which seems to grow up as fast as cut down, its great distance from any center that we would call civilization and still greater distance from a source of supplies of oil well materials—all were forgotten in the joy of discovery with which we contemplated this little hill from whose base flowed oil in various directions. We felt that we knew, and we did know, that we were in an oil region which would produce in unlimited quantities that for which the world had the greatest need—oil fuel.[12]

Doheny acquired for $325,000 the three-hundred-thousand-acre "El Tulillo" hacienda at the convergence of the states of Tamaulipas, Veracruz, and San Luis Potosí and organized the Mexican Petroleum Company with an initial capital of $10 million. William L. Clayton, the American ambassador, introduced him to Díaz, who encouraged his prospecting with tax and tariff concessions. A decade later the counselor for the embassy lived in the home of the oil magnate's representative in Mexico, and Doheny's representative subsequently resided in the counselor's house.[13] The web of relationships thickened when Mrs. Doheny donated property on which a new chancery for the embassy was built.

Doheny benefited from the services of Ezequiel Ordóñez, who—unlike most Mexican geologists—believed that oil existed in Mexico. They made their initial strike on May 14, 1901, at a depth of 525 feet, in Ebano, a jungle-infested region of San Luis Potosí.[14] When Ferrocarril Central refused to purchase the fifty-barrel-per-day (bpd) output of this well—Pierce may have feared Doheny as a potential competitor[15]—Doheny set up a firm to provide asphalt for Mexico City, Guadalajara, Morelia, Tampico, Puebla, and Chihuahua.

By December 1903 Doheny's company had sunk about nineteen wells, and he was disappointed by the poor results after spending $1.5 million. At this point Ordóñez intervened, proposing that a well be drilled in the vicinity of massive oil oozings near the volcanic obtrusion, Cerro de la Pez, where Doheny had first seen the hill of oil. La Pez 1 came in on Easter Sunday, April 3, 1904, yielding fifteen hundred bpd from a depth of 1,650 feet. This was Mexico's first important oilfield, and as output rose sharply, the company signed long-term supply contracts with the Central and other railroads. It also agreed to sell two million barrels a year for five years to Standard Oil and shipped significant quantities to the east coast of the United States. The Mexican Petroleum Company found the Chijol reservoir, with geological characteristics similar to Ebano, a few miles northeast of Ebano in 1909.[16]

Doheny and his associates created the Huasteca Petroleum Company to explore outside the area of their original holdings. They first turned their attention south of Tampico, near the zone where Cecil Rhodes's company had unsuccessfully prospected, developing the Juan Casiano tract. After drilling two moderate producers, the company brought in Juan Casiano 7 in September 1910. By 1913 this well furnished nine million barrels annually or one third of total national production. La Huasteca was in a position to use much of this oil because it had prudently invested millions of dollars in the construction of roads, pipelines, and pumping stations.

Doheny's company was also responsible for an even more abundant gusher, Cerro Azul, which was completed in February 1916. Lying at the culmination of the "Golden Lane" or "southern fields," a series of reservoirs that stretched thirty miles from Dos Bocas to El Chapopote on the Tuxpan River, it was "the largest well the world had seen" with a measured flow of 260,858 bpd.[17]

La Huasteca owned its property in fee simple and enjoyed no concession from the state, although President Díaz had helped Doheny acquire access to the land of a recalcitrant owner. Still, Díaz urged Doheny, who thanks to oil properties in Mexico and elsewhere had become the largest oil magnate after John D. Rockefeller,[18] to allow the government first refusal should he decide to sell his holdings lest they fall into the hands of a powerful monopoly such as Standard Oil. Standard Oil of Indiana, which signed the purchase over to Standard Oil of New Jersey, did acquire the Doheny family's interest in La Huasteca and the Mexican Petroleum Company, but this was not until 1925—fourteen years after Díaz's political demise.

The third petroleum baron was the Yorkshireman Pearson, who had earned an international reputation in the construction business. Among his feats were the Dover naval base, the tunnel under the Thames River in London, the first dam in the Aswan region of Egypt, and two subway tunnels under the Hudson River.[19] Díaz paid him to undertake the drainage of Lake Texcoco, rebuild the port of Veracruz, and reconstruct the trans-Tehuantepec railway, with harbor terminals at both ends. Pearson supplied half the capital for the Tehuantepec Railroad Company, a government-organized enterprise, and supervised all construction and repair work. His efforts were rewarded with 37 percent of the system's net profits for five years and an even larger percentage thereafter. Díaz and Pearson inaugurated the new line by riding its entire 170 miles with their sons.[20]

In the course of completing these projects, Pearson's engineers informed him of the presence of oil in Mexico. Pearson found himself in Texas shortly after the great Spindle Top well spewed a jet-black stream of oil two thousand feet over Beaumont in 1901, marking the Lone Star State as an important depository of hydrocarbons. He immediately cabled his associates to buy and lease land and by 1906 controlled eight hundred thousand to nine hundred thousand acres. Knowing that geology in parts of Mexico was similar to that of Texas, he hired the engineer who had brought in the Spindle Top and began exploring near San Cristóbal on the Isthmus of Tehuantepec. His goal was the establishment of the country's first integrated oil firm. The engineer's discoveries led him to build a refinery, storage center, and pipeline at Minatitlán to take care of the antici-pated output from the San Cristóbal wells. Low productivity, however, forced him to purchase supplies from other producers. Undaunted, Pearson sought to shatter the Waters-Pierce kerosene monopoly in a confrontation that became known as the "great Mexican oil war." He absorbed severe losses on the way to garnering nearly one quarter of the retail market.[21]

The Díaz administration granted Pearson vast concessions on government-owned property in five states in order to prevent Standard Oil from securing a monopoly. All told, he invested $25 million in exploration before bringing in San Diego 2 in May 1908 with an initial daily yield of twenty-five hundred barrels, opening the Dos Bocas pool. This discovery was eclipsed when San Diego 3 blew in on July 4, 1908. So great was the pressure on this well, located seventy miles south of Tampico on the shore of the Laguna de Tamiahua, that it destroyed the derrick and drillpipe. Worse, the well caught fire, sending a one-thousand-foot column of flame into the air and burning uncontrollably for fifty-eight days. Estimates of the oil output varied from fifteen thousand to eighty thousand bpd. Vividly described in newspapers throughout the world, the "blowout" aroused interest in Mexico's oil wealth. Even though the disaster virtually exhausted the local pool, two subsequent wells produced at a rate of several hundred barrels.[22]

Pearson, who may have "garnered larger profits from Mexico than any other man, either during or since the Spanish conquest,"[23] set up the Compañía Petróleo "El Aguila" in 1908 to operate his leases in the Tampico district. Everything Pearson touched turned to black gold for his second major well, Potrero del Llano 4, completed about twenty miles west of Tuxpan in December 1910, was a gusher. Despite tremendous hydrostatic and gas pressure, the work crew avoided a fire as it struggled for two months to cap a flow estimated at one hundred thousand bpd. Potrero del Llano produced ninety-three million barrels of oil in twenty-four years and was the most prolific producer in the Golden Lane. The discovery of this field launched an oil boom in Mexico which attracted attention comparable to that lavished on the Persian Gulf a half century later. Pearson recorded his third key discovery in 1913 when the Aguila company opened the South Chinampa–North Amtalán field on the Los Naranjos lease several miles south of La Huasteca's Juan Casiano wells.

Many of Díaz's advisers favored British over American investments. Still, the president was anxious to balance these interests to prevent the domination of either. He therefore encouraged the work of men of both countries by sponsoring helpful legislation and assisting land acquisition. For example, the Petroleum Law of 1901, which authorized the government to grant concessions in public lands, enabled Pearson and Doheny to operate in extensive areas. The November 25, 1909, mining law reaffirmed the subsoil rights of the surface owner. This measure became the basis for the companies' later resistance to the nationalization of Mexico's oil.[24] Díaz also permitted the unfettered exploration of petroleum and petroleum products, duty-free importation of drilling and refining equipment, and the exemption of capital stock from virtually all federal taxes.

Fierce competition with Doheny notwithstanding, Pearson's El Aguila company—which boasted Díaz's son and his finance minister, José Y. Limantour, the foremost *científico,* as board members—became Mexico's preeminent producer following the Potrero del Llano strike. Winston Churchill, then first lord of the Admiralty, contributed mightily to its success when in 1912, over strenuous opposition, he converted the Royal Navy from coal to oil so that its vessels could achieve sufficient speed to outmaneuver the German battle fleet. Thus the First Battle Division glided into combat with the kaiser's Kriegsflotte propelled by Mexican bunker oil, which met fully 75 percent of British needs. That Pearson was a backbencher of the then-ruling Liberal party helped El Aguila land the supply contract.[25] Henry Ford also nourished the demand for Mexico's petroleum when he transformed the automobile from an upper-class luxury to a convenience affordable by millions of families. Too, the lowering of import duties by the United States enhanced the attractiveness of Mexican crude to major fuel oil consumers who had often been inconvenienced by undependable supplies from the Gulf Coast fields of Texas and Louisiana.[26]

Even though Doheny and Pearson, knighted in 1911 as Viscount Cowdray, came to dominate the petroleum industry, some 160 American and European firms, many of which were small independents, sought to profit from the newly found wealth. The Southern Pacific Railroad, which had been vigorously prospecting for oil since 1907, discovered the important Pánuco field near Tampico in 1910. With a cumulative production of 276 million barrels to January 1, 1935, the deposit proved to be the most important of the northern fields. Emboldened by its success, the railroad chartered the East Coast Oil Company in early 1911 to handle its petroleum activities in the Tampico region. The same year, it opened the productive Topila pool twelve miles east of Pánuco. The Royal Dutch Shell group entered the country in 1912 and soon formed the Mexican Dutch Company "La Corona," whose Pánuco 5, completed in January 1914, proved one of the most spectacular wells in the northern fields. Royal Dutch Shell acquired control of El Aguila from Pearson in mid-1919 and gradually integrated it with La Corona.[27] The leases of Penn Mex yielded 59 million barrels from eighty-seven producing wells, the most productive of which was Alamo 2 near the southern end of the Golden Lane.

Gulf Oil secured valuable leases in both the northern and southern fields and recorded a peak production of nearly 25 million barrels in 1922. The Texas Company commenced activities in the Tampico region around 1911 and achieved a maximum output of 9.8 million barrels nine years later. Standard Oil of New Jersey did not begin operations in Mexico as a producer until 1917, when it bought the Transcontinental Petroleum Company, owned by American, British, and Mexican interests. Transcontinental boasted only a single active well at the time of the purchase; however, massive investment by Jersey Standard raised the company's output to 50,000 bpd by mid-1918. The lure of profits attracted venture capital into pipelines, refineries, railroads, docks, and tankers.[28] "From the Doheny fields north and west of Tampico to the Cowdray fields almost two hundred miles south near Papantla and Poza Rica, the foreigners owned or controlled virtually every acre of potentially productive oil land."[29] Map 1 shows the principal deposits.

Production soared from 10,000 barrels in 1901 to 3.9 million in 1908 to 55 million in 1917 to 193.4 million in 1921. This rapid increase took place despite a populist social revolution which began in 1910, when Francisco I. Madero, a well-to-do intellectual and spiritualist, challenged the domestic and foreign policies of Díaz and helped overthrow him the following year. Some American oil companies allegedly backed the insurgency because they perceived a decidedly pro-British cast to policies effected by Díaz and his lieutenants. Specifically, Standard Oil is believed to have offered Madero and his forces a

MAP 1 *Principal Oil Deposits*

loan of between $500,000 and $1 million for "certain concessions" under a new government.[30] Only circumstantial proof buttresses this contention, which also holds that Pearson contributed to the Porfirista cause. For more than a decade, much of Mexico resembled a Tolstoyan battlefield festooned with conniving politicians, stacked corpses, penurious peasants, invading forces, and disheveled armies marching off in different directions.

That the "Golden Age" of Mexican oil occurred during this upheaval can be explained by the remote location of the coastal fields, the poor roads between them and the country's central combat theaters, the companies' use of bribery and their own brutal "white guards" or *guardias blancas* to secure the tropical petroleum zone, and the constant threat to combatants that American forces would intervene if the area's sanctity were violated. After all, U.S. Marines occupied Veracruz for seven months following the bloody seizure of the port on April 21, 1914, ostensibly to prevent the unloading of German arms.[31] As once explained by Philander C. Knox, President Taft's secretary of state, U.S. naval vessels patrolled the Gulf Coast to keep the Mexicans "in a salutary equilibrium, between a dangerous and exaggerated apprehension and a proper degree of wholesome fear."[32] In addition, Gen. Manuel Peláez staged an uprising against the constitutional government of Venustiano Carranza, who emerged as president in 1914, and seized control of a large portion of the oil-producing regions. At the height of the civil war, the oil companies paid the rebellious general, with whom they closely cooperated, $15,000 per month to protect their properties.[33] Other commanders also received danegeld from the companies.

Ironically, output began declining in 1922 after the next chief executive, Alvaro Obregón, had made great strides toward restoring peace. Ruinous exploitation had diminished the productivity of the wells, and operating costs had begun to rise. In addition, the companies began fretting over their status under a viable government that might collect taxes and listen sympathetically to labor demands. Unless designed to advance their interests, government interference was anathema to these corporations, especially the American firms which—except on public lands—faced few regulations at home. At this time there were no taxes levied on oil production in the United States, and only fifteen state governments imposed a gasoline tax in 1921.[34]

As a result cheaper fields enticed many investors to Venezuela, where President Juan Vicente Gómez, who seemed almost as pliant and cooperative as Díaz, promised huge reserves and immediate profits. (Daily production in this Andean republic expanded from 6,100 barrels in 1922 to 515,200 in 1938.)[35] Some medium-sized firms left Mexico, while others shut down refineries, dismantled pipelines, and closed terminals. More than half of the workers employed by foreign companies were discharged.[36] The rate of well-drilling increased two and one-half times between 1921 and 1926, and the number of wells drilled in the 1924–1927 period was five times greater than the number completed before the production peak in 1921. But the majority of wells opened at this time were dry, compared with less than 38 percent of the wells abandoned at the termination of drilling before 1921.[37]

This downward slide in output continued until the early 1930s when El Aguila, controlled for a decade by Royal Dutch Shell, discovered the rich Poza Rica deposits between Tampico and Veracruz, approximately twenty-five miles south of the old producing fields of the Golden Lane. Until the opening of the Reforma area near Villahermosa in the early 1970s, Poza Rica was Mexico's premier producer—with a cumulative output of 941 million barrels to December 31, 1968, and an estimated ultimate production of 2.8 billion barrels. Although companies such as La Corona opened new pools in the Tampico area, both the northern fields and the Golden Lane suffered depletion or serious water encroachment. The decline in their production was partially offset in later years by Poza Rica and discoveries in the Isthmus of Tehuantepec (Tonalá and Nuevo Teapa, 1928; El Burro, 1930; El Plan, 1931; and Cuichapa, 1935).[38] Nonetheless, daily yields plunged from 193,398 bpd in 1921 to 32,805 in 1932, rising slightly to 38,506 in 1938. The companies contended that the country possessed promising new reservoirs. Yet a medley of factors—the debilitation of existing fields, low market prices, domestic taxes, U.S. tariffs, the financial depression in the United States, and uncertainty over ownership rights—militated against large investments in exploration and development.[39] The companies turned their attention to Venezuela and the Middle East.

Expropriation

For years the firms delighted their American and European shareholders with attractive dividend checks. Yet they won few friends in Mexico because they treated their holdings as foreign enclaves, sought the protection of their home governments, and arrogantly embroiled themselves in the host country's politics. These corporations claimed not simply to hold ten- or twenty-year concessions but to own the mineral deposits which they exploited. Their proprietary view, embedded in the 1884 mineral law and subsequent Porfirian statutes, offended Mexico's social-reformist Constitution of 1917. Strongly influenced by the Hispanic concept of property, Article 27 of the charter described as "inalienable and imprescriptible" the nation's ownership of subsoil deposits, including "petroleum and all hydrocarbons, solid, liquid, or gaseous." Although the Mexican Supreme Court ruled ten years later that companies holding concessions before 1917 enjoyed perpetual rights, the question of ownership constantly agitated relations between the corporations and the Mexican government.

Article 123 of the 1917 fundamental law stipulated that workers should enjoy an eight-hour day, one day of rest each week, full compensation for overtime, cash payment for their labor, healthy working conditions, social benefits, and the right to organize. It created the Federal Board of Conciliation and Arbitration to resolve intractable labor-management disputes. While conditions improved somewhat in the 1930s, the petroleum workers endured physically exhausting work, squalid living conditions, and harsh discipline. Their salaries were well above the national average, but the stark contrast between the earnings and benefits enjoyed by foreign managers and professionals and those of Mexican employees sharpened the sense of deprivation.

Lázaro Cárdenas, then a young, restless cavalry officer, received a first-hand impression of such discrimination when assigned to Tampico for three years during the mid-1920s. He found separate dining rooms for Mexicans with whom he ordered his officers, invited to share the foreigners' table, to eat. Foreigners received twice the pay and better housing than Mexicans, even when they performed identical tasks. Moreover, the oil companies showed scant interest in the social conditions of the people whose subsoil they exploited. One company turned down a request by peasants that a hydrant be installed on a water line crossing their village. This forced the *campesinos* to continue carrying water from the nearest river. Although there were no funds for a discharge tap, the company had money to spend on a shiny new Packard that it offered to give Cárdenas. The young colonel adamantly refused a gift which he considered tainted and continued to navigate the area's rut-filled roads in his broken-down old Hudson.[40] Later, as governor of Michoacán, he demonstrated his contempt of industrialists when—in defiance of the president—he ordered the expropriation of plants that had closed their doors rather than implement labor regulations.[41]

Conditions changed in Mexico with Cárdenas's election as president in 1934 on a platform committed to implementing the social goals of the Mexican Constitution. By this time the oil industry had ten thousand workers who belonged to nineteen separate labor organizations. Encouraged by the chief executive's sympathy for the working class, the *petroleros* forged a single national union on August 15, 1935. The Union of Oil Workers of the Mexican Republic (Sindicato de Trabajadores Petroleros de la República Mexicana—STPRM) acted to protect and improve the economic status of its members. It affiliated with the Confederation of Mexican Workers (Confederación de Trabajadores Mexicanos—CTM), the rapidly expanding labor federation, and convened a "general assembly" on July 20, 1936, to draft the industry's first collective-bargaining contract.

The workers' foremost demand, an increase of 26,329,393 pesos in wage and welfare benefits, set off a chain of events that led directly to the expropriation of the private component of the oil industry. The companies' counteroffer, approximately half the union request, gave rise to litigation and mutual recrimination. The Board of Conciliation and Arbitration upheld the contract demands, specified additional retroactive payments, and imposed restrictions on the number of "confidential employees" (*empleados de confianza*), who served at management's pleasure. The Supreme Court, in ruling on a petition for an injunction (*amparo*), fully sustained the findings of the board. As a result the oil trusts expressed a willingness to increase remuneration by 26 million pesos but balked at controls over personnel matters. According to the government, the foreigners sought to put pressure on the Mexican authorities by manipulating the exchange rate. Distinguished scholars believe, however, that the doubt in the reliability of the president's word expressed by the companies during the negotiations was "the decisive factor in galvanizing national pride."[42]

The dispute between the workers and the oil trusts culminated at 10:00 P.M. on March 18, 1938, when the husky-voiced Cárdenas announced by radio the expropriation of the property of seventeen American and European corporations. He did not, however, take the properties of several large companies— Mexican Gulf and Ohio-Mex, for example—which enjoyed amicable relations with their workers and which had not been the object of strikes. The labor conflict provided its catalyst and justification, but a fundamental cause motivated the takeover: the desire of a progressive president to assert the sovereignty of a proud country which the firms had treated as a fiefdom for the last six decades.

As he said in the speech, later christened "the Declaration of Mexico's Economic Independence":

It is evident that the problem which the oil companies have placed before the executive power of the nation by their refusal to obey the decree of the highest judicial tribunal is not the simple one of executing the judgment of a court, but rather it is an acute situation which drastically demands a solution. The social interests of the laboring classes of all the industries of the country demand it. It is to the public interest of Mexicans and even of those aliens who live in the Republic and who need peace first and afterwards petroleum with which to continue their productive activities. It is the sovereignty of the nation which is thwarted through the maneuvers of foreign capitalists who, forgetting that they have formed themselves into Mexican companies, now attempt to elude the mandates and avoid the obligations placed upon them by the authorities of this country.[43]

An overwhelming response met the decree. At the call of the Confederation of Mexican Workers, some two hundred thousand people thronged the Plaza de la Constitución in front of the National Palace on March 22. The demonstrators, hoisting banners such as "They shall not scoff at Mexican laws," screamed themselves hoarse as Cárdenas greeted them from a central balcony. Nearby hung the bell that Father Hidalgo had rung 128 years before in the distant town of Dolores to proclaim Mexico's fight for political independence from Spain.

The call for public collections to support Cárdenas's pledge that "Mexico will honor her foreign debt" revealed the popularity of this bold act. As the late Howard Cline observed: "State governors, high Church officials, patriotic grand dames, peasants, students—all the numberless and picturesque types of Mexicans—pitched in what they had, including money, jewels, even homely domestic objects, chickens, turkeys, and pigs."[44]

How should the newly expropriated oil industry be run? For the Oil Workers' Union, the answer was simple: Turn it over to the workers just as was done the year before with the nationalized railroads. Cárdenas rejected this approach because rail service had gone from bad to worse under the union. Instead he established a public corporation operated jointly by labor and government, with the latter boasting a majority of the appointments to the nine-member board of directors. Thus on June 7, 1938, Petróleos Mexicanos was formed for the attainment of "social" rather than "speculative" ends.

In early 1940 PEMEX absorbed the Distribuidora de Petróleos Mexicanos, formed at the time of expropriation to handle domestic and foreign marketing, and the Control de la Administración, established fifteen years earlier to regulate oil prices at home and to compete with private firms in the production and refining of hydrocarbons. A desire to eliminate duplication and reduce costs inspired the fusion of these three agencies, which brought 90 percent of the industry within the domain of the state enterprise.

The oil companies responded aggressively to the expropriation decree, whose legality was upheld by the Mexican Supreme Court on December 2, 1939. Mexico appeared vulnerable to pressure because its petroleum, 60 percent of which was exported, earned a significant quantity of the country's foreign exchange. The trusts summoned home key administrative and technical personnel, a few of whom elected to stay behind and work for the nationalized firms. They orchestrated a boycott to deprive Mexico of both tankers to carry its products and tetraethyl lead required for producing high-octane gasoline. They also engineered a secondary boycott to prevent sales of badly needed American machinery to the expropriated industry and hampered oil shipments to Europe by slapping liens on Mexico's "stolen" cargoes.

A malicious publicity campaign complemented these economic reprisals. The foreign corporations insisted that Cárdenas's move was "confiscation," not expropriation, inasmuch as Mexico had neither the will nor wherewithal to pay for the seized property which, including remaining subsoil reserves, was generously valued at between $400 and $500 million. It was alleged that the Mexicans dumped unmarketed oil into the sea, that the Cárdenas government had leased seaports to the Japanese, and that the Nazis enjoyed the use of Mexican airfields. The *Lamp*, a publication of Standard Oil of New Jersey, carried some of the most vicious statements and caricatures concerning Mexico. It was outdone, however, by the *Atlantic Monthly,* whose special July 1938 issue bearing the headline "The Atlantic Presents Trouble Below the Border" reeked of the oil companies who surely subsidized it.

The articles depicted the country as overflowing with brigands, Communists, and thieves hell-bent on exploding the Good Neighbor Policy. The Spanish-speaking neighbor, beset by "racial degeneration," appeared as a debased country crammed with contemptible people. Every Mexican portrayed looked as if he had just stepped out of a Pancho Villa conclave. One especially virulent cartoon showed Mexico doing at home what Hitler had done in Austria, Japan in China, and Mussolini in Spain. According to Josephus Daniels, the American ambassador to Mexico from 1933 to 1942, the issue was a treatise in "misrepresentation and slander and hate."[45] Meanwhile, the British government severed relations with Mexico over the seizure of El Aguila, an act deemed arbitrary and illegal by the Foreign Office.

Several factors frustrated this drive to bring Mexico to its knees. To begin with, the oil workers and the Mexican people showed an indomitable will to resist foreign pressures. In addition, the Roosevelt administration as principal architect of the Good Neighbor Policy rejected military intervention on behalf of the oil trusts. As FDR expressed it at Monterrey on April 20, 1943: "We know that the day of the exploitation of the resources and people of one country for the benefit of any group in another country is definitely over."[46] Although Secretary of State Cordell Hull increasingly pressed Mexico to assure prompt, fair payment for oil holdings as well as for farms seized earlier under the agrarian reform, he never challenged the right of a sovereign state to expropriate property. Washington might have devastated the depressed, reform-riven economy by not only suspending purchases—as it did—but by ceasing to buy silver. This precious metal provided employment for one hundred thousand men, generated 10 percent of the government's revenue, provided 17 percent of the income of the national railways, and earned more foreign exchange than any other export. Altruistic concerns aside, the United States resisted this move because a bankrupt Mexico could ill afford American goods and services, and precipitate action might have provoked Cárdenas to take over the silver industry, in which 80 percent of the mine owners were Americans.[47] Silver, like black gold, boasted a powerful lobby on Capitol Hill.

The prospect of a European war further encouraged American moderation. In the face of the boycott and propaganda campaign, Mexico had reluctantly begun selling and bartering oil to the Axis powers. Such activities coincided with the noisy emergence of the Fascist-oriented Hispanidad and Sinarquista movements in Mexico. To curry better relations with its strategically situated southern neighbor, the State Department quietly responded to an inquiry in mid-1939 that it had no opposition to Mexico's sale of oil in the United States. As a result the Mexicans landed a major contract with a New York buyer, giving them a new foothold in the American market.

Finally, the Sinclair Oil Company broke the united front of expropriated firms in 1940 and negotiated a separate settlement with the Mexican government. The reason for this action may have been a belief, which proved correct, that the first company to come to terms would receive a better cash settlement.

While most of the other corporations continued to posture and bluster, it became evident that the U.S. government would not pull their economic chestnuts out of the fire, and that Mexico could find outlets for its oil. Consequently a Mexican-American general agreement was signed on November 19, 1941. This accord stipulated inter alia that Mexico would pay $40 million to the expropriated American firms. President Roosevelt encouraged the renewal of diplomatic ties between Britain and Mexico a few months after Pearl Harbor. Nonetheless, the two countries did not settle outstanding claims over El Aguila until 1948.

More than two years of diplomatic pressure and official scoldings gave way to a "special relationship" as the Roosevelt administration concentrated on the war effort. In return for acting as an ally, the Mexican government obtained an assured market for silver and other items, major credits to stabilize the peso, Export-Import Bank loans for road construction, and a comprehensive trade treaty. The war years marked the period of closest economic, political, and military ties ever enjoyed between the two countries.

Organizing the Industry

The struggle for control and administration of the newly nationalized industry proved as difficult as the conflict with the foreign firms. The Oil Workers played a crucial role at the time of expropriation. Although responses varied from firm to firm, the foreign companies withdrew the overwhelming majority of their personnel shortly after General Cárdenas's decree. The union immediately filled this vacuum, taking over key posts as they were vacated by employees of the nationalized companies. In addition, the STPRM dominated the petroleum sector through administrative councils (*consejos de administración*) made up of key officers in each local functioning in their respective areas.

An esprit de corps characterized the activities of the *petroleros*. Reports abounded of union personnel in management positions who worked exceedingly long and irregular hours, volunteered for difficult and dangerous assignments, and performed their duties with makeshift machinery because of the foreign companies' reluctance to repair or replace equipment during the tempestuous period before March 18, 1938. Many of the new managers reportedly received little or no pay for their herculean efforts.

Thus they developed a proprietary view of the industry which fostered a commitment to syndicalism: that is, a belief that the workers employed in the petroleum sector should have the right to operate it through their labor organization. As a precedent they cited the 1937 nationalization of the railroads. As has been mentioned, Cárdenas rejected this approach and instead established a public corporation operated jointly by labor and government but controlled by the latter.

Labor did not accept gracefully Cárdenas's rejection of its syndicalist goals. The union, angered at not receiving "its industry" and extremely combative because of its conflict with the oil companies, resisted all efforts to organize Petróleos Mexicanos in an efficient, businesslike manner—a move that would have necessitated firing a significant percentage of the bloated work force. So ubiquitous was the redundancy and wasted effort that a dramatic reorganization was required to avoid complete collapse. The continuing expansion of PEMEX's employment rolls induced the president to force the badly needed reorganization.

Between 1938 and 1946, the Oil Workers' Union and the government locked horns over who would be master of the industry. Among the most salient points of contention was union opposition to the following government-supported policies:

1. Restructuring the industry on a nationwide basis, eliminating duplication of functions carried out by the expropriated firms—e.g., advertising, sales, engineering, accounting, and medical services.
2. Reorganizing the Oil Workers' Union to reflect the new structure of the industry.
3. Conferring upon PEMEX the right to appoint "confidential personnel" to key positions.
4. Authorizing PEMEX to eliminate positions it deemed unnecessary, limit the number of permanent employees, and reduce the number of temporary employees to no more than 10 percent of the permanent work force.

5. Holding the line on pay increases, reducing administrative salaries to equitable levels, and limiting such benefits as allowances for housing.
6. Emphasizing merit over seniority in making promotions.
7. Demanding greater efficiency and output from workers.
8. Allowing management a free hand in reassigning workers to other parts of the country.[48]

The union rested its case for improved salaries and benefits on the Board of Conciliation and Arbitration award of 1937. In its view the legitimacy of this award was enhanced after expropriation when the government announced plans to implement it gradually, a move later rescinded on financial grounds.

The government employed a variety of arguments to justify its position. Cárdenas, who stressed his sympathy for socialism over syndicalism, insisted that the petroleum industry operate for the benefit of all Mexicans, not simply those men and women who worked in it. His supporters also observed that even before expropriation, the *petroleros* enjoyed higher salaries and better benefits than any other major group of workers in the country. Moreover, the nation's honor required payment of an indemnity to the affected companies, and a portion of each year's production should be set aside for this purpose. The government also observed that the industry had changed dramatically, and that a new organization should reflect this change. It reiterated that imminent financial problems prevented its acceding to many of labor's bread-and-butter demands. Such difficulties, which made impossible the implementation of the 1937 award, were exacerbated by the sharp increase in the number of employees from 15,895 in April 1938 to 19,316 in January 1940.

Vicente Cortés Herrera, the first director general of PEMEX, pointed out that in the twenty-two months following expropriation, the monopoly's budget had grown by more than twenty-six million pesos—the amount in dispute with the foreign companies—but that the status of individual workers had not improved because of the growth of the work force. He called this increase unjustified because output had fallen, indicating declining efficiency. He implored the STPRM to "save the industry," then in a "disastrous" state because of labor's inexperience, irresponsibility, lack of discipline, and wholesale thefts. He noted that workers in Minatitlán caused the Edeleaunau plant to shut down when they balked at complying with reorganization plans fashioned by management. Workers almost precipitated the closing of the Arbol Grande refinery near Tampico by refusing to allow materials to be shifted to another processing center on the "infantile pretext" that local workers would be harmed. Meanwhile, a large quantity of oil ran into the sea from an unnamed plant because the employees would neither repair a defective pipe nor stop the pumps. The union leaders, livid over these highly publicized allegations, demanded the director general's removal.[49]

Although Efraín Buenrostro succeeded Cortés Herrera as head of PEMEX in December 1940, relations remained tense. However, a collective contract signed between the state firm and the STPRM on May 17, 1942, appeared to restore harmony. It conferred greater administrative leeway with respect to filling vacancies, while the petroleum workers received a forty-four-hour week, higher holiday pay, and special benefits for union officials.

No sooner had the ink dried on this accord than new disputes erupted over PEMEX's hiring and firing practices. Tensions mounted until August 1943, when workers in Ciudad Madero called a twenty-four-hour strike in retaliation for three hundred alleged violations of the collective agreement. Growing inflation fueled the unrest as prices in Mexico City increased approximately 50 percent between 1941 and 1943, thereby renewing union demands for implementation of the 1937 decree.

Faced with possible chaos in the industry, President Manuel Avila Camacho instructed PEMEX to implement the wage scales contained in the award. He also decreed a 10 percent boost in wages for workers receiving less than 5.52 pesos per day. The result was that the portion of PEMEX's budget devoted to compensation shot up from 90 million pesos in 1943 to 123 million in 1944, an extremely burdensome increase.[50]

The agreement ushered in a year of calm before the whole scene of confrontation, mutual recriminations, and work stoppages was repeated during negotiation of the 1946 collective contract. An impasse prompted the union to call a general strike for November 28, 1946. After Buenrostro resigned the STPRM postponed its *huelga* until the newly elected president, Miguel Alemán Valdés, could name another head of the state oil company.

Thus ended the first postexpropriation stage in the history of Mexico's oil industry. A number of difficulties beset PEMEX in 1946. Current expenditures, notably for personnel, had devoured so much of the company's revenues that it was unable to invest in exploration, much less meet ordinary tax obligations. Foreign-exchange earnings dwindled because of the contraction of the export market, and sales abroad were further inhibited by a scarcity of tankers. Even had additional funds been available, it would have been extremely difficult to obtain critically needed parts and machinery due to wartime shortages and restrictions. The capital-starved monopoly's record of discoveries was poor through the mid-1940s, and 60 percent of the limited drilling that took place occurred in the Poza Rica and Isthmus fields. Specifically, PEMEX extended the producing area of Poza Rica into the Mecatepec hacienda in 1941 and found deeper deposits in the isthmian El Plan field in

1943. "PEMEX inaugurated an exploration campaign also, which was admirably conceived to meet long-range requirements." The first new discovery—the Misión oil and gas field near the Texas border—was recorded in 1945.[51] The rapid draining of reserves threatened to exhaust Mexico's supplies within a generation. Moreover, the industry—geared for export—had yet to integrate itself with the local economy. Pipelines ran down to the Gulf instead of to the potential industrial, residential, and governmental customers in major inland cities. In addition, much of Mexico's crude was low grade, and the country lacked the refining capability to process the needed products—a situation that existed for decades. In the twenty years after expropriation, the value of oil imports exceeded that of exports by almost $120 million, even though the volume of exports was 95,231,000 barrels greater than that of imports.[52] Worst of all, PEMEX teemed with nepotism, corruption, and inefficiency.

The picture was not completely negative. In seizing control of its economic birthright, Mexico had mobilized popular sentiment and gained the attention of the world by an expropriation unparalleled since the 1917 Russian Revolution. Ingenuity and care compensated for the lack of new equipment or modern technology. Companies such as El Aguila and La Huasteca had halted exploratory work four years before expropriation. The absence of new reserves notwithstanding, PEMEX drilling crews succeeded in squeezing oil from existing reserves, and crude output in 1946 was the highest it had been since America's entry into World War II. The *New York Times* reported sources to the effect that the "industry had been running with a fair degree of efficiency," considering the wartime obstacles it faced.[53] Furthermore, there had been an unbroken rise in domestic oil sales since 1938, and—most important of all to nationalists—the industry had slowly begun to shift from an export orientation to a supplier of the domestic market. Like a sick man, PEMEX reeled and staggered, yet somehow managed to avoid collapsing as it lurched forward. That it remained on its feet at all was a minor miracle. But the time had come for serious treatment, if not extensive surgery.

In summary, it should be noted that expropriation and the events that preceded and followed it placed an indelible imprint on the development of Mexico's petroleum sector. *First,* despite Cárdenas's failure to deliver the industry into its hands, the Oil Workers' Union secured excellent pay and attractive benefits for its workers compared to those in other areas of the economy. It also boasted tremendous influence—especially in the hiring of PEMEX personnel—as a result of Mexico's "closed-shop" law and the crucial role that the syndicate played in keeping oil flowing immediately after expropriation. *Second,* PEMEX's "Generation of '38"—those dedicated engineers, technicians, and managers who went to work for the newly formed public body—consistently advocated conservation of Mexico's national patrimony lest output again plummet, as it did following the 1921 production peak, or the industry once more become the focus of foreign aspirations.

Third, Petróleos Mexicanos, in accord with its original aims, is expected to fulfill a vital social mission; that is, to furnish subsidized fuel to mines, factories, farms, villages, and myriad state agencies, including the railroads, the military, and the state electric-power company. The social mission also entails paying high wages, conferring generous social benefits, and building schools, roads, bridges, and other facilities in *zonas petroleras*. *Fourth,* the bullying and conniving of the American, British, and Dutch companies during this period convinced Mexican leaders as well as the public at large that foreign capital should play an emphatically restricted role in Mexico's emancipated petroleum sector.

2

Forging a National Oil Company

The Bermúdez Years

It was suggested that President Alemán ask Lázaro Cárdenas, the hero of expropriation, to take charge of the oil company that he had created.[1] Admirers of the beloved general insisted that he alone could heal the ailing industry. His detractors saw poetic justice served in giving him such an apparently hopeless task. Petróleos Mexicanos still suffered from dwindling reserves, decrepit plants and equipment, grossly inadequate transport and pipeline facilities, a chronic shortage of Mexican technicians, continuing foreign pressures, intractable labor problems, low prices, and the challenge of hammering the foreign firms into a single corporation at the nation's service. Exacerbating the situation was the country's appetite for energy: Pent up during the war years, it grew voraciously under Alemán, whose foremost priority was industrial progress.

Alemán ignored the gratuitous advice concerning the appointment and selected instead Antonio J. Bermúdez, a wealthy, energetic, and honest businessman-turned-politician from Chihuahua. Bermúdez had amassed a $6-million fortune as the agent for Waterfill Whiskey before winning election in 1942 as mayor of Ciudad Juárez, a "wild-west" town opposite El Paso renowned for gambling, prostitution, and graft. The new mayor waded into the morass, discharging crooked politicians, closing down brothels, and banishing gangsters. Thanks to this housecleaning, Juárez acquired an air of modest, if short-lived, respectability. Bermúdez next became the treasurer of Chihuahua and, in 1946, won a seat in the federal senate with the endorsement of Alemán and organized labor. The president decided that he was just the kind of man to whip the petroleum monopoly into shape.

No sooner had Bermúdez settled into office than he and the union clashed over a scheme to standardize wage scales which, eight years after expropriation, still reflected the variations among the private companies. The new director general rejected a union plan to raise salaries fourteen million pesos annually, proposing instead to limit the increase to nine million. The STPRM's response took the form of illegal twenty-four-hour strikes at the Poza Rica and Azcapotzalco refineries.

Bermúdez responded with dispatch in the face of this challenge to his and the government's authority. He fired the fifty syndicate leaders, including the entire executive committee, who had called the walkout. At the same time, he petitioned the Federal Board of Conciliation and Arbitration to annul the violated collective-labor contracts and initiated proceedings for a conflict-of-economic-interest order. Under such an order, the board would render a decision after considering a report, prepared by a special commission of experts, on the monopoly's ability to meet the demands.[2]

On December 21, 1946, Alemán threw his weight behind the PEMEX chief. It was his duty to act, he said, because the illegal union action was a threat to the well-being of the entire country. He disallowed the strikes, mobilized army units near the fields, and ordered machine-gun-bearing soldiers in jeeps to escort gasoline trucks through the streets of the capital. This bold action elicited the support of powerful elements of organized labor who believed in socialism, not syndicalism, and resented the petroleum workers having used their status as heroes in the "battle against imperialism" to obtain an average salary of twenty-four pesos a day, approximately ten to twelve times the national rate.[3] Both the Mexican Regional Workers' Confederation and the Industrial Syndicate of Mining and Metal Workers sided with the government. Even the CTM openly criticized the intransigence of the Oil Workers. Unfortunately, no intrepid pollster was on the scene to measure the popular temper. But a leading scholar reported that public opinion was "disgusted with the odorous conditions of their national oil industry."[4]

Then as now, Mexican workers understood and respected strong personal leadership. Bermúdez and the president showed the nation who was in charge and inspired the Oil Workers to replace confrontations with conversations. The director general tried to convince the union that Petróleos Mexicanos represented the Mexican people, not Wall Street barons. He not only proved a formidable negotiator but used his friendship with Vicente Lombardo Toledano, once the head of the CTM and for years a powerful voice within the labor movement, to manipulate the STPRM leaders. This combination of external and internal pressure enabled Bermúdez to hammer out an agreement for resuscitating the industry. PEMEX gained the right to cut fat from its payrolls, transfer workers more freely, and contract with private firms for projects unsuited for union labor. This last item, contained in Clause 39 of the collective contract (later Clause 36), ended the STPRM's exclusive right to serve as contractor for pipelines, refineries, offshore drilling, roads, schools, or any other project which PEMEX itself elected not to execute. It also gave impetus to corruption, as will be discussed in chapter 4. For its part the STPRM won an acceptable wage standardization plan, a 15-percent wage increase for workers in hardship areas, and the pledge of a fixed sum to cover the expenses and allowances of union officials. The two-year accord created four labor-management commissions to continue discussions of unresolved grievances between the two parties.[5]

Bermúdez's determination and vigor earned the admiration of the workers and the union's current leader considers him PEMEX's finest director general.[6] Tales of his generosity enlivened lunchroom gossip among the *petroleros*. He put his $9,600 annual salary in a special account to be used for schools and other PEMEX-sponsored social projects.[7] And when he learned of the hardship suffered by the family of a man who had perished in a refinery fire, he saw to it that the mortgage on the family's house was retired, that the widow obtained a job with the state firm, and that her daughter received a scholarship.[8] He cemented relations with administrators and workers alike by making regular weekend visits to PEMEX facilities. Each Saturday a company plane took him to a different part of the country for an inspection, a meeting with local administrators, an informal meal, a visit to PEMEX schools, clinics, and housing developments, and an opportunity to fraternize with the men.

Mexico City cartoonists delighted in depicting Bermúdez as a whip-cracking Simon Legree. In fact the workers, who came to adore their chief, fared quite well during his twelve years in office. Initial efforts to limit the size of PEMEX's labor force succeeded for three years before its inexorable growth resumed; the number of employees leaped from 28,822 in 1947 to 45,532 in 1958.[9] During the same period, total compensation—salaries and benefits— shot up from 250 million pesos to 1,098 million, figures uncorrected for inflation. PEMEX continued to spend freely on housing, schools, clinics, and recreational facilities for the workers and their families. The union, which got a cut on virtually every contract negotiated by the monopoly, developed a more centralized structure within the expanded and consolidated industry.

In any case Alemán and Bermúdez brought order out of chaotic labor relations. The conclusion of this battle allowed the director general to concentrate on his goal of putting PEMEX on a "businesslike basis." "We have more oil than Venezuela," he boasted to a *Time* magazine correspondent, "and it is our job to get it up. . . . When I get through, every man in PEMEX is going to be on his toes, anxious to win a good record and improve the organization."[10]

Bermúdez believed that above all a good record entailed locating oil, and he launched the most extensive exploration program since 1930. He increased outlays on geological studies and exploratory drilling fourteenfold. PEMEX had drilled 37 exploratory wells between 1938 and 1946. It drilled 758 under the hard-driving director general. The number of developmental wells soared from 221 to 2,478 during the same period, and the average distance drilled each year for wildcat and developmental wells shot up by nearly a factor of ten under

Bermúdez (1,399,594.2 feet) compared with the 1938–1946 period (145,763 feet).

All told, PEMEX discovered some sixty-five new oil deposits. These included the Santa Agueda and Ezequiel Ordóñez fields, which accounted for 70 percent of total yields in a "new Golden Lane" one hundred miles south of Tampico near the Gulf of Mexico. Other noteworthy fields were Moralillo (fifty miles northwest of Poza Rica, 1948), Angostura (Veracruz, 1953), Nuevo Progreso (Poza Rica, 1955), and San Andrés (Poza Rica, 1956). Production in the last-named, situated fifteen miles southeast of Poza Rica, benefited from increased proficiency in deep drilling techniques as oil flowed from Jurassic limestone at a depth of more than ten thousand feet. PEMEX opened three other Jurassic wells in this vicinity in 1957, with El Hallazgo being the most fecund. Discovery of the Constituciones field north of Tampico in 1956 and the Tres Hermanos field near the old Golden Lane in 1959 further expanded supplies from the Jurassic layer, which provided 22 percent of national production in 1959 and was destined to be the principal source of future output. Three of the most bountiful new oil fields in the Isthmus—La Venta, Ogarrio, and Magallanes—came on line between 1954 and 1957 near the old El Plan and Tonalá fields in the eastern portion of the Sabinas Basin. As a result proven reserves of crude oil, natural gas, and condensates tripled from 1.4 billion barrels (1947) to 4.1 billion (1958) as average crude output climbed from 156,487 bpd to 275,730 bpd. Table 1 presents production figures from 1938 to 1976.

Extraordinarily impressive were achievements in capturing natural gas, a substance that had to be burned off or "flared" because PEMEX lacked the capital with which to construct complex and expensive facilities for gathering, treating, storing, and transporting it. For example, while railcars, trucks, ships, and pipelines can move petroleum, gas can only be transported economically through mains. While "dry" gas—that which is free of liquefiable hydrocarbons—can be placed directly into these ducts, a good deal of natural gas, "sour" with hydrogen sulfide, must be "sweetened" in processing plants before shipping. Bermúdez reported that "untold billions" of cubic feet were burned before PEMEX installed the integrated system required to deliver it to population centers.

Petróleos Mexicanos discovered thirty natural gas fields during the Bermúdez years as reserves rose from the equivalent of 372 million barrels of oil to 1,558 million and annual output rose from 32.9 to 262.6 billion cubic feet. Three areas embraced 90 percent of Mexico's known reserves in 1958. Reynosa, in the Burgos Basin of the Northeast just below the United States–Mexican border, contained one of the largest gas deposits in the Western Hemisphere. Next in importance were the José Colomo fields, opened in the

TABLE 1
Mexico's Crude and Natural Gas Production, 1938–1976

Director-General	Period	Daily Crude Production in Period[1] (barrels)			Daily Natural Gas Production in Period (Mcf)		
		Beginning	End	% Change	Beginning	End	% Change
Vicente Cortés Herrera and Efraín Buenrostro	1938–1946	106,351	135,707	27.6	66,009	71,377	8.6
Antonio J. Bermúdez	1946–1952	135,707	215,592	58.9	71,377	255,537	258.0
Antonio J. Bermúdez	1952–1958	215,592	275,730	27.9	255,537	719,524	181.6
Pascual Gutiérrez Roldán	1958–1964	275,730	353,835	28.3	719,524	1,325,105	84.2
Jesús Reyes Heroles	1964–1970	353,835	486,573	37.5	1,325,105	1,821,932	37.5
Antonio Dovalí Jaime	1970–1976	486,573	894,000	83.7	1,821,932	2,108,672	15.7

Source: Petróleos Mexicanos, *Anuario estadístico 1976* (Mexico City: Petróleos Mexicanos, 1976), pp. 34, 36.

1. Includes petroleum, condensates, and absorption liquids.

Macuspana Basin of eastern Tabasco in 1951, and the Veracruz Basin in the central Gulf Coast, which gave rise to a modest level of production.

These vast quantities allowed PEMEX to sign a contract in 1955 with the Texas Eastern Transmission Corporation to export up to 200 million cubic feet per day of dry natural gas. The agreement called for a minimum of 14.2 cents per thousand cubic feet (Mcf), with an ascending scale each year and the right, after five years, to revise the price consonant with market conditions. The negotiated price exceeded that obtained by any U.S. gas producer in the border area. The contract would generate between $116 million and $232 million during its twenty-year life, and Mexican gas would flow as far north as Boston. Credits derived from the deal helped finance such projects as the complex at Ciudad PEMEX, gas-treating plants in Reynosa, the lube-oil plant at Minatitlán, the catalytic cracking facility at Azcapotzalco, and the new Ciudad Madero refinery.[11]

Bermúdez served not only under Alemán but also under his hand-picked successor, Adolfo Ruiz Cortines, elected in 1952. He effusively praised the former for being "very knowledgeable about the industry" and exhibiting an "extraordinary dynamism." He went so far as to label the 1946–1952 period a continuation of the "Golden Epoch of Petróleos Mexicanos" that began with the expropriation and was characterized by an unusual "spirit" and a "heroic" commitment to keep the industry growing despite severe adversities.[12] Yet the greatest advances came under the less flamboyant Ruiz Cortines, known by associates as *"El Estudiante"* because of his penchant for reflection and meticulous study before reaching decisions. A markedly higher percentage of wildcats (68.7), development wells (65.7), total feet drilled (72.6), and oil and gas reserves discovered (68.5) took place during the second *sexenio* that Bermúdez held his post. Nonetheless, he criticized the meditative chief executive for having "braked the rhythm of development" by continuing a policy of low prices and subsidies to other areas of the economy.[13] (Alemán had granted a modest but controversial price increase in 1949.) These subsidies, which accelerated economic growth in a country that relied on oil and gas for 90 percent of its energy needs, amounted to 2,800 million pesos during Bermúdez's tenure, while investments totaled 7.640 million.[14] Subventions and anemic prices prevented PEMEX from reaching a 400,000-bpd level in 1954, as the optimistic director general had once predicted.

In addition to its impact on production, the uneconomic pricing system retarded the construction of such vitally needed facilities as a pipeline network to bring oil and gas from the coastal plain over the mountains to the cities of the central plateau as well as refineries to satisfy the surging industrial demand—increasingly met by imports—for high-grade lubricants and other refined products. Only prompt action could prevent Mexico's extreme dependence on foreign supplies. In all fairness to Ruiz Cortines, who was by nature parsimonious, Alemán had spent public monies with so much abandon that a policy of economic retrenchment was unavoidable.

So ubiquitous was the red ink that Petróleos Mexicanos had stopped publishing a financial balance sheet of its operations in 1947. Still, Bermúdez moved aggressively to strengthen the country's oil infrastructure. A financial expert would not have dared the risks that he took. Although toward the end of his term it was possible to obtain five- and seven-year credits, in the mid-1940s Bermúdez could only secure short-term financing. Thus he had the audacity to finance long-term projects such as refineries with short-term credits.[15]

Completion of the Reynosa refinery in 1950 proved a notable achievement because, despite a modest capacity (4,000 bpd), it had been designed and constructed by Mexican technical experts, thousands of whom received training during the Bermúdez years. PEMEX later enlarged its maximum volume to 10,000 bpd and installed an 1,800-bpd thermal cracking unit. The 30,000-bpd Ing. Antonio M. Amor refinery at Salamanca, the country's geographic center, also opened in 1950. More ambitious was a new plant at Minatitlán, to serve the Pacific coast; it boasted a primary distillation capacity of 85,000 bpd and a catalytic cracking capability of 12,000 bpd. The erection of these three facilities raised to eight the number of refineries owned and operated by Petróleos Mexicanos. Private companies had constructed five before expropriation: Minatitlán (El Aguila—Veracruz); Azcapotzalco (El Aguila—Mexico City); Ciudad Madero (El Aguila—Tamaulipas); Arbol Grande (Pierce—Tamaulipas); and Mata Redonda (La Huasteca—Veracruz). PEMEX had finished or begun construction of two others before Bermúdez joined the monopoly: Poza Rica (Veracruz) and 18 de Marzo (Azcapotzalco, D.F.). At the beginning of 1947, the state company could refine 153,000 bpd; this potential had more than doubled to 328,000 bpd by the end of 1958.[16]

Processing complexes sprang up at Poza Rica, Reynosa, and Ciudad PEMEX to handle the rapidly increasing volume of natural gas, while the monopoly constructed 977 miles of gas pipelines to carry the fuel to urban areas. The first main, inaugurated in 1950, extended 162 miles between Poza Rica and Mexico City. Even more important was the new Reynosa-Monterrey gasoducto, completed in 1958, which assured year-round supplies to northern Mexico's industrial center. Construction had just begun on the most challenging project thus far contemplated—a 24-inch, 663-mile trunk gas line from Ciudad PEMEX to Salamanca by way of Mexico City—when Bermúdez left office.

Bermúdez saw the monopoly's potential as the nucleus for a cluster of related industries. He persuaded private entrepreneurs to manufacture the pipe, gauges, compressors, and other items used in the petroleum sector. Petróleos

Mexicanos, totally dependent on imports in 1947, obtained half its required supplies from Mexican industry a decade later.[17]

Even more controversial than his early forays into finance was Bermúdez's decision—perhaps forced by the probusiness Alemán—to allow foreigners to reenter Mexico's oil sector. In March 1949 the director general signed a contract with a U.S. firm, the Compañía Independiente México-Americana (CIMA). He later concluded fifteen other agreements, although twelve of the interested parties opted out of what they believed would be uneconomical deals. The only contracts remaining in force were those of CIMA, Pauley Noreste, Sharmex, and the Isthmus Development Company. These companies would receive no payment for dry holes, but 50 percent of the revenues derived from productive wells would be set aside to cover the expense of drilling these wells, with an additional 15 percent earmarked as compensation for risks assumed. To circumvent the legal prohibition on oil exploitation by foreign interests, PEMEX agreed to supervise carefully each contractor's work and take charge of all completed wells.[18]

The wisdom of these arrangements was placed in question by their addition to yields. By late 1958 the foreigners accounted for only 2 percent of production, although their contribution to marginal output was somewhat higher. The first discovery pursuant to these contracts was Xicalango, a gas well drilled by CIMA near the coast of Campeche. CIMA was also responsible for finding Santa Ana, a field located in the Gulf that was the first deposit in Mexico developed exclusively by offshore operations. The private firms recorded a 27.2-percent success rate for wildcats compared to 31.3 percent for PEMEX; their percentage of productive developmental wells (70.6) was also inferior to that of the monopoly (74.7 percent).

It was said that oil, not blood, coursed through the veins of the indefatigable director general who developed such a passion for the industry which bore his indelible stamp. As the *New York Times* expressed it: "Senator Bermúdez' role in building up PEMEX to the status of a modern, progressive industry is one of the few things about PEMEX that no one disputes whether they are for or against the nationalized oil company."[19] He calmed the troubled waters of labor relations even as the union grew stronger and more graft-ridden. He created a highly effective, professionally trained, and loyal team that catalyzed the most successful period of exploration until the 1970s and earned an international reputation for PEMEX. He pioneered the development of Mexico's abundant natural gas deposits and in 1953 began the exportation of this important fuel. At the same time, he preached the doctrine of conservation; namely, that "future petroleum reserves are worth more than the money received for them now."[20] Despite a rear-guard action by private corporations, he launched an ambitious investment program which included a doubling of the monopoly's refining capacity. Above all, he forged a congeries of run-down, capital-starved, technically deficient companies into a modern respectable entity to address the needs of the world's sixth-largest energy consumer. Ironically, President Ruiz Cortines took the politically difficult step of increasing prices—a move urged for years by Bermúdez to advance PEMEX's program—only a few days before they both left office. On his last day as president, he also had Article 27 of the Constitution amended to prevent foreign interests like CIMA from entering the country's oil sector; Congress passed the implementing legislation in August 1959.

An Engineer-Banker-Industrialist Takes Over

Upon assuming the presidency in 1958, Adolfo López Mateos appointed Bermúdez ambassador to Iran and named in his place at PEMEX Pascual Gutiérrez Roldán. Gutiérrez Roldán, who described himself as an "engineer by profession, a banker by trade, and an industrialist by chance,"[21] was born in Mazatlán, Sinaloa, in 1903. He secured degrees in both agriculture and economics at Mexico's National Autonomous University (Universidad Nacional Autónoma de México—UNAM) before earning a Ph.D. in economics in the United States. He held a series of responsible posts dealing with banking and credit in the Department of Treasury and other government agencies before beginning the highly successful El Roble Savings Bank, financed by funds from Spanish émigrés. President Ruiz Cortines placed him at the head of the national steel industry, where he gained recognition for efficient and skillful management. After naming Gutiérrez Roldán director general, López Mateos put PEMEX, along with the Federal Energy Commission, the state steel industry, and other agencies, under the jurisdiction of a newly created ministry, National Patrimony, to assure a rational and better-coordinated development of the country's natural resources. The minister, Eduardo Bustamante, was an industrial lawyer with whom Gutiérrez Roldán had worked closely when both men served as specialists in Treasury. Bustamante gave firm backing and considerable leeway to his good friend's management of the oil monopoly.

A series of thirteen articles in *Novedades,* a Mexico City newspaper, and the *News,* its English-language edition, dispelled any doubts that López Mateos would make changes in the industry.[22] Written between January 9 and January 21, 1959, by Antonio Vargas MacDonald, whose public-relations firm had a contract with PEMEX, they focused on waste, inefficiency, and labor-racketeering within the state enterprise. The specific abuses cited included (1) theft of oil and equipment, (2) chicanery in leasing service stations, (3) neglect and improper maintenance of machinery, (4) unnecessary and costly ship repairs in the United States, (5) wrongful estimates on contracts, (6) flagrant use of company automobiles for nonofficial activities, (7) overprescribing of drugs by PEMEX physicians in collusion with pharmaceutical firms, and (8) the establishment of "independent" businesses by administrators who then enriched themselves at PEMEX's expense. Vargas inveighed against the creation within Petróleos Mexicanos of baronies in which superintendents controlled the workers and local population as though they were vassals. He singled out Poza Rica, the nation's most productive oil district, where Jaime J. Merino had for sixteen years ruled like a medieval lord. Merino called upon union allies to keep workers in line, kept order with ruthless guards, manipulated the politicians of the region, and dominated its economy. More will be said about him in chapter 4.

Vargas leveled his sharpest attacks at such labor abuses as featherbedding, payroll-padding, payment for phony overtime, negligence in the performance of duties, the feigning of illness, and—the "most vicious" practice of all—*vendeplazas,* the selling of jobs. Unscrupulous STPRM leaders extorted between $450 and $1,200 from "hundreds" of supernumerary workers as payments for positions in the industry. As a consequence the once glorious oil company had become a hotbed of black-market labor operations. The writer concluded that the survival of PEMEX was imperiled, and that its failure would have horrific consequences for the national economy.

Pedro Vivanco García, a Merino cohort and the STPRM leader from Poza Rica who later became the union's national secretary general, took bristling exception to these accusations which, he contended, sprang from "gross ignorance" and "malicious intent." The culprit, in his opinion, was not the union but PEMEX itself. He charged that the firm had purchased "millions upon millions" worth of drilling, storage, and pumping equipment that wound up on the scrap heap without having been used. Thanks to generous kickbacks to PEMEX employees from ship-repair firms, the company unnecessarily dispatched its tankers to the United States for overhauling. Vargas, according to Vivanco, was merely "an agent of financial interests" intent on "preparing public opinion" to accept the return of foreign ownership of Mexican oil.[23]

Vargas dismissed these fulminations as nonsense, reiterated his commitment to shielding the industry from the grasp of foreigners, and accused the labor chieftain of "creating ghosts and then fighting them."[24] Few doubted that Vargas was an agent. However, his principal was neither Standard Oil nor Royal Dutch Shell but the president of the Republic, who wanted to prepare the Mexican people for drastic changes in PEMEX, long deemed untouchable. López Mateos sent Vargas, who had worked in his election campaign, to help Gutiérrez Roldán at PEMEX. Vargas's good friend Ramón Beteta, publisher of *Novedades,* was "anxious" to run the series, which received international coverage because of its simultaneous appearance in English. A major Mexican newspaper would never have published such intense, prolonged, and mordant criticism of a national institution without approval from the highest political level. Less enthusiastic than López Mateos about the venture was Lázaro Cárdenas, who sent word through intermediaries that he disapproved of the articles.[25]

Gutiérrez Roldán's steel firm had supplied PEMEX, and he knew from first-hand experience the company's poor payments record. Still, he was astonished to find upon replacing Bermúdez that there was insufficient money available to meet the next month's payroll. He clearly had his work cut out if he was to put PEMEX on a businesslike footing. Losses in 1958 alone equaled $9.2 million.

The new director general began by lopping off "deadwood" from the more than forty-five thousand employees, with a view to cutting $20 million from the $100 million payroll. He got rid of a number of political appointees who owed their jobs to friends in high places. Poza Rica received special attention because of the sharp practices, violence, and inefficiency that suffused the oil business there. With the president's approval, Gutiérrez Roldán politely thanked the district baron, Jaime J. Merino, for his past services, assured him a comfortable salary, and sent him packing to a PEMEX sales office in Los Angeles. He then reorganized the Poza Rica operation.[26] In the six years that Gutiérrez Roldán presided over the company, the work force grew only from 45,695 (1959) to 50,372 (1964), or 10.2 percent, less than half the increase in production (22.1 percent).

The chief executive's ties to the union's new leadership help explain the relative tranquillity of labor-management relations at this time. Joaquín Hernández Galicia, known as "La Quina," replaced Pedro Vivanco García as leader of the STPRM, becoming secretary general for two years beginning in 1962. His career will be discussed in chapter 4. It need only be noted here that La Quina and López Mateos were political allies: The strongman from Ciudad Madero backed the politician's election and kept the *petroleros* in line behind him, though he had not attained the power he was to enjoy in the 1970s; in return, the *jefe máximo* publicly praised and supported Hernández Galicia and even offered to make him a senator from his home state (Tamaulipas)—a post which La Quina declined because of union responsibilities.[27]

There was, however, no love lost between the syndicate and Gutiérrez Roldán. He eliminated the political role of Vivanco García and other corrupt union officials at Poza Rica; clamped a lid on employment; and slowed the rate of contributions to social projects for the workers and their families. STPRM members also disliked his style. In contrast to the ubiquitous, unpretentious Bermúdez, Gutiérrez Roldán seemed arrogant and aloof. Early in his six-year term, he made one inspection trip to the fields; subsequent visits were political or social in nature. The director general's reputation sprang in part from a reluctance to mix with the men, a well-known appreciation of classical music, and his eagerness to decorate his office with paintings rather than dreary pictures of oil rigs and meandering pipelines. Retirees still joke of his golden anklet, which they claimed to have seen when Gutiérrez Roldán climbed a ladder during a tour of a facility.[28] Still, the business-minded PEMEX chief claimed good relations with La Quina, "who always kept his word with me."[29]

Mexican presidents since Cárdenas have relished direct involvement in hydrocarbon matters, and López Mateos was no exception. He and Gutiérrez Roldán regularly conversed over the telephone system which linked their offices. The director general often stayed late after Tuesday cabinet meetings "to talk for hours" about the industry. Whether it was buying a crane to start a port at Pajaritos (López Mateos approved) or honoring Fidel Castro's request to buy three hundred thousand barrels of heavy oil (the president emphatically disapproved), López Mateos liked to be consulted.[30]

Extremely adept at raising money, Gutiérrez Roldán quickly devoted himself to improving the position of a firm chronically short of operating capital and investment funds. He was particularly anxious to convert PEMEX's indebtedness from short- to long-term obligations. In 1959 he obtained $40 million in loans and credits from four private U.S. banks to finance a major gas pipeline. Other American financial institutions provided a ten-year, $50-million credit, backed by natural gas exports. He also found willing lenders in Europe where prosperity had followed reconstruction to swell the available funds of lending agencies. Banks within the European Economic Community furnished at least $35 million. In 1963 the French government granted a $100-million loan for expansion of the nascent petrochemical industry. The only condition imposed by the young finance minister, Valéry Giscard d'Estaing, was that Gutiérrez Roldán "include the private sector in your plans."[31]

López Mateos excused PEMEX from tax payments in 1958. He also agreed to capitalize the firm's outstanding public debt (1,750 million pesos) in ninety-nine-year securities that the company would retire from current revenues.[32] Past payments by the firm had actually been a sham because they often came on the heels of huge government subsidies that proved a drain on the federal budget. On January 1, 1960, Petróleos Mexicanos began paying to the federal treasury a "single tax" of 12 percent on its gross income, an amount greater than that paid by all other private and state enterprises.[33]

Gutiérrez Roldán knew that PEMEX itself had to generate additional revenues and applauded the increase in product prices instituted before Bermúdez left office. Even then Mexican prices remained among the lowest in the world. Revenues grew 23 percent during Gutiérrez Roldán's first year on the job.[34] An annual financial statement began appearing for the first time since 1947, and its red ink finally gave way to black. Total net assets, which stood at 6,958 million pesos in 1958, climbed to 12,579 million pesos in 1963. During the same period, the monopoly's debt fell from 3,839 to 3,110 million pesos.[35]

The director general put his new resources to work. He considered his foremost accomplishment to be the construction of a 456-mile gas pipeline from Ciudad PEMEX to Mexico City.[36] This 100-million-peso project helped expand the capital's daily consumption of natural gas tenfold to 233 million cubic feet in 1963. Under Gutiérrez Roldán a network of distribution lines grew like a web across the North and Gulf Coast, supplying natural gas to two dozen cities. Also significant was the laying of a 357-mile products line from Minatitlán to Azcapotzalco which allowed refineries in each locality to coordinate their production activities. Space limitations prevent the description of every project, but suffice it to point out that between 1958 and 1963, pipeline capacity for crude, natural gas, and refined products increased 58 percent in length and 75 percent in volume.[37]

Gutiérrez Roldán expanded refining capacity by constructing a 140,000-bpd facility at Ciudad Madero and increasing the output of the Minatitlán, Salamanca, and Poza Rica plants. Meanwhile, he closed down old, inefficient installations at Mata Redonda and Arbol Grande. The total daily volume of products increased from 287,474 (1959) to 346,000 (1964).

Gutiérrez Roldán excited a protracted controversy over his drilling strategy. A doubling of resources enabled him to expand the number of wells completed (3,752) and feet drilled (28.9 million) during his *sexenio* compared to those of the previous one (2,148 and 12.2 million) (see table 2). Yet in contrast to his predecessor, he concentrated almost exclusively (84.8 percent) on developmental rather than exploratory activities. Even so, there was only a modest increase in crude output, from 275,730 bpd (1958) to 353,835 (1964). Petroleum engineers who have retired from PEMEX and wish to remain anonymous attribute the slight production gain to poor field management, overdrilling, and the sinking of "political wells." Such wells, which invariably turned out to be dry, were drilled at sites for which there was little or no scientific evidence of the presence of hydrocarbons. However, they provided drilling sites for contractors eager to collect fees. According to these well-informed sources, the private firms receiving lucrative contracts gave money to selected company personnel to show their appreciation for the work and to assure a renewal of the accord. Every prospective well represented a possible new contract and another deposit in a Swiss bank account. One of the industry's best-known men, and a key figure in PEMEX's production department, acquired the sobriquet of "Mr. Ten Percent" during this period for the putative size of payoffs received.

TABLE 2
Drilling Activities in Mexico, 1938–1976

Director General	Period	No. of Exploration Wells		No. of Developmental Wells		Total No. of Wells	% of Total No. Drilled by Contractors	Avg. No. of Wells Drilled per Year	Avg. No. of Feet Drilled per Year	Total Amount Spent on Exploratory Drilling in Period (thousands of pesos)
		Prod.	Unprod.	Prod.	Unprod.					
Vicente Cortés Herrera and Efraín Buenrostro	1938–1946	8	29	151	70	258	0	28.7	145,763.0	N.A.
Antonio J. Bermúdez	1946–1952	83	154	542	309	1,088	36.9	181.3	768,006.4	681,979.0
Antonio J. Bermúdez	1952–1958	154	367	1,309	318	2,148	23.5	358.0	2,031,182.0	1,152,906.6
Pascual Gutiérrez Roldán	1958–1964	125	445	2,570	612	3,752	48.9	625.3	4,832,114.5	3,031,091.2
Jesús Reyes Heroles	1964–1970	225	617	1,487	471	2,800	7.5	466.7	3,710,670.8	2,455,789.0
Antonio Dovalí Jaime	1970–1976	143	498	1,416	370	2,427	5.2	411.2	3,509,741.1	4,237,057.0

Sources: The following volumes published by PEMEX: *Anuario estadístico 1975* (Mexico City: Petróleos Mexicanos, n.d.), pp. 30–33; *Anuario estadístico 1976* (Mexico City: Petróleos Mexicanos, 1976), *passim*; and *Memoria de labores 1976* (Mexico City: Petróleos Mexicanos, 1976), p. 78.

When López Mateos mentioned that he heard Gutiérrez Roldán favored friends and partners in the distribution of contracts, the assertive director general responded: "Do you expect me to give them to my enemies!"[38] Even López Mateos's personal pilot established a drilling company that did business with PEMEX.

Gutiérrez Roldán justified the use of outside contractors on the dubious grounds that they were only one third as expensive to employ as PEMEX personnel. Bermúdez, highly critical of his successor, claimed that in fact the cost per hole was substantially higher because the contractors, who also had to hire costly union labor, received generous advances from PEMEX which often underwrote their equipment purchases. These firms, sometimes newly organized, lacked skilled professionals, technicians, and blue-collar workers, so they often hired men away from the state enterprise which had invested heavily in their training.[39]

Natural gas reserves expanded 41.7 percent during the *sexenio,* but neglect of exploration and surging consumption restricted the increase in crude reserves to 5 percent. Failure to maintain the rhythm of exploration begun the previous decade consigned Mexico to the status of an importing nation in 1968.

Bermúdez also excoriated Gutiérrez Roldán for having involved the monopoly in peripheral activities that detracted from exploration. For example, the director general became a passionate and forceful advocate of a proposal, broached by López Mateos in a January 1960 meeting with President Juscelino Kubitschek of Brazil, to construct a deepwater canal across the 150-mile Isthmus of Tehuantepec. The idea, first advanced by Hernán Cortés in the fifteenth century, had been seriously explored by an American, Robert W. Shufeldt, and a Mexican, Fernández Leal, in 1871. Beginning in the late 1940s, the prospect of a "second Panama Canal" captured the imagination of a number of engineers and industrialists. When the concept received Gutiérrez Roldán's imprimatur, specialists within PEMEX went to work immediately to chart a suitable route. The monopoly also awarded a contract to an engineer who had completed many hydrological works for the government.[40] The project quickly folded when critics, including Lázaro Cárdenas, ever vigilant of Petróleos Mexicanos, voiced unalterable opposition on the grounds that the waterway would split the country into two parts.[41] High construction costs and the possible use of nuclear explosives to cut a channel combined to kill the idea of a canal that gained currency in 1964.[42]

Criticism also attended Gutiérrez Roldán's initiatives in petrochemicals. His predecessors had talked about the desirability of turning out such products, which greatly enhance the value of petroleum. Yet it remained for this shrewd engineer-banker-industrialist to provide the impetus. PEMEX began with sulfur production near its refineries, then built small detergent plants, and gradually extended its activities to provide the basic materials for four product groups: fertilizers, plastics and artificial fibers, synthetic rubber, and benzene ring compounds known as aromatics. Plants sprang up across the country, from the arid sun-drenched flats of the North to the steamy jungles of Veracruz. Minatitlán, Pajaritos, Salamanca, Ciudad Madero, and Reynosa were the principal production centers.

Opposition arose not to the industry itself; everyone agreed that it was a worthwhile, potentially lucrative venture. The objections were to an eagerness, denied by Gutiérrez Roldán, to allow foreign investment in primary (basic compounds) as well as secondary petrochemicals (derivatives of basic compounds). Bermúdez bluntly charged him with "attempting to deliver the basic petrochemical industry to private interests, including foreigners."[43] He claimed that Gutiérrez Roldán was attempting to conclude a deal with Dow Chemical and that he actually signed a contract, in violation of the law, with foreign investors for a plant to produce polyethylene, used in making plastics, film, containers, and insulation.[44] PEMEX officials reportedly delivered the documents relating to the project to General Cárdenas, who simply held on to them for several months until a new president and director general assumed office.[45] Another source alleged the PEMEX chief was chairman of the board of directors of Poli-Rey, S.A., a company that planned to produce polyethylene.[46]

Whatever Gutiérrez Roldán's intentions may have been, a chorus of outcries assured that basic petrochemicals would be reserved, in accordance with Mexican law, to Petróleos Mexicanos. In 1963 nearly 205,000 tons of petrochemicals were sold in Mexico; the following year sales totaled 277,473 tons at a value of 222.8 million pesos. The number of such petrochemical products increased from an original six to thirty-four in 1979.

Cortés Herrera and Buenrostro shepherded PEMEX through the difficult postexpropriation and war years. Bermúdez made peace with labor, galvanized PEMEX into a unified, technically proficient company, spurred the development of natural gas, and sharply boosted its reserves. For his part Gutiérrez Roldán put the firm's finances in order, greatly expanded its pipeline system, and established what has become an extremely important petrochemical sector. But the 1958–1964 period also saw corruption, present for years, spread like wildfire. Peculation molded a drilling strategy, distinguished by generous contracts awarded private firms, that led to a stagnation in reserves. As a result Mexico would again import hydrocarbons by the end of the decade.

A Resurgence of Nationalism

The election of Gustavo Díaz Ordaz brought a change in the leadership of PEMEX as Jesús Reyes Heroles, a lawyer and politician, succeeded Gutiérrez Roldán, a financier and entrepreneur. Born in Tuxpan, Veracruz, Reyes

Heroles entered politics soon after finishing law school at UNAM and graduate studies at the University of La Plata and the University of Buenos Aires. In 1949, at the age of twenty-eight, he became auxiliary private secretary to the president of the Revolutionary party's (PRI) National Executive Committee, a post that he himself would hold from 1972 to 1974.[47] He served on the Federal Board of Conciliation and Arbitration, worked in his party's Institute for the Study of Economic and Social Problems (IEPES), directed economic studies for the national railroads of Mexico, and was technical subdirector general of the Mexican Social Security Institute. He also found time to advise Ruiz Cortines and López Mateos, serve as a federal deputy from the state of Veracruz, offer classes at the university, and write a half-dozen books on economic and political questions. Even though he had no knowledge of the oil sector, he set about to master its workings. Associates report that he would spend eighteen or twenty hours each day poring over reports, immersing himself in pertinent data, receiving briefings, and visiting installations.

In strong contrast to his predecessor at Petróleos Mexicanos, Reyes Heroles was a militant supporter of the Mexican Revolution, a fervent nationalist, a committed reformer, and a firm believer that PEMEX's director general should be a politician. In an extremely candid statement made on March 18, 1967, he said: "I do not know what requirements a director of Petróleos Mexicanos should fulfill and I am sure I do not meet any of them. But I do feel that there is one that is absolutely indispensable and which I feel I do meet: the Director of Petróleos Mexicanos should have complete faith and conviction in the Mexican Revolution."[48]

Fidelity to the revolution meant a public-service role for PEMEX as opposed to a profit-seeking one, and adherence to a conservationist doctrine, which he labeled an "iron law of crudes." As Reyes Heroles said in his first annual report: "The variations in prices, the mechanics ruling distribution, make it desirable, neither to depend on the exports of low-priced crude with a fluctuating demand, nor on the imports of crudes, exposing ourselves to phenomena equally unfavorable, whatever be their sense." Self-sufficiency was the key. To import large volumes gave external forces undue influence over a nation's destiny; to export major quantities risked a return to the excessive waste of the 1921–1924 period, "which was not a boom, but a fever, an overexploitation, a frenzy."[49]

Consistent with this outlook, he encouraged wildcat drilling, so neglected in the previous *sexenio*. The percentage of exploratory wells increased from 15.2 to 30.1 as 958 million of the 2,041 million pesos earmarked for drilling were devoted to this purpose. The cost per well mounted sharply as PEMEX crews sank four wells deeper than thirteen thousand feet and six offshore in this period. The quality of exploration improved thanks to the installations of processing centers for seismological information in Tampico, Poza Rica, Coatzacoalcos, and Reynosa. Advances in exploration equipment and new seismological techniques paved the way for the Reforma discovery in the early 1970s. All told, the company discovered seventeen new fields and twenty-three extensions of existing ones.

Proven holdings increased only 6.5 percent—from 5.23 billion to 5.57 billion barrels—under Reyes Heroles. Bermúdez deprecated the results with respect to crude reserves and called the natural gas situation "disastrous" inasmuch as PEMEX, which achieved a rise of over 40 percent during the previous six years, witnessed a decline equivalent to 22.9 million barrels of oil from 1964 to 1970.[50] The picture brightened when PEMEX discovered the Arenque field in the Gulf of Mexico, eighteen miles east of Tampico. Even more significant was the opening of a Marine Golden Lane lying off the coast of Veracruz roughly between Tampico in the North and Poza Rica in the South, embracing the Arrecife Medio, Isla de Lobos, Tiburón, Esturión, and Atun fields. Even though the offshore Golden Lane never proved as productive as its onshore counterpart—output has declined from 60,911 bpd (1970) to approximately 40,000 bpd (1979)—it turned out to be the largest strike in four decades.[51] Too, gas manifestations in El Gato 1, a well in Coahuila, alerted experts to the presence of prolific gas supplies in the Sabinas Basin. Despite the scant progress in proven reserves, the director general boasted a 67-percent increase in proven and probable holdings.[52] Although crude production expanded from 362,030 to 486,573 bpd, it failed to keep pace with demand, then accelerating at 7 percent annually. In 1968, as we have seen, Mexico again became a net importer of hydrocarbons. Even though prices were low, Reyes Heroles dismissed as "uninformed" those who favored the importation of cheap crude. "To change the Mexican policy of seeking self-sufficiency . . . to one that emphasizes 'complementary' imports of crude," he said, "would be to change from a policy of independence to one of dependence."[53]

To promote self-sufficiency, he encouraged the output of refined products and petrochemicals. Our interest, he insisted, is "not to obtain more primary or natural yield of a larger amount of crudes, but to reprocess, and to again refine residues which can give us more valuable products. Not simply more barrels in order to obtain the natural yield of crudes, but greater yield per barrel. For every peso derived from highly refined gasolines, over 50 could be obtained from petrochemical products."[54] Domestically produced petrochemicals obviated the need for 712 million pesos worth of imports between 1959 and 1964 and 3,999 million between 1965 and 1970. Reyes Heroles predicted that after mid-1971, with the completion of plants to produce ethyl benzene, polyethylene, and acetaldehyde, Mexico would no longer have to purchase basic petrochemicals abroad.[55]

The director general's nationalism also inspired support for a special facility to give Mexico its own research and development capability and furnish technological services to the petroleum sector. Greatly needed was an institute to do things that PEMEX did not do properly, such as work on project designs for exploration installations that company engineers simply did not have the time to perfect. In a speech to the Mexican people, Reyes Heroles bemoaned PEMEX's technical and scientific deficiencies which meant, for example, that the firm did not even know the precise chemical composition of its own crude. The Mexican Petroleum Institute (Instituto Mexicano del Petróleo—IMP), strongly endorsed by Reyes Heroles and Ing. Francisco Inguanzo Suárez, the key man in PEMEX's Production Department of which he was subdirector between 1969 and 1976, was established in an August 23, 1965, presidential decree and began operations on March 18, 1966. Its first director was Ing. Javier Barros Searra, who had served as minister of public works under López Mateos. The IMP was charged with (1) performing hydrocarbon-related research that accentuated applied science, (2) training the industry's personnel, and (3) forming a bridge between technical experts and the petroleum sector.[56]

The institute grew in subsequent years from a fledgling operation with a small technical staff to a sophisticated, internationally respected research and development center employing three thousand men and women, twelve hundred of whom are professionals. In contrast to the frequent criticism directed at PEMEX, the IMP has consistently won accolades for its achievements. Friction has occasionally developed between the institute and PEMEX. One such dispute broke out over the responsibility for reservoir engineering, which PEMEX engineers felt better suited to accomplish because of its close relationship to daily operations. They believed the IMP to be more attentive to theoretical, long-range concerns. Still, the two agencies have generally worked hand in glove, with the institute tailoring its research and training programs to the state monopoly's operational requirements. It has justified its establishment by doing on its modern, attractive Mexico City campus what formerly had to be done under contract with private, often foreign, firms. The IMP regularly and efficiently provides project design and engineering services for plants and facilities, laboratory analyses, and special studies in secondary recovery as well as rendering scientific support in seismological, geochemical, and petrographic undertakings. It trains scientists and technicians from foreign countries with which Mexico exchanges research findings; keeps the industry abreast of technological advances in other countries; publishes a leading technical journal; conducts hundreds of courses to improve the skills and advance the careers of PEMEX employees; and has ninety patents to its credit. Investments on industrial plants at home and abroad which employ processes patented by the institute exceed the sum of seven billion pesos.[57]

Reyes Heroles also built up the PEMEX fleet, which began operations in 1938 with only one tanker and 41 smaller ships. Soon thereafter it acquired several used vessels, many of which already operated in Mexican waters. During World War II, 5 tankers were sunk, including the *Faja de Oro* and the *Potrero del Llano*. Anxious that the fleet meet the industry's needs with ships bearing the green, white, and red Mexican flag, Reyes Heroles expanded its size from 191 to 211 vessels and thereby increased its capacity from 191,670 tons (1964) to 264,294 tons (1970). Equally important, he oversaw the modernization of the fleet, which in 1978 numbered 213 vessels with a 476,618-ton capacity.

If Reyes Heroles did not mastermind the strident nationalism of the last half of the *sexenio,* he contributed mightily to it. The always tense relations between private enterprise and the "revolutionary" government perceptibly improved during Díaz Ordaz's first three years in office. The director general had urged that there be "complementarity and not conflict between the private and the public sector" and that communication and coordination be stressed.[58] The ostensibly favorable climate saw the return of the traditional two-to-one ratio between private investment and that of the state, whereas it had almost been one-to-one under López Mateos.

Then the sparks began to fly. The increasingly prosperous business community apparently felt that greater wealth entitled it to a larger role in decision-making. This precipitated a reaction from the president of the PRI, who declared in early 1968: "There is a minority which tends to concentrate economic power in its hands. This is unjust, but the injustice would be disastrously aggravated if that minority were also to monopolize political power."[59] Reyes Heroles reiterated this message at the ceremony commemorating the thirtieth anniversary of the expropriation. He brusquely warned the private sector that it should

> reinvest, in lieu of squandering, pay taxes satisfactorily, comply with social legislation, and forget "company unions," make its own decisions in its own businesses and not operate as an errand boy for foreign capital, choose limited, steady profits; act as an articulate portion of Mexican society, and not as a pressure group; remember that property, in Mexico, is subject to social function[s], and finally, understand the solidarity which unites and eliminates selfishness, which isolates.[60]

The presence on the speaker's platform of Díaz Ordaz, next to whom was seated the redoubtable General Cárdenas, gave official sanction to this reprimand.

The government did not confine its criticism to domestic businessmen. At a public gathering in Mérida, Díaz Ordaz employed a time-proven revolutionary idiom by denouncing foreign "monopolies" in Yucatán and vowing to take action against them. A London newsletter reported that the government apparently incited the textile workers to call a strike against their employers.[61]

Reyes Heroles extended this onslaught to the petroleum field. Soon after taking office, he commenced negotiations to terminate voluntarily the risk contracts concluded by Bermúdez with CIMA and other U.S. companies. On June 5, 1969, PEMEX rescinded the offshore, onshore, and sales contracts with CIMA, which was paid $18 million for its rights, goods, equipment, data, and "anticipated" proven reserves.[62] Later it rescinded contracts with Sharmex, the Isthmus Development Company, and Pauley Noreste. "In this manner, President Díaz Ordaz liberated without any species of limitation 3,858 square kilometers of National Territory for the exclusive utilization of Petróleos Mexicanos, for the benefit of the nation."[63] The director general argued that PEMEX's technical and financial progress obviated the need for foreign involvement in the industry. The presence of a drill ship flying the Mexican flag at Gaviota 1, an offshore well, symbolized the company's new status.

Reyes Heroles also announced that at the president's "expressed indications," he would eliminate another situation that obscured "absolute nationalization." By this he meant the recision of twenty-two agreements concluded between 1960 and 1962 that allowed independent producers, some of whom were foreign, to drill for oil. PEMEX paid 7.5 million pesos as part of this settlement.[64] Finally, PEMEX took complete control of the production of basic petrochemicals in a move that allegedly "undid" much of the work of Gutiérrez Roldán.[65]

Reyes Heroles's interest in reform is well known. Most recently, as secretary of the interior (1976–1979), he successfully pressed for legislation to alter the electoral system and expand the number of seats to assure representation for newly recognized minority parties in Mexico's Chamber of Deputies. As head of PEMEX he quickly divested the monopoly of "Mr. Ten Percent," who had profited so greatly from conferring private contracts. He also pensioned off corrupt union leaders, warning that he would "cry at their funeral" ["llegaría a llorar sobre sus tumbas"] if they refused retirement. In his first major address, he reminded the union that discipline was a prerequisite to liberty. "Irresponsibility, unjustified absences and lack of uprightness, contradict the moral dignity of work, as well as the dignity of legitimate rest, to which all of us have a right, after our day's work," he said. Later references in the same speech seemed to cast doubt on the capacity and honesty of the leadership of the labor movement.[66] In comments to the press, the director general also criticized the practice of selling of jobs. These statements infuriated La Quina, who—when asked to evaluate Reyes Heroles's performance—made a spitting sound.[67] The labor chieftain encouraged Rafael Cárdenas Lomelí, then the STPRM's secretary general, to persuade the head of PEMEX that the union knew corruption existed and was prepared to clean its own house without his sullying the reputation of all labor leaders.[68] Cárdenas Lomelí proved too weak either to confront Reyes Heroles or to combat effectively the corruption within his organization.

Reyes Heroles found an ally for his cause in Cárdenas Lomelí's successor, Samuel Terrazas Zozaya, a labor activist from Poza Rica who served as secretary general from 1968 to 1970. Growing protests by transitory workers pressured to pay two thousand to three thousand pesos for ninety-day employment encouraged the two men to begin their clean-up campaign with an attack on *vendeplazas*. At first it was a "terrible struggle" against Hernández Galicia and his entourage, who finally withdrew in the face of the strong backing that Reyes Heroles gave Terrazas, only to reemerge after both adversaries had left office.[69] The STPRM's top post rotates every three years among the largest locals in the northern, central, and southern zones. As it happened, a representative of Ciudad Madero's Local 1 (northern zone) was next in line to assume leadership. La Quina wasted no time in placing Salvador Barragán Camacho, his closest associate, in this position. They soon undid all that Terrazas and Reyes Heroles had attempted to accomplish.[70] Reyes Heroles would surely head any "enemies list" kept by La Quina.

Reyes Heroles contributed notably to the workers' well being. He improved PEMEX's poor safety record, extended health coverage to transitory workers, and increased medical and hospital services (including an expansion of the kinds of specialized treatment available to all employees). The director general's previous service in the Mexican Social Security Institute made him sensitive to the occupational safety and health requirements of the nation's labor force.

Reform also appeared in the widespread introduction of computers and the exercise of central control over twenty-seven warehouses so that PEMEX could determine the size and composition of its inventory, eliminate old or surplus materials, and avoid unnecessary purchases. This innovation saved the company over one billion pesos,[71] although hundreds of millions of pesos are still lost annually as obsolete parts gather dust in remote storage centers. The new data-processing equipment enhanced efficiency with respect to personnel management and the payroll (PEMEX began distributing paychecks every fourteen days, on Thursdays, instead of once monthly).

"Stubborn reality is the grave of the perfectionist's efforts," Reyes Heroles once said.[72] As head of Petróleos Mexicanos, he proved himself a realist, pragmatist, and astute politician. Despite a genuine interest in reform, he used the enterprise as a source of patronage, contributing to the 33-percent expansion in personnel between 1965 (53,973) and 1970 (71,062). More serious, in the view of career professionals, was a readiness to place political appointees in three of the five subdirectorships, the company's key management posts, as well as in other high-level positions, some of which were newly created. For example, he established an Office of Industrial Security (Gerencia de Seguridad Industrial) whose head, the son of former labor minister Salomón González Blanco, was, in organizational terms, on a par with that of the Office of Production (Gerencia de Explotación), the heart of the company with nearly half its personnel. Assistants to executives also gained appointments on political grounds. Some of the inexperienced appointees proved effective; many, including a number who still work in Petróleos Mexicanos, did not.

The self-serving pragmatism of the six previous years gave way to aggressive nationalism under Reyes Heroles. The professional politician returned PEMEX to its revolutionary path by increasing exploratory drilling, creating the Mexican Petroleum Institute, enlarging PEMEX's fleet, cancelling foreign contracts, and asserting control of basic petrochemicals. Ironically, during this *sexenio* Mexico again became a net importer of hydrocarbons: an odious symbol of dependence. Reyes Heroles greatly expanded the firm's work force, with more political appointees than PEMEX had ever seen, while seeking to extirpate the most conspicuously venal practices of labor chiefs. This latter effort, which proved temporary, produced a nadir in labor-management relations even though Reyes Heroles advanced the physical welfare of the *petroleros*.

Mexico Regains Exporter Status

Luis Echeverría Alvarez won the 1970 presidential election and replaced Reyes Heroles, with whom he had attended law school although the two men were not close, as director general. He later named his former classmate to the presidency of the PRI. To fill his place at Petróleos Mexicanos, Echeverría appointed Antonio Dovalí Jaime, a sixty-five-year-old engineer from Zacatecas who was a foremost authority on bridge construction, a subject that he had taught at UNAM's National School of Engineering and the National Military College. He had previously served as director general of railroad construction in the Ministry of Public Works, subsecretary of public works, and director of construction for the Chihuahua-Pacific Railroad. His closest connection to the oil industry had come between 1966 and 1970, when he served as director of the Mexican Petroleum Institute. Former president Alemán strongly recommended Dovalí Jaime for the PEMEX post.

The new director general consistently appeared as a low-keyed, apolitical professional in comparison to Echeverría, an unremittingly volatile and controversial chief executive. Normalcy, not nostrums, characterized the enterprise during the *sexenio,* when Ing. Francisco Inguanzo, the experienced head of production and a scrupulously honest member of the "Generation of '38," helped to fill the painfully obvious vacuum in leadership that beset the firm.

Echeverría, consistent with his sometimes wild optimism, set ambitious goals for the company at the outset of his administration. Specifically, he announced an eighteen-billion-peso investment plan to (1) locate new reserves, (2) raise oil and gas production over 8 percent annually, (3) attain self-sufficiency in crude, and (4) quadruple the output of petrochemicals.[73] PEMEX managed to achieve the first three goals, although it fell short of the objective in petrochemicals.

Even without a dynamic leader at the head of the state oil company, Mexico's energy fortunes soared in 1972 with oil discoveries in the Villahermosa area. Pearson's El Aguila company had sunk shallow wells in this region between 1911 and 1913 but had failed to locate commercial quantities of crude. The imperative to find new reserves in the 1960s led PEMEX, employing modern deep-drilling procedures and seismological techniques such as the "stacking" of data from signals generated by induced explosions or vibrations, to intensify its activities in the Southeast. The success of this effort was evident on May 15, 1972, when the company announced two important discoveries in Chiapas state: Sitio Grande 1, about three miles south-southeast of the town of Reforma, and Cactus 1, about two miles north-northeast of the same village. The first well tested 1,720 bpd of oil and 3,800 Mcf of gas from a 524-foot-thick section of fractured and permeable carbonate structure; the second yielded 2,550 bpd and 5,800 Mcf.[74]

These discoveries, revealing oil over 13,775 feet below the thick tropical vegetation that blankets much of the states of Tabasco and Chiapas, alerted engineers to the presence of the prolific Reforma trend, often compared favorably to Alaska's Prudhoe Bay.

PEMEX carefully mapped the area, employing various seismological methods, and recorded 125 to 150 potential oil structures, many of which proved to be rich producers. Reforma's output of oil and gas rose from 10 million barrels (1973) to 62 million (1974) to 118 million (1975) to 164 million (1976), when daily production averaged 451,000 barrels. The decision by OPEC to boost the price of oil fourfold gave an impetus to production activities in a nation whose oil bill shot up 140.7 percent to 3,600 million pesos between 1973 and 1974. By 1976 the Villahermosa area boasted eighty-one producing wells. The exceptional productivity of the wells in Chiapas and Tabasco led to the area's designation as "little Kuwait." While the nation's older wells, some of which date from the 1940s, yield an average of 120 bpd, Sitio Grande, Cactus, Níspero, and Río Nuevo in Chiapas were providing 2,808 bpd, and

Samaria, Cunduacán, and Iride in Tabasco were furnishing 9,672 bpd by
1976.[75] The Reforma trend contributed to a rise in daily gas production from
1,822,000 Mcf (1970) to 2,114,000 Mcf (1976).

The new oil province enabled Mexico to regain its status as a net exporter on
September 17, 1974, thereby freeing the nation from the "ominous consequences of a burdensome and servile dependence with regard to the supply of
indispensable oil."[76] During the first half of 1974, Mexico imported 33,884
bpd of crude; by the end of the year it was exporting 60,000 bpd, a figure which
had doubled by mid-1975.

Three other events marked 1974 as a red-letter year: On January 8 Dovalí
Jaime announced a 36,000-million-peso development program for the last three
years of his term; the first increase in PEMEX's domestic prices in fifteen years
took effect; and annual yields of crude and liquids climbed to 238.3 million
barrels to surpass, for the first time in the state enterprise's history, the record
1921 output. Also encouraging was exploratory work in Campeche Bay. The
success of Chac 1, a well offshore from Ciudad del Carmen, offered the
prospect of another vast new production zone similar to the Poza Rica fields
and the promising Reforma trend. This topic will be discussed in chapters
3 and 9.

In 1971 Dovalí Jaime had forecast that proven reserves would total 9.1
billion barrels in five years. Despite the vertiginous rise in crude output,
reserves increased only from 5.6 billion to 6.3 billion barrels during his tenure.
Once again the conservationists within PEMEX imposed their cautious line: It
was imperative to husband reserves for future generations and to guard against
improper exploitation that would damage the country's reservoirs.

The surge in output necessitated new processing plants and pipelines. The
most important facility, begun under Reyes Heroles but completed during
Dovalí Jaime's term, was the 150,000-bpd Tula refinery, which cost 3,400
million pesos. Work also began on refineries at Salina Cruz, Oaxaca
(170,000), and Cadereyta, Nuevo León (235,000), and on a 205,000-bpd
increase in the capacity of the Minatitlán refinery. This work, to be completed
by the end of the decade, was designed to assure the nation self-sufficiency in
gasolines, fuel oils, lubricants, and other products as well as surpluses for
export. PEMEX's capacity for primary distillation and fractioning of absorption liquids rose from 592,000 bpd in 1971 to 760,000 in 1974 to 968,500 in
1976.

Dovalí Jaime initiated construction of a petrochemical complex at La Cangrejera in Veracruz state. Its nineteen plants and their projected 2.8 million-
ton-per-day capacity would give Mexico undisputed leadership in petrochemicals within Latin America. It was anticipated that by the end of 1976, Mexico
would eliminate imports of ammonia, methanol, acetaldehyde, ethylene oxide,

polyethylene, vinyl chloride, acrylonitrile, and other products vital to agricultural activities and the secondary petrochemical industry.[77] Despite a 70-percent increase in installed capacity, the projected date for self-sufficiency in basic petrochemicals was later postponed to the end of the decade. Nonetheless, Mexico produced 13,678 million tons with a commercial value of 16,861 million pesos during the first five years of Dovalí Jaime's administration.[78]

PEMEX also made great headway in pipeline construction to improve efficiency in handling large volumes of crude oil and gas. The most ambitious of the new facilities was a 30-inch, 372-mile pipeline that arched across the Gulf plain from Cárdenas to Poza Rica. It made possible the conveyance of crude from the Southeast to the Salamanca, Azcapotzalco, and Tula refineries, thereby reducing the burden on PEMEX vessels operating between Pajaritos and Tuxpan.[79] Among other projects completed were a second pipeline from Ciudad PEMEX to Mexico City (and from there to Salamanca), a Salamanca-Guadalajara product line, an additional Minatitlán–Salina Cruz pipeline, and a crude line between Ciudad Madero and Cadereyta. Mains to gather and handle the expanding output spread like steel ribbons across Chiapas and Tabasco.[80]

Unlike his predecessor, Dovalí Jaime avoided confrontations with the Oil Workers' Union, to which he was all too ready to capitulate. The president occasionally talked about the problems of vice and the need to change "not only the political structure, but the [country's] mental structure." He cautioned that "we must overcome old modes of thinking and ways of acting that have already given their best fruits."[81] Nevertheless, it was business as usual—or, better said, "corruption as usual"—within labor's ranks. The Manpower Budget Committee established to monitor and control the growth of the work force was clearly for cosmetic purposes as the number of PEMEX employees jumped by over 30 percent, from 71,737 in 1970 to 94,501 in 1976. Dovalí Jaime attributed the large and growing employment level to the company's "completely integrated" character—a "virtually unique instance within the world petroleum industry."[82] The surging output in Tabasco and Chiapas offered a more appealing but still unconvincing justification. As table 3 shows, salaries and fringe benefits—housing, health care, recreation, education, consumer stores, etc.—grew at a phenomenal rate during this period. The timid PEMEX chief readily acceded to the demands of STPRM representatives who made the threat, dismissed by many as posturing, to paralyze the industry with strikes.

The issue of number of workers and their compensation paled in comparison to another change accomplished during the *sexenio*. In response to the union's demand, Echeverría agreed to require most professionals, technicians, and other white-collar employees to join the STPRM. The new ruling affected

TABLE 3
Employment, Salaries, and Benefits, 1958–1976

Director General	Period	Avg. No. of Regular Workers (Planta) in Period		Avg. No. of Temporary Workers (Transitorios) in Period		Total No. of Workers			Salaries Paid (millions of pesos)		Fringe Benefits (millions of pesos)		Total Compensation (millions of pesos)		
		Beginning	End	Beginning	End	Beginning	End	% Change	Beginning	End	Beginning	End	Beginning	End	% Change
Pascual Gutiérrez Roldán	1958–1964	28,668	33,472	16,864	16,900	45,532	50,732	10.63	545,712	846,580	552,003	1,005,656	1,197,745	1,852,236	68.73
Jesús Reyes Heroles	1964–1970	33,472	43,728	16,900	28,009	50,372	71,737	42.41	846,580	1,981,988	1,005,656	1,859,205	1,852,236	3,841,193	107.38
Antonio Dovalí Jaime	1970–1976	43,728	50,634	28,009	43,866	71,737	94,501	31.73	1,981,988	6,488,604	1,859,205	4,802,513	3,841,193	11,291,117	193.95

Source: Petróleos Mexicanos, *Anuario estadístico 1976* (Mexico City: Petróleos Mexicanos, 1976), p. 35.

engineers, seismologists, geologists, geophysicists, section chiefs, and many others who once served as *empleados de confianza*. This step has led to a decline in output by these skilled employees, according to the former president of the Mexican Institute of Chemical Engineers.[83] The influence of the STPRM in the running of the industry will grow even stronger as these individuals advance to important management positions.

Dovalí Jaime signed the unionization order "under protest" at about 3 o'clock in the morning on March 18, 1976, after hours of discussion that included the labor minister and secretary general of the union. However, had he stood firm against the president on this matter, many career professionals were prepared to offer their resignations en masse as an act of solidarity. The director general rejected the suggestion, offered by key lieutenants in PEMEX, that he confront Echeverría with his own letter of resignation, as well as those of a half-dozen subdirectors and managers, in an effort to block the move. Especially outspoken was Francisco Inguanzo, subdirector of production, who had strongly urged the president, when both were in Cancún in early March to brief Marshal Tito on Mexico's oil prospects, to abandon the syndicalization scheme. Distaste for the union's behavior may have motivated cryptic references to "internal blemishes" and "administrative maladjustments" found in his final report to the nation.

Dovalí Jaime did respond to other advice from subordinates, and he evinced leadership in another area. He was the first director general to acknowledge publicly his company's "responsibility as one of the major contributors to environmental pollution."[84] PEMEX's products clearly encouraged the 14-percent annual growth in the number of automobiles; at the same time its refineries, processing centers, and other installations spewed pollutants into the air. By 1970 Mexico City earned distinction as one of the most smog-infested places in the world; nearly five thousand tons of carbon monoxide, hydrocarbons, nitrogen oxides, sulfur dioxide, and other compounds poured into its air each day. The Azcapotzalco refining center, located in the northwestern portion of the city, contributed sixty-seven tons of hydrocarbons and thirty tons of sulfur dioxide daily. In the interest of the capital's air quality, Dovalí Jaime decided to obtain additional capacity by enlarging several other refineries and constructing new ones instead of expanding the Azcapotzalco facility.[85] He also tried to discourage the construction of so many thermoelectric plants, for these not only contribute greatly to air pollution but also consume enormous quantities of fuel oil and gas.

At the urging of Ing. Hector Lara Sosa, subdirector of industrial production, Petróleos Mexicanos created the Committee for Environmental Protection in 1971 to determine the extent of the problem nationwide, raise the company's consciousness of it, and bring pollution within lawful limits. As will be

discussed later, PEMEX has consistently subordinated environmental protec
tion to production. Yet Dovalí Jaime ordered the installation of oil collectors
and separators to prevent water contamination, improved the Salamanca and
Minatitlán refineries to achieve the same goal, and placed special equipment at
three marine terminals (Pajaritos, Salina Cruz, and Tuxpan) to accomplish
primary treatment of the ballast water removed from tankers.[86]

The director general implored the nation to use renewable energy supplies
more prudently. He urged industry to economize by operating more efficiently
and repairing defective equipment. He called for better coordination of public
transportation and improved maintenance of automobiles and suggested reductions in public and domestic lighting. "In brief, the whole population of the
country bears unavoidable responsibility for this change in attitude and mentality, which, in making us conscious of the wealth we possess, will lead to its
better use."[87]

The increase in domestic petroleum prices helped to ease PEMEX's day-
to-day financial burden. But growth in output attracted the interest of foreign
financial institutions. By March 18, 1976, Dovalí Jaime could note with
satisfaction that PEMEX enjoyed "ample credit" from domestic and foreign
sources.[88] Wall Street appeared especially bullish on Mexico. David Rockefeller, president of the Chase Manhattan Bank, described the country as "sensa-
tional, one of the best international credits in the world." Such tributes may
explain why PEMEX's first foreign bond issue—$20 million on the London
capital market—was oversubscribed.

Fortuitously, Dovalí Jaime's administration represented a watershed in the
evolution of Petróleos Mexicanos. Before the early 1970s, it had been a
struggling firm whose conservationist doctrine had become an article of faith,
except for the Gutiérrez Roldán period. Growing domestic demand, the need to
resume hydrocarbon imports, spiraling international prices, and economic
problems at home coincided with the development of the Reforma fields. A
sharp increase in union salaries, benefits, and membership paralleled the rise in
reserves and production. Pressures mounted to forsake both cautious husbanding of the country's hydrocarbon wealth and an emphasis on selling products in
favor of massive shipments of crude to earn badly needed foreign exchange.
Mexico's oil industry stood on the threshold of a new era.

3

The Oil Policy of José López Portillo

López Portillo Takes Office

As millions of Americans celebrated their nation's Bicentennial on July 4, 1976, Mexican voters took part in their own patriotic ritual: the election of a president. As predicted, José López Portillo overcame three "independent" challengers, including a Communist candidate, to secure victory for the Revolutionary party, PRI, which has captured every presidential contest since its founding in 1929. He received 94.4 percent of the votes cast. Still, a high abstention rate marred the results for the fifty-six-year-old lawyer, scholar, political scientist, and former finance minister, who owed his nomination to outgoing President Echeverría. Nearly three out of every ten eligible citizens (28.6 percent) defied a compulsory voting law and failed to exercise their franchise. Detractors of the regime claimed that a large number of spoiled ballots were also cast. Mass absenteeism plagued the July 1, 1979, congressional elections as well; over 50 percent of the 27.9 million registered voters stayed home. Although López Portillo attributed this phenomenon to "satisfaction with the government's leadership of the country," for the first time since its formation, the official party failed to win a number of votes in a national contest greater than the number of those who abstained.[1] Table 4 shows the percentage of abstentions in recent elections.

The public also registers its distrust of politics, parties, and politicians through responses to opinion surveys. In May 1979 a pollster for the *Los Angeles Times* asked a cross-section of Mexican adults who is "most to blame" for the nation's failure to achieve "the progress it deserves." The respondents placed "bad government" (35 percent) at the head of the list, followed by "foreign exploitation" (19 percent) and "corruption in the bureaucracy" (17 percent).[2]

Such answers spring in part from disenchantment with a system dominated by a powerful political machine. The PRI, virtually synonymous with the government, appears more committed either to advancing the economic and political interests of its leaders or coopting and suppressing adversaries than to uplifting the masses, who live in abject poverty. The absenteeism, vote spoliation, and critical responses also reflect concern over conditions in a country where the economy had registered an "economic miracle" for twenty-five years after the end of World War II. In this period the Gross Domestic Product shot up 6 or 7 percent annually.

As López Portillo prepared to assume the presidency, he witnessed the worst economic conditions in Mexico since the Great Depression. The population was growing faster than the economy. Mexico's then sixty-three million inhabitants made it the world's tenth largest nation, and the number of people was expected to double by 2000, when Mexico City—now third behind Tokyo and Shanghai—will become the planet's foremost metropolis. This steep rise (3.2 percent in 1976) meant that every other Mexican was under fifteen years of age. Soon these youngsters would be elbowing their way into a labor force half of whose members already lacked work or were marginally employed. It also meant that U.S. cities would continue to attract millions of "undocumented" emigrants from Mexico, where 17 percent of the people earn less than $75 a year, and three out of ten cannot afford a minimum balanced diet.[3]

Echeverría, elected president in 1970, had pledged to free his country from dependence and to stimulate economic growth. He attempted to create jobs, boost incomes, and breathe life into the egalitarian goals of the 1910 revolution by encouraging public investment in everything from steel-making to sky-scraper construction. He placed particular emphasis on infrastructure projects requiring heavy capital-goods imports. Mexico had—and still has—one of the world's lowest taxation levels, and Mexican investors, alarmed by the chief executive's leftist rhetoric, spurned government securities. As a result the

TABLE 4
Abstentions in Mexican Elections, 1961–1979

Year	Election	% Abstentions
1961	Congressional	31.5
1964	Presidential	20.0
1964	Congressional	33.3
1967	Congressional	37.4
1970	Presidential	36.0
1970	Congressional	35.0
1973	Congressional	36.2
1976	Presidential	28.6
1976	Congressional	38.1
1979	Congressional	50.0

Sources: Proceso, June 25, 1979, p. 24 (1961, 1964, 1967, 1970, 1973, and 1976 congressional); *Visión,* July 28, 1979, p. 14 (1979 congressional—estimated); the *New York Times,* July 7, 1964, p. 12 (1964 presidential—estimated); *Facts on File,* July 18, 1970, p. 544 (1970 presidential—estimated); and *Facts on File,* July 10, 1976, p. 498 (1976 presidential).

development program was financed largely by an expansionary monetary policy and external borrowing.

The upshot was that the public debt rose fivefold to $20 billion under the mercurial Echeverría, while private-sector obligations surpassed $7 billion. Debt service as a percentage of exports climbed from 23.6 percent in 1970 to 32.3 percent in 1976, and the annual interest on the external public debt rose from $217 million to $1.07 billion. The burden of these loans, over 50 percent of them made by American financial institutions, becomes apparent when one remembers that the nation's GDP was only $79 billion in 1976. With the possible exception of Brazil, Mexico remained the most debt-ridden populous country in the world. A confluence of factors—the size of the debt, Mexico's propensity to import more than it exported, and double-digit inflation—brought a halt to private investment, massive capital flight, a prodigious flow of funds into real estate and other currencies, and two devaluations of the vaunted peso in late 1976. On the eve of López Portillo's December 1 inauguration, wild tales filled the capital that the generally quiescent armed forces would stage a coup d'état to prevent his taking office. These rumors proved false. Yet never in recent memory was the country more beholden to international financial institutions and the United States and its banks.

Oil to the Rescue

A communication dispatched to López Portillo during the campaign proved as disquieting as the worsening social and economic conditions described above. In this document a group of petroleum engineers warned that reserves were such that Mexico, which had begun selling oil abroad in late 1974 after importing it for six years, would become an importer again before his term ended in 1982.[4]

Petroleum could be the solution to Mexico's problems. But if this report were true, there might be no way for the country to regain its economic momentum. It faced the prospect of defaulting on loans, watching inflation corrode its badly shaken currency, and having its political institutions undermined—of becoming, in short, a Spanish-speaking Weimar Republic, dependent on foreign largesse.

López Portillo, an extremely forceful and determined man, recoiled from this prospect. He believed that the problem might lie less with meager reserves than with the cautious, conservationist policy of Petróleos Mexicanos, many of whose executives wished to husband reserves for future generations rather than embark on a vigorous program of export promotion. It also seemed curious to him that reserves had increased only 12 percent in the last *sexenio*, from 5.57 to

6.34 billion barrels, while production had nearly doubled thanks to the new discoveries in Chiapas and Tabasco. This phenomenon was explained largely by rising economic consumption. To gain a fresh assessment of his country's hydrocarbon wealth, the PRI standard bearer turned to Jorge Díaz Serrano, a friend since boyhood whom he described as "honest, efficient and patriotic."

Díaz Serrano had accumulated a fortune in the petroleum industry. Born in the border town of Nogales, Sonora, in 1921, he won a scholarship to study mechanical engineering at the National Polytechnic Institute in Mexico City. When Díaz Serrano was a student, Cárdenas seized control of the oil industry in a vividly remembered move: "I was more interested in sports than politics, but it was a very emotional moment for everyone. We all went to demonstrate in support of the measure. The companies seemed all-powerful and we were filled with a great sense of patriotism when the president had the courage to confront them. It seemed like our economic independence, our true independence."[5]

Díaz Serrano later worked for the government's national irrigation commission before studying at the University of Maryland under the auspices of the U.S. Department of Commerce's Inter-American Training Administration. He was subsequently employed in a Pennsylvania tool factory. Upon returning to Mexico in 1946, the ambitious young engineer spent a decade with Fairbanks Morse de México, a subsidiary of an American engineering firm. He began as a machinery salesman, spending a great deal of time in the Southeast, where his entrepreneurial talents flourished. "Fairbanks Morse sold machinery but didn't install it, so I set up a company to install the equipment. Then I'd be asked for machinery we didn't make, so I set up another company to build machinery."[6]

Díaz Serrano developed many contacts with PEMEX, the major customer for his oilfield devices. When associated with Dresser Industries, he manufactured mud pumps, drilling masts, and substructures for the monopoly. Once contract drilling was authorized, he formed companies to take advantage of this lucrative opportunity and had as many as seven rigs in operation at one time. He persuaded PEMEX to use inland barge drilling because of the high cost of rebuilding equipment-scarred roads in Tabasco and Veracruz after every rainy season. From 1969 to 1973, he represented General Motors in Mexico, specializing in sales of locomotives and electric generating plants. In 1965 Díaz Serrano founded the Golden Lane Drilling Company, with offices in Houston and Galveston, to sink wells off Texas, California, and Mexico. He was also owner of or partner in Electrificación Industrial, S.A.; Servicios Petróleos EISA; Perforaciones Marinas del Golfo, S.A. (PERMARGO); Dragados, S.A.; and Cia. Golfo de Campeche, S.A.[7] In addition, he held interests in a Texas bank and fishing fleet. His fluent English and familiarity with the United

States has led one specialist on Mexico to say, "Díaz Serrano is practically a Texan."[8]

Candidate López Portillo named his friend as coordinator in charge of analyzing industrial matters with particular responsibility for energy. Díaz Serrano moved at once to evaluate official reserve estimates, especially those of the Reforma trend, which appeared so promising. He began by meeting with friends in the monopoly's Exploration Department in order to discuss particular wells and fields, secure drilling logs, and obtain core and reservoir information. Díaz Serrano seldom if ever visited Petróleos Mexicanos during the campaign. Instead Jesús Chavarría García, a key official in the Production Department, joined other PEMEX personnel, oilmen from the private sector, and party representatives in the Institute for the Study of Economic and Social Problems, the official party's planning and research organization which worked closely with Díaz Serrano. In a move typical of Mexico's political system, Chavarría's assignment to IEPES from his PEMEX post in Coatzacoalcos was made without the knowledge of his superior, the manager for production,[9] presaging the differences that erupted between the old and new teams at the monopoly.

The Reforma reservoirs—like those of Iran, the North Sea, and the Alaska Slope—occur in carbonated rock formations, which are extremely difficult to analyze. Consequently, Díaz Serrano sought the help of reservoir engineers familiar with the foreign fields as well as the advice of University of California scientists. He placed less reliance on the traditional method of drilling "step-out wells," spaced an appropriate distance apart, depending on the characteristics of the reservoir, in a widening circle around a discovery well until the boundaries of the deposit had been delineated. Instead he claimed to employ new techniques in quantification, certification, and development that sprang from work in the North Sea and Alaska; namely, identifying proven reserves "with the evidence of only two or three key wells per deposit and with high quality modern seismological work."[10] The more he learned from reevaluating data found in PEMEX's headquarters, the firmer became his conviction that Mexico had larger reserves than had been announced.

A wide barrier reef that dates from the Jurassic-Cretaceous period 130 million years ago surrounds Mexico's southern zone. The reef, which gave rise to a large atoll-like island with an immense brackish lagoon in the center, served as a collector of rich layers of marine life whose decomposition would eventually yield oil and gas. Dissimilar in shape from the current land mass, the reef resembles the shallow shelf called the Yucatán platform which juts from the coastline more than 62 miles into the Gulf of Mexico. Over millions of years sediments buried the reef, and horizontal and vertical geological movements produced a complex fault network and folded together the reef sediments with a deeper blanket of salt. This process resulted in sealed structures in which temperature and pressure converted the organic material into hydrocarbons. The reef, composed of highly faulted sedimentary rocks, sandstones, and clays, may extend from the Papaloapan Basin southwest of Veracruz some 186 miles east through Chiapas and Tabasco into the Gulf, where it circles the Yucatán peninsula before winding onshore through Belize and Guatemala.[11]

The reservoirs in the Reforma, characterized by large pay thickness, extreme porosity, high uniformity, and low viscosity, have been called an "oilman's dream."[12] Subsequent drilling showed that a number of separate fields actually formed a single field of 58 square miles with an average productive thickness of over 3,300 feet. As the monopoly's southern zone exploration manager expressed it: "The best way to visualize a typical Reforma or Bay of Campeche field is to imagine a sealed and pressurized tank full of coarse gravel, with a bottom layer of water and then full of oil to the top. The vertical and horizontal flow channels mean the individual wells can sustain high production rates with little if any pressure drop when aided by water-flooding."[13]

Thus in meetings of López Portillo's advisers, Díaz Serrano increasingly attacked the reserve figures advanced by Dovalí Jaime, then director general of PEMEX. The engineer-turned-counselor insisted that oil could be the motor of the country's economic resurgence. As a result of their investigations, Díaz Serrano and his collaborators concluded that Mexico's proven holdings of crude oil, natural gas, and natural gas liquids should be increased from 6,338 million to 11,160 million barrels. Francisco Viniegra, PEMEX's exploration manager, alleged that the company kept the true reserve level from the president lest he or corrupt officials misuse increased earnings. "We were afraid of Echeverría," he said. "Let's face it, he would have given the oil to Cuba and other Communist countries."[14] Viniegra's charge, denied by his superiors in the state firm, has yet to be substantiated. In any case, soon after the inauguration López Portillo announced the new figure, which included neither supplies in the Campeche continental shelf nor nascent reservoirs not yet brought into production.

López Portillo selected Díaz Serrano to head PEMEX during his administration. Before taking the assignment, he reportedly liquidated his financial interests in the several petroleum firms with which he had been associated. López Portillo's faith in his appointee was evident when Heberto Castillo, leader of the opposition Mexican Workers' party, asked him in August 1978 to investigate charges that the oilman had profited from contracts with the government agency. "Jorge Díaz Serrano has been my friend since we were children," the chief executive replied, "and as friends I know he wouldn't deceive me."[15] Once in his new post, Díaz Serrano became a frequent traveling companion of López Portillo and his key energy adviser. This closeness and the salience of oil and gas to Mexico have nourished speculation that Díaz Serrano will become a candidate for president in the 1982 election, despite his lack of a political base.

The new director general not only had to verify his reserve figure, but he had to demonstrate that the $10-billion corporation—the largest in Latin America—could overcome a bloated bureaucracy and ubiquitous corruption to sharply expand output. He brought the same discipline to the firm that he had exercised over his own life when he conquered a serious drinking problem in 1968 and deemphasized money-making for cultural pursuits four years later. He diplomatically denied that a full-scale house-cleaning ensued when he took charge on December 1, 1976. Still, he placed men loyal to him in charge of the firm's major sections (exploration, production, industrial production, commercial, technical-administration, and financial), which are shown in figure 1.

An American contractor, who asked to remain anonymous, observed that under Dovalí Jaime, the failure to meet a goal on time could be ironed out over a relaxed Mexican *almuerzo*. The new team, however, demands action and stands ready to take contracts away from firms unable to fulfill promises. Lengthy lunches have given way to crisp office visits as PEMEX executives seek to avoid public fraternization with foreigners. When bottlenecks emerge the monopoly's operational people identify the area of difficulty, assign a group of young technicians to dissect the problem, and then move with dispatch to resolve it. But the need to pay off key company and union officials in return for economic opportunities is as great, if not greater, than at any time in the industry's history.

The designation of new subdirectors prompted or at least coincided with a decision by Ing. Francisco Inguanzo Suárez, the most prestigious and respected member of the "Generation of '38," to retire (he had offered his resignation on at least two other occasions). "Maestro Inguanzo" began his career as a petroleum engineer with the British firm El Aguila while in his early twenties. He joined PEMEX soon after its formation and rose steadily through the ranks until his appointment in 1969 as subdirector of primary production, with responsibility for both exploration and production. In this position he helped furnish the leadership so conspicuously absent in PEMEX during the Dovalí Jaime years. Inguanzo, who also taught for thirty years at UNAM and the National Polytechnic Institute, enjoyed universal admiration for integrity, modesty, efficiency, and unrivaled knowledge of the industry. Also retiring after the change of administrations was Walter Friedeberg Merzbach, a highly regarded engineer who spent thirty-three years with Petróleos Mexicanos and was manager for production between 1969 and 1976; Francisco Viniegra

FIGURE 1—*PEMEX Organizational Chart*

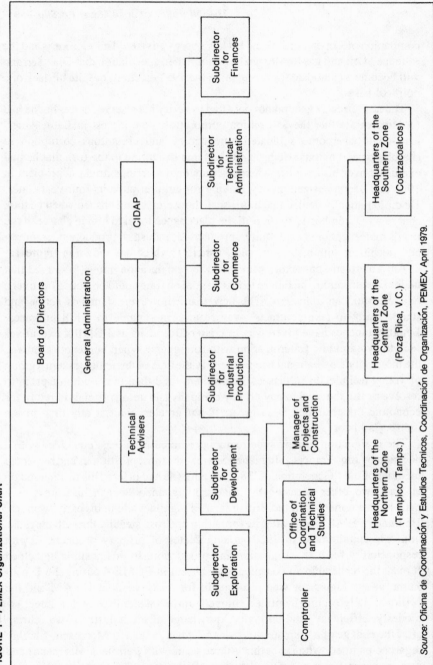

Source: Oficina de Coordinación y Estudios Tecnicos, Coordinación de Organización, PEMEX, April 1979.

Osorio, a PEMEX geologist for thirty-five years and exploration manager under Dovalí Jaime; Antonio Echeverría Castellot, a petroleum engineer and former assistant manager for production who had served the company since 1944; Héctor Lara Sosa, a chemical engineer who had held the post of subdirector for industrial production (refining and petrochemicals) before Díaz Serrano took office; and Mario Hernández Samaniego, an engineer employed by PEMEX for thirty years who held the post of refining manager.

With his men at the helm, Díaz Serrano next sought to apply modern business techniques to improve coordination within the company. As he expressed it in understated fashion:

> Actually there's been no big shake up. Things were going fine when I became Director General and we had excellent geologists and engineers (though not as many as we want). There was, however, a lack of communication between departments. The geologists and the geophysicists were working separately, each on their own, and they were not making a great effort to cooperate with one another. Similarly, exploration was not communicating with production; and production, in turn, was isolated from refineries. Even sales was going its way without communication with refineries.[16]

Hyperbole more than accuracy may have characterized this assessment. Although occasional disputes had occurred between personnel responsible for production and refining, the Production and Exploration departments had cooperated well under Dovalí Jaime. Moreover, open channels of communication existed between geologists and geophysicists, whose activities were coordinated by the exploration manager.[17]

Still, to remedy this "lack of communication," the director general began convening regular meetings of all the subdirectors and managers. He thereby made certain that his administrators kept abreast of major problems and that they clearly understood "our over-all goals and . . . their responsibilities."[18] The transformation of the firm is far from complete. "There are still many corners of inefficiency, some corruption and low productivity that have resisted the broom of change being wielded by the company's director general."[19] Nonetheless, PEMEX's public-relations specialists emphasize the strong leadership and contagious enthusiasm of Díaz Serrano to evoke comparisons with the Bermúdez years. In fact, divisions that beset the oil company because of opposition to the accelerated production program advocated by the director general and the omnipresent corruption, discussed in the next chapter, suggest similarities with the administration of Gutiérrez Roldán.

The Six-Year Plan

López Portillo wasted little time in emphasizing the pivotal role of Petróleos Mexicanos and the oil sector. Soon after taking office, he released a six-year program which called for production to climb from 894,219 bpd in 1976 to 2.25 million in 1982, when half of the output would be exported to earn 124 billion pesos per year.[20] Natural gas production is expected to double to 3.6 billion cubic feet per day. The program's entire capital requirements were placed at 310,000 million pesos, or approximately $15.5 billion, more than five times the sum invested in the industry between 1971 and 1976. The pressing need to earn foreign exchange meant that the lion's share—8 percent for exploration and 46 percent for development—was earmarked for production. These funds would finance the drilling of 3,476 new wells—1,324 wildcats and 2,152 development wells—with the major effort (477 development wells) in Tabasco and Chiapas, which embraced two thirds of Mexico's proven reserves. Clearly the Reforma, one of the world's major deposits, would furnish the largest amount of oil and gas.[21] But the program contemplated an expansion of the production base by (1) opening new fields in such established areas as Chiapas, Tabasco, Cotaxtla, and Nuevo Laredo, (2) intensifying offshore efforts, particularly in Campeche Sound, and (3) stepping up geological and seismological work in the Central Plateau, Coahuila, Chihuahua, and the Chiapas Mountains.

Despite a surfeit of refining capacity worldwide, 15 percent of investment was designated for refineries, with a view to nearly doubling PEMEX's capability from 968,500 bpd (1976) to 1,670,000 bpd (1982). The program called for improving and expanding existing installations through technology provided by the Mexican Petroleum Institute. It also specified the completion of new refining complexes at Salina Cruz, to meet the demand on the Pacific coast, and at Cadereyta, to serve the North and Northeast.

Meanwhile, the program anticipated a trebling of petrochemical production to fifteen or eighteen million tons per year to enable Mexico to reach self-sufficiency in 1979 and thereafter earn foreign exchange from shipments abroad. The 55 billion pesos (17 percent of the budget) devoted to this sector would increase the number of plants from 60 to 115, with the greatest amount of construction concentrated in petrochemical centers in La Cangrejera and Allende, Veracruz.

As PEMEX expands its product line from thirty-eight to forty-four items, the state firm will also become one of the world's predominant suppliers of ammonia. It is anticipated that production of this vital chemical will grow from four thousand to thirteen thousand tons per day thanks to the construction of seven new plants at Cosoleacaque, Allende, Salina Cruz, and Salamanca.

Sharp increases in output should also occur in ethylene, propylene derivatives, and aromatics, produced from the use of natural gas liquids as feedstocks.

Even with the director general's tireless efforts and promotional skills, can PEMEX carry out the bold plan? Will it be necessary to invite in foreign companies to achieve the program's production goals? Anticipating that the monopoly would need substantial outside help, agents of international companies crowded into Díaz Serrano's office. Their briefcases bulged with "offers of assistance and proposals for joint ventures," according to the *New York Times.*[22]

Díaz Serrano rejected most of the overtures. "We're being offered risk contracts," he told a reporter at the time, "but there is no risk. We know where the oil is to be found."[23] He described as "unthinkable" auctioning off development zones or blocks to private firms. None would be allowed a stake in production. Instead the director general assigned PEMEX's own engineers the chief responsibility for exploration and production. In mid-1977 he contracted with Halliburton's giant subsidiary, Brown & Root, to coordinate the engineering work, construct required land-based facilities, and oversee the purchasing of production platforms, pipelines, and gathering mains to initiate large-scale operations in the Campeche Sound. Unwilling to rely exclusively on foreign firms for this undertaking, PEMEX in early 1978 acquired an additional five offshore rigs to be manned by Mexican crews with foreign advisers. In an unusual display of pragmatism, two of these rigs will be operated by the Texas-based Rowan companies (49 percent equity) in a joint venture with Mexican private investors.[24] Tenneco is also heavily involved in offshore activity, while BS&B de México, S.A., the local subsidiary of an American firm, has won contracts totaling $30 million for the design and construction of gas treatment plants. By January 1978 PEMEX had signed ten agreements with domestic and foreign private drilling companies for equipment and technical assistance; ten more agreements were being negotiated at that time. In late 1979 the monopoly ordered six prefabricated drilling-rig packages for use in the Gulf of Campeche from Lan-Dermott, a joint venture of Empresas Lanzagorta of Mexico and J. Ray McDermott & Company of New Orleans.[25]

Petróleos Mexicanos has modified the six-year program several times since its publication. The company has consistently surpassed production targets as average daily output exceeded projections in 1977 (953,000) and early 1980 (2.1 million). It has raised anticipated investment to 1982 to nearly $20 billion.[26] In September 1978 López Portillo disclosed that his country would recover 2.25 million barrels of crude oil and natural gas liquids in 1980, two years ahead of schedule. Subsequently, a ranking Ministry of Patrimony official stated that output would reach 2.5 million bpd by December 1980.[27] In his March 18, 1980, report to the nation, Díaz Serrano announced that daily output had reached 2.07 million barrels, a figure that rose to 2.3 million by September 1.[28]

Also of interest is the promotion of other energy sources to diminish Mexico's dependence on oil and gas. Greater attention has been given to coal deposits, most of which are in Coahuila; the National Polytechnic Institute in Mexico City has carried out extensive research on the uses of solar power; and a BWR nuclear reactor, designed by General Electric, is under construction in Laguna Verde, Veracruz, with the first of the two units scheduled to begin operation in 1982. Table 5 depicts 1976 consumption patterns and those predicted for the year 2000.

Nevertheless, PEMEX has accelerated its exploration. This has contributed to a major increase in official calculations of proven reserves and estimates of probable and potential holdings between 1976 and 1980. Even more important was the introduction of a new accounting system. Mexico used to consider only "onstream" oil and gas obtained through primary recovery—that is, hydrocarbons from deposits exploited without gas or water injection—as part of its proven reserves. On December 31, 1978, PEMEX adopted the industry-accepted definition of "proven" reserves as quantities capable of recovery through primary and secondary techniques "with reasonable certainty . . . under existing economic and operating conditions."[29] This definition embraces holdings identified by existing facilities as well as the drilled portion of a reservoir and the adjacent area deemed productive as a result of "available geological and engineering data." The conversion helped boost reserves from sixteen billion to over forty billion barrels. By early 1980 PEMEX claimed over fifty billion barrels of proven holdings, forty billion barrels of probable reserves, and two hundred billion barrels of potential deposits.[30]

PEMEX has not changed its calculation of "probable" holdings derived from "an estimate of reserves taking into consideration known geology, previ-

TABLE 5
Mexican Energy Consumption, 1976 and 2000

Energy Source	% 1976	% 2000 (Projected)
Oil and gas	86	72
Hydroelectricity	7	5
Coal	6	12
Nuclear	0	9
Other	1	2
Total	100	100

Source: Data supplied by Ing. Juan Eibenshutz, executive secretary, National Energy Commission; see *Proceso*, December 4, 1976, p. 6.

ous experience with similar types of reservoirs, and seismic data if available." "Potential" reserves are essentially hypothetical and consist of "undiscovered resources that may reasonably be expected to exist . . . under known geologic conditions."[31] Table 6 summarizes the figures disseminated by the state company.

A number of factors bolster the optimism concerning holdings in Mexico, where hydrocarbon areas can be classified as follows: the southern zone (the Reforma area, Campeche Sound, and the Isthmus of Tehuantepec, comprising Isthmus Salinas, Macuspana Basin, and Tabasco); the central Gulf Coast (the Tampico-Misantla Basin—comprising Ebano-Pánuco, the Golden Lane, and Poza Rica—the Veracruz Basin, and Chicontepec); the northeastern basins (Burgos and Sabinas); and other areas.[32] Map 2 shows the location of principal onshore and offshore fields as well as the route of a gas pipeline linking the Reforma fields and northern Mexico.

Southern Zone

PEMEX continues to find new oil in the southern zone, which accounted for over 75 percent of the nation's hydrocarbon production in 1980 (compared to 56 percent in 1974). The Reforma area, which embraces 2,700 square miles, yields over one million bpd from twenty-five fields with daily production averaging six thousand barrels per well. The area also produces 1.8 billion cfd (cubic feet per day) of gas. PEMEX had not delineated the area fully by 1980,

TABLE 6
Official Mexican Oil and Gas Reserves
(Billions of Barrels)

Date Released	Proven Reserves	% Increase	Probable Reserves	% Increase	Potential Reserves[1]	% Increase
1938	1.240					
March 18, 1955	2.609	110.40				
January 10, 1961	4.787	83.48				
December 31, 1965	5.078	6.07				
December 31, 1975	6.338	4.81				
December 31, 1976	11.160	76.08			120	
December 31, 1977	16.000	43.36	31.000		120	0
December 31, 1978	40.194	151.21	44.612	43.90	200	67
December 1, 1979	45.803	13.95	45.000	0.87	200	0
March 18, 1980	50.022	10.97	40.432	-10.15	200	0
September 1, 1980	60.100	20.14	38.000	-6.01	250	25

Sources: U.S. Congress, Senate, *Mexico: The Promise and Problems of Petroleum,* Report prepared for Committee on Energy and Natural Resources, 96th Cong., 1st sess., 1979, Pub. No. 96–2, p. 17; *Daily Report* (Latin America), September 5, 1979, p. M-4. Also used were various reports of the director general.

1. Potential reserves include both proven and probable holdings.

MAP 2 *Principal Fields and Gas Pipeline Route.*

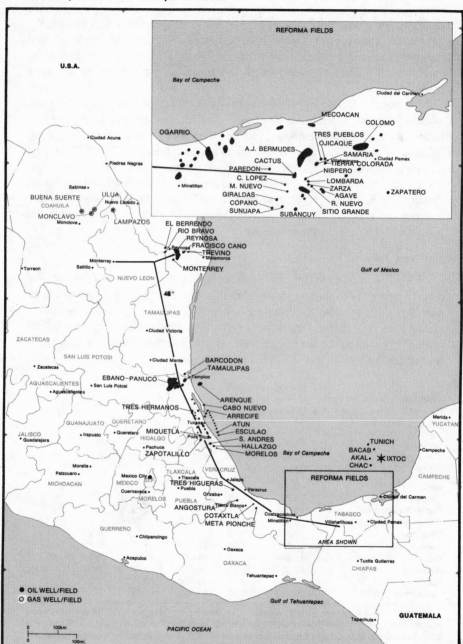

Courtesy of T. J. Stewart-Gordon, international editor of *World Oil*.

but it had identified three principal reservoirs. The northernmost, which extends over fifty-eight square miles and has an average thickness of oil-impregnated limestone of 1,475 feet, has been named for the late director general Antonio J. Bermúdez. Over six hundred thousand bpd flow from this important complex, which is composed of the Cunduacán, Samaria, and Iride fields located in Tabasco state. One well in the Samaria block penetrated more than 6,500 feet of oil-saturated carbonate reservoir.

To the south, in northern Chiapas, is the Cactus reservoir, embracing five fields: Cactus, Sitio Grande, Río Nuevo, Mundo Nuevo, and Nispero. Officials claim that this reservoir covers 50 square miles and boasts an average pay thickness of 1,400 feet. East of Cactus lies the Agave reservoir, discovered in 1977, whose size has yet to be ascertained. Company experts have calculated the thickness of its pay zone as nearly 4,000 feet in some places. PEMEX has announced the discovery of seven other reservoirs of similar dimension which had not been included in proven reserves as of late 1979.[33] One of the most promising of these fields, Iris-Giraldas, covers 4,800 square miles and contains reserves estimated at 1.5 billion barrels.[34] PEMEX believes that fields similar to those of the Reforma will be discovered further to the northeast.

One economist estimated that the cost of locating, developing, and producing oil from the Southeast (Reforma and Campeche) would range between $1.00 and $2.50 per barrel in mid-1978 prices compared with 35 to 50 cents for Saudi Arabia, $6.50 for the most expensive North Sea crude, and $13.00 plus for the highest-cost stripper wells (yielding less than ten barrels per day) in the United States.[35]

Díaz Serrano has announced that the Campeche Sound area, off the west coast of Yucatán, is "the greatest offshore field ever." Less is known about Campeche than Reforma, to which it appears to be linked, but PEMEX has mapped over two hundred seismic structures in the Gulf, "all . . . with surprisingly gentle slopes, and thus quite larger than those of Reforma."[36] The fields of Poza Rica, which cover just over 50 square miles and contain an oil column of 330 feet, produce 114,000 bpd; over one million bpd flow from the developed portion of Reforma, which extends over 138 square miles. Campeche is twice the size of Reforma and boasts a pay thickness of more than 2,100 feet. Its potential is awesome, and by mid-1980, Petróleos Mexicanos had fixed its proven reserves at 14.4 billion barrels.

Akal, the largest field in the prodigious Cantarell complex, where at least thirteen wells were onstream in January 1980, embraces the Akal 3 and Akal 74 wells which, respectively, produce 60,000 and 50,000 bpd.[37] Other areas explored by PEMEX are Ku, Maloob, Kutz, Bacab, Chac, Abkatún, and Ixtoc—all named for Mayan gods. The size of the deposits and the possibility of recovering oil from Jurassic as well as Tertiary and Cretaceous strata led one

PEMEX official to state: "We seem to be facing another startling development. It's like the nurse coming out of the delivery room to tell the anxious man he's the proud father of a baby . . . only to come back a while later to tell him it's twins.[38]

Díaz Serrano insists that it costs far less to lift oil from Campeche than from the North Sea. Production in the former takes place at 230 feet on the average; in the latter it occurs at 500 feet. This means that North Sea platforms must weigh up to thirty thousand and forty thousand tons compared with three thousand in the Gulf. The Campeche fields are nearer to shore than the North Sea fields, which reduces outlays on the 105-mile, 36-inch pipeline that links them to the terminal at Dos Bocas.[39] In addition, the Gulf boasts milder weather conditions than those generally found in the North Sea.

The state enterprise anticipates installing twenty-three drilling platforms, each with twelve slots, to develop the complex. PEMEX set a production target of 200,000 bpd in 1979, rising to 360,000 bpd in 1980, for the offshore fields, whose high-density oil has a low gas-to-oil ratio (GOR).[40] The GOR for most fields (Abkatún and Ixtoc are higher) is 250 to 500 cubic feet of gas to one barrel of oil compared to 7,000-to-one in the Agave fields not yet onstream.[41] However, on August 23, 1980, the director general stated that PEMEX was already lifting 807,000 bpd of crude from Campeche's fields, with output expected to reach 1.3 million bpd by the end of the year.[42] In fact, output from Campeche has enabled PEMEX to lift more crude at a time when it has failed to reach production goals in Reforma. Chapter 9, which focuses on the Ixtoc 1 oil spill, provides additional information about the Campeche area.

Most of the hydrocarbons have probably been found in the Tertiary basins of the Isthmus of Tehuantepec, where production has taken place for many years. The Isthmus Salinas Basin, geologically similar to the Louisiana Gulf Coast, is immediately west of the Reforma area. Filisola was the first accumulation to be discovered in this area, in the early 1920s. The most significant fields here are El Plan, Cinco Presidentes, Magallanes, La Venta, Cuichapa, and Ogarrio. East of the Reforma area lies the Macuspana Basin, where José Colomo and Chilapilla near Ciudad PEMEX are still productive fields. Mecoacán and Tupilco are the most notable accumulations in the Tabasco region.

Central Gulf Coast

The Tampico-Misantla Basin furnished most of Mexico's oil before the discoveries in the Reforma area. The Ebano-Pánuco group of fields contains the nation's oldest producers. Development of the Golden Lane fields began in 1910 and proved to be Mexico's most prolific source of oil for nearly twenty-five years. Cerro Azul 4 attracted world attention for its initial 260,000-bpd yield, and the first Dos Bocas well experienced a blowout that released 200,000 bpd for two months. The output of the Golden Lane has declined in recent years, but as mentioned in chapter 2, PEMEX in 1963 discovered a marine Golden Lane of which Atun proved the largest accumulation. The Poza Rica field about thirty miles south of Tuxpan, discovered in 1930, was Mexico's most noteworthy producer until PEMEX struck oil near Villahermosa.

The Veracruz Basin is west of the Reforma-Campeche reef trend. The productive potential of this area appears extremely limited; however, the Miralejos well in the Papaloapan River Basin near Cotaxtla, Veracruz, produces gas and condensates in geological formations similar to those of the Copite and Mata Pionche fields, with which it joins to form a productive strip of more than fifteen miles.

PEMEX is much more enthusiastic about the 1,275-square-mile Chicontepec area located between Tampico and Poza Rica on the Gulf coastal plain. Díaz Serrano informed participants in the 1978 annual meeting of the American Petroleum Institute, held in Chicago, that the Chicontepec field was "one of the bigger hydrocarbon accumulations in the Western hemisphere." It could contain up to 100 billion barrels of crude and 40 trillion cubic feet of natural gas, he said. The director general reported that the basin had been known to have oil but, it was thought, in small deposits. New technology, however, "led us to detect an erosion canyon filled with tertiary formations . . . with producing intervals of an average thickness of 200 feet. The oil beds vary in depth between 3,000 and 6,000 feet, but because of low porosity and permeability they will have to be developed by the drilling of up to 16,000 wells over 13 years."[43] The initial development plans call for bringing thirty rigs a year into the area for four years. Also needed will be 217 miles of new paved road, 1,553 miles of gravel roads, and a railroad spur to service the zone. The gas production network alone will necessitate some 17,000 miles of pipe. Approximately 212 production batteries will have to be installed to reach full production of 740,000 bpd of oil and 1,200 Mcf of gas.[44]

Northeastern Basins

Burgos is the Mexican portion of a basin that extends into southern Texas and Louisiana. It has never produced as much as 1 percent of the country's oil, but it contains extremely important gas fields, including Reynosa, on the Texas border, and Brasil, southeast of Reynosa. Treviño, a few miles to the east, is the basin's most notable oil deposit.

PEMEX has verified the presence of prodigious new gas fields in the Sabinas Basin of Nuevo León (Lampazos) and Coahuila (Monclova and Ulúa), thus permitting the commercial production of hydrocarbons outside the Gulf coastal plain and the Gulf of Mexico for the first time in the country's history. The productivity of the wells in the 15,440-square-mile basin makes the area extremely attractive. Sabinas wells average about 3,000 Mcf per day compared to 1,000-Mcf yields of wells in the Reynosa district to the northeast. Moreover, Monclova, Mexico's steel-making capital, lies in the center of the basin, and the industrial city of Monterrey is nearby. By late 1979 PEMEX had identified sixty-five interesting structures in the basin, thirty-five of which appear to be more than eighteen miles long.[45]

Other Areas

Serious work is just beginning in Durango, while exploration efforts continue in Chihuahua, Sonora, Nuevo León, and Baja California, where the Bombas well confirms the occurrence of hydrocarbons on this western peninsula. Still, as of late 1980, Baja California had proved a major disappointment: The two offshore holes turned out to be dry, while the onshore wells lacked commercial quantities of gas. PEMEX exploration crews are also hard at work in the "Portal del Balsas," a section of Mexico that formed a water bridge between the Atlantic and Pacific oceans during the Jurassic and Cretaceous periods. This area embraces part or all of the states of Michoacán, Jalisco, Colima, and Nayarit. Despite the above-mentioned activities, less than one fifth of the country's geologically promising areas had been explored by the end of 1980.

The foregoing exploratory and developmental work has led PEMEX officials and foreigners to trumpet the prospect of Mexico's becoming the world's foremost producer. On March 21, 1978, Dr. Bernardo F. Grossling, a scientist at the U.S. Geological Survey, regaled congressmen with an "astonishing" 340-billion-barrel estimate of Mexican holdings.[46] No less enthusiastic was William D. Metz, who wrote in *Science* magazine that PEMEX "exploration . . . is turning up oil fields so immense that they could overturn the conventional wisdom about world oil supplies and significantly alter the geopolitics of energy." He claimed that by early 1978, "it was clear that Mexico was at least the equal of another Kuwait," whose reserves hover around 60 billion barrels. Metz went on to quote Lawrence Goldmuntz, a Washington energy consultant, who argued that the Mexican fields "may be equivalent to the entire Middle East." Richard B. Mancke, an economist at the Fletcher School of Law and Diplomacy, has lent his voice to the chorus of optimism. In his opinion the continued drilling successes in the Southeast provide "some basis for inferring that Mexico may actually have sufficient reserves to supplant Saudi Arabia as the world's richest oil-producing country."[47] Should PEMEX continue to maintain a twenty-to-one ratio of proven reserves to annual output, it could produce "an enormous 13.5 million bpd of crude and natural gas equivalents," he contends. Mancke's more conservative estimates fix production at 4.7 million bpd in 1985, rising to 7.2 million bpd in 1990.[48]

But questions have emerged as to the extent of Mexico's deposits. A study completed by the Subdirection of Primary Production just before Díaz Serrano assumed office placed potential reserves at 60 billion barrels. Within a matter of weeks, PEMEX announced a figure twice as high. Of the 200 billion potential barrels reported in 1978 (see table 6), PEMEX's former manager of exploration, Francisco Viniegra, told reporters for the *Los Angeles Times:* "It's impossible. I know the geology of Mexico and it's not there." A. A. Meyerhoff, a Tulsa-based geologist who has worked for the monopoly, termed the number "utter nonsense," a judgment corroborated by Mexican specialists in Washington.[49]

The proven holdings have also sparked questions. To begin with, Mexico, unlike Persian Gulf producers which flare a large percentage of their natural gas, includes this vital fuel in its reserve calculations. By so doing it automatically expands the amount by one third. Moreover, PEMEX considers 5,000 cubic feet of gas the equivalent of one barrel of oil, while the accepted industry practice is to use 6,000 cubic feet. Hence in making conversions, the firm overstates its oil holdings by 20 percent. But this accounting procedure does not explain how the new team could announce a doubling of reserves so soon after taking office in 1976. The new estimate sprang from desk calculations; that is, no new field work was carried out. Before Díaz Serrano took over, PEMEX neither announced probable and potential reserves nor included oil derived from secondary recovery, such as water injection, in proven holdings until (1) completion of a careful engineering study of the reservoir, (2) design of the appropriate recovery method, and (3) rigorous calculation of the recovery factor.[50] Apparently the new director general estimates the secondary recovery factor for certain deposits rather than following these rigorous verification procedures. For example, the state firm may have concluded that 40 percent—not 10 percent—of the volume in a given well could be obtained by using water injection or some other secondary recovery method. PEMEX may also have extrapolated from known holdings, assuming that a given formation would yield as much oil and gas as an adjacent one already under production. Two key PEMEX engineers, now retired, told me in mid-1979 of the tendency—in violation of industry standards—to mix probable reserves with those rigorously certified.

Also wary of this practice is James W. Watson, senior vice president of DeGolyer and MacNaughton, a Dallas consulting firm that certified proven reserves of 14.7 billion barrels in 1977. "We've never substantiated that 40 billion figure," he stated. When asked if Mexico's proven holdings might be as high as 28 billion, Watson exclaimed, "Oh, no. Oh, no." He also said, "We've reminded [the Mexicans] over and over about the danger of losing credibility. We've been trying to hold them back."[51] When Petróleos Mexicanos sought to have its higher reserve figures verified by the Mexican Petroleum Institute, the IMP's engineers in the Production Department refused because PEMEX declined to supply sufficient data on which a conclusion could be based. Agustín Straffon Arteaga, director of the IMP, finally confirmed PEMEX's holdings in a personal act based on political, not scientific, considerations.

The vaunted Chicontepec fields exemplify PEMEX's legerdemain. The internal company study completed before Díaz Serrano was appointed showed that these deposits, marred by low permeability, pressure, and porosity, embraced 7 billion probable barrels. Two years later, as we have noted, the director general touted them as "one of the bigger hydrocarbon accumulations in the Western hemisphere" and suggested that Chicontepec might contain up to 100 billion barrels, a volume later scaled down to 17.6 billion proven barrels. To obtain even half this amount will require over a dozen years and the sinking of more wells than the monopoly has drilled since its creation. The need to construct roads, rail spurs, gathering lines, and production facilities further complicates the picture.

The ready accessibility of hydrocarbons in other areas, notably Campeche, the Reforma, and the Gulf of Sabinas, makes it difficult to understand why PEMEX, whose funds are abundant but not unlimited, has earmarked Chicontepec for immediate attention. The explanation may lie more with politics than economics. It will take 23,300 workers to develop the field; approximately 150,000 people will depend on oil production for their livelihoods; and the area may be able to support a total population of 350,000.[52] Thus by moving ahead with Chicontepec, Díaz Serrano can assure the president of his agency's support for a vaunted industrial development plan released in March 1979. Such a commitment may enable PEMEX to fend off bureaucratic rivals who would like to see a larger percentage of petroleum revenues channeled to other areas of the economy.

Several reasons may have inspired Díaz Serrano to advance the highest reserve figure possible. Domestically, the prospect of massive deposits strengthens the monopoly's position with respect to bureaucratic competitors, bolsters the pride of a highly nationalistic people who for four decades have viewed the oil industry as a sacred symbol of national dignity, enhances public confidence in an economy beset by surging inflation and massive unemployment, and lofts the star of Díaz Serrano, who clearly enjoys public office. The oil discoveries may even have contributed to the recovery of the Mexico City stock market in 1978.

Large reserves also justify the rapid exploitation of individual wells—a tendency which alarms petroleum engineers conversant with the excesses of earlier periods in the nation's history. For example, they cite the potential damage to the Sitio Grande field as a result of imprudent management. An accelerated flow of oil gives rise to more contracts that provide opportunities for sub rosa "commissions" in an industry where graft and corruption flourish. Companies founded by Díaz Serrano hold lucrative contracts with PEMEX, and critics doubt his claim to have severed financial ties with them. One such firm, PERMARGO, drilled the exploratory well Ixtoc 1 that poured over three million barrels of crude into the Gulf of Mexico in what scientists have called the worst oil spill in the history of the industry.[53] As will be discussed in the next chapter, the venal Oil Workers' Union, itself a major contractor, has profited from the boom conditions. The rapid development of the Reforma trend combined with high production rates has excited worries about exhausting these fields.[54]

The possession of major reserves raises a country's standing within the community of nations, improves its position with the World Bank and other financial agencies, and increases its attractiveness to private foreign lenders, from whom Mexico has succeeded in extracting lower interest rates. Mexican officials no doubt anticipated that the inflated figures would bolster their leverage in negotiations with the United States on issues such as trade, illegal immigration, and the price of natural gas.

Constraints

Although PEMEX executives now exude optimism, there are a number of constraints—political, economic, and administrative—on Mexico's ability to produce oil and gas. No doubt Mexico could raise output even faster and earn foreign exchange in greater amounts by allowing the international oil companies to participate in production. Construction of an oil pipeline from the Reforma fields to the Texas border might complement such a move.

But memories of the 1938 expropriation, revived in stentorian speeches every March 18, are strong. Mexicans would view the presence of the "Seven Sisters" as a desecration of the national patrimony, an indication that their country was becoming more, not less, dependent on the "Colossus of the North." Any official daring to tender an invitation to these transnationals would commit political suicide, for the greatest act of apostasy in contemporary Mexico is "selling out to imperialism."

Detractors have already accused Díaz Serrano of this crime. Most vocal and articulate has been Heberto Castillo, a civil engineer who is leader of the small Mexican Workers' party and a columnist for *Proceso,* an influential weekly magazine. He asserts that Díaz Serrano, allegedly mired in conflict of interest, has too readily agreed to export large amounts of hydrocarbons. Thus, he warns, Mexico will be importing oil and gas by 1992.[55]

Castillo has organized a National Front for the Protection of Natural Resources, composed of a score of small Marxist organizations, to protest the "massive and indiscriminate" exploitation of the country's mineral wealth. Sympathetic to this effort are the National Center of Communication, the Unity of Leftist Communists, the Socialist Workers' party, the Revolutionary Socialist party, the Communist party of Mexico, the "Democratic Tendency" of the Electrical Workers' Union, and an assortment of university groups.

Castillo and his chief ally, Demetrio Vallejo, a sixty-nine-year-old former railroad union leader, have difficulty drawing even a few thousand supporters to demonstrations. Still, they are sufficiently visible, outspoken, and well prepared that the government does not want to hand them issues with which to inflame the passions and attract the loyalty of their countrymen who are disaffected with PRI and the authoritarian, manipulative politics it practices. "Díaz Serrano is convinced that Mexico can climb out of its underdevelopment by pumping out its oil," Castillo said. "We've been trying to show public opinion that this isn't so."[56] As will be discussed in chapter 8, the government is vulnerable to charges of offending the goals of the revolution as a result of amendments to the Constitution, secured in 1977, to permit automatic expropriation of lands required for such PEMEX operations as the construction of a gas pipeline from the Reforma fields to the U.S. border.

A political reform promulgated by López Portillo's administration assures a louder voice and legislative forum to critics of governmental policy. The Communist party of Mexico, which secured over 7.6 percent of the ballots cast in the July 1979 congressional elections, left the legal wilderness for the officially sanctioned status accorded parties obtaining at least 1.5 percent of the total vote. The leftist Socialist Workers' party and the conservative Mexican Democratic party also gained permanent registration. One hundred of the four hundred seats in the newly expanded Chamber of Deputies have been guaranteed to the combined opposition parties, which also enjoy greater access to the mass media. The objectives of the new law may have been (1) to provide a legal outlet for detractors to articulate their beliefs, lest they contemplate violence, (2) to coopt dissidents (by a regime that has converted cooptation from an art form to an exact science), and (3) to encourage flexibility, if not democracy, within the PRI and the government by subjecting both to searching examination by opposition spokesmen. The replacement of Jesús Reyes Heroles, minister of the interior and architect of the reform, by Enrique Olivares Santana, a pragmatic stalwart of the Revolutionary party, appears to have diminished hope of accomplishing the third goal, if—in fact—it was ever a serious one.[57]

Castillo neither sought nor gained official recognition of his Mexican Workers' party because he viewed the reform as cosmetic. "You go to Congress if you want to make a speech," he said. "But the revolutionary struggle takes place outside Congress, in factories, on farms, in cities. It is a mistake for the left to become absorbed by Congress."[58] Nevertheless, the results of the 1979 congressional contests found 104 opposition representatives in the Chamber of Deputies. While their presence did not break the official party's monopoly on power, it has enlivened parliamentary proceedings in a heretofore calcified body. For example, in September 1979 the deputies heatedly questioned Díaz Serrano on a wide gamut of sensitive subjects related to the petroleum industry. The director general's appearance before the legislators will be discussed in chapter 9.

Some politicians may be willing to work through the system, but peasants fighting for their land are not. For years Indians in Tabasco state have resisted the intrusion of exploration and drilling crews. The situation worsened in December 1978 when Chontales Indians blocked roads to ten wells near Jalpa and Cunduacán, located some twenty miles from Villahermosa. Later they took up arms against workers at the Salta de Méndez and Cuytan wells. Meanwhile, acts of vandalism and sabotage were reported against wells in the Samaria and Iride fields. All told, the *campesinos* blocked six oil wells and fifty-five roads. These actions, which caused the loss of nearly one million barrels of crude, apparently followed the monopoly's failure to pay indemnities for damages to arable land and other property. The prospect of "chaos" in the nation's richest oil zone led PEMEX to summon the army and file charges with judicial authorities because of "actions against the country's economy." Company officials, who claimed that the Chontales had been incited by "unofficial agitators" who "are trying to practice extortion" by demanding excessive sums of money or the payment of indemnities for nonexistent damage, later asked the postponement of military intervention while a peaceful resolution was attempted.[59]

Officialdom's thin-skinned sensitivity to criticism combined with the precept that "Mexico's oil belongs to Mexicans" has other consequences. For example, it makes it difficult to raise the domestic price of hydrocarbons: Oil sold for $6.00 a barrel in 1978 while consumers purchased gas in Monterrey for less than 35 cents per Mcf. Extremely low prices feed a rapidly expanding domestic demand that diminishes the volume of oil and gas available for exports. The six-year plan contemplates a 7-percent annual increase in consumption at home, but Mexico's population is growing faster than that of any other major country, and the nation's appetite for energy seems insatiable, expanding steadily each year.

Petróleos Mexicanos also faces physical and economic constraints as it attempts to accelerate production. Large volumes of natural gas are "associated" with oil in the Reforma wells. To produce 2 million bpd of oil, the company will have to handle approximately 3.5 billion cfd of gas, large amounts of which require "sweetening"—i.e., the removal of hydrogen sulfide, carbon dioxide, and other corrosive chemicals. The gas must also be "stripped" of liquid hydrocarbons. Absorption plants and cryogenic units now have a capacity of 3.1 billion cfd, far above the level of domestic consumption. The monopoly will either have to find new customers—chapter 8 focuses on a deal with the United States—or reinject and flare gas to keep up oil output.

A shortage of rigs presents another problem. PEMEX, with 178 land and 30 offshore rigs at work in mid-1980, was the world's largest drilling company. An expanded capability is contingent upon the prompt delivery of another 34 rigs ordered in 1978. But enlarging the scope of operations gives rise to headaches along with oil. Breakdowns are frequent, and supply difficulties persist. The crews have often found themselves waiting up to two weeks for the delivery of replacement parts. Many rig operators have limited onshore experience and virtually none in marine drilling. After all, Mexico had only 3 offshore rigs before the Campeche strikes.[60] The result has been a failure to achieve drilling targets. In 1977 PEMEX announced plans to sink 497 wells, but it completed only 307, with an average depth of 9,790 feet. In 1978 the firm drilled only 306 of 450 projected wells as the average depth declined to 9,755 feet. The 1979 program called for 409 wells, 359 onshore and 50 offshore.[61] But as new rigs begin work in the shallow Chicontepec area, the average yearly number of wells completed—only 2 per rig were drilled in the difficult Chiapas and Tabasco fields—will climb.[62] Moreover, the state firm has shown its willingness to contract with private corporations, foreign and domestic, for equipment and technicians.

The absence of modern port and storage facilities combines with an inadequate transportation system to introduce another serious bottleneck. While PEMEX has exceeded its production goals even with fewer wells, exports have fallen short of expectations. In early 1979 the firm anticipated that shipments abroad would average 682,000 bpd, a figure later revised upward to 778,000 bpd. Yet the level attained was approximately 540,000 bpd. Problems encountered early in the year caused reductions in contracted export volumes of as much as 40 percent; by December this percentage had fallen to 15. A major factor in the shortfall was congestion and bad weather at the Pajaritos export terminal, where shortages of marine diesel fuel reportedly delayed tanker departures by up to five days. A temporary facility was set up at the Campeche maritime fields to alleviate loading problems at Pajaritos, and modernization and expansion of this Gulf port as well as Salina Cruz on the Pacific Ocean were well underway in 1980.

Completion of Latin America's largest deepwater facility at Dos Bocas near

Villahermosa will improve conditions even more. Construction crews had cleared approximately thirty-five hundred acres by November 1979; nevertheless, building a tanker terminal is time-consuming, and the project will not be finished until late 1981 or early 1982. Meanwhile, storage capacity is severely limited. PEMEX officials admit that even then conditions will have to be ideal before the monopoly can export half of its production as planned.[63]

The poor condition of the heavily traveled arteries of the Southeast forces most trucks to operate at markedly reduced speeds. Others roar ahead with impunity, leading to accidents, breakdowns, and damage to vehicles. An American journalist, who compared Villahermosa's main street to "a chariot scene out of Ben Hur," reported ten to fifteen serious crashes a day as drivers openly violate the law. "They hit people and drive away," one citizen said, "and the *sindicato* . . . bosses transfer them to other districts so they'll be safe from the police."[64]

International agencies have evinced misgivings about the direction of Mexico's development. Under an agreement concluded in late 1976, the International Monetary Fund (IMF) limited all public-sector borrowing from abroad to $3 billion in 1977, 1978, and 1979 because of the country's notoriously high foreign debt. Each year the IMF also negotiates a limitation on new indebtedness in terms of a percentage of Gross Domestic Product. These restrictions, at first ominous, were relaxed in 1978 as the IMF determined that PEMEX would have to borrow large sums to finance an increasingly ambitious development program deemed crucial to long-term economic growth. American, Japanese, and European bankers have competed to make loans to the state oil firm. More serious has been the World Bank's concern over the ability of the Mexican economy to absorb the spending contemplated in its six-year program, which the international agency reportedly believes should be scaled down by 25 percent.[65] Rapid oil development could trigger a sharply higher rate of inflation, further distorting the nation's economy and threatening the stability of the political system. López Portillo seemed to recognize the danger when he stated: "The capacity for monetary digestion is like that of a human body. You can't eat more than you can digest or you become ill. It's the same way with the economy."[66] More will be said in chapter 5 about digesting oil revenues.

The Oil Workers' Union, spokesman for approximately 92 percent of the nearly one hundred thousand men and women employed in the industry, also represents a constraint on production. Its unique status and peculiar practices entitle it to an entire chapter.

PEMEX activity is not seriously impeded by one obstacle greatly affecting energy production in the United States: environmental protection. The company's annual *Memoria de labores* extols the importance of the ecological system, although this subject is invariably treated last. Areas such as Poza Rica and La Venta have suffered irremediable damage. Exemplary of a current problem is the harm done to Villahermosa, Tabasco, in the heart of the Reforma petroleum zone. Foreigners have often besieged this fly-specked old city situated near the spot where Cortés first set foot on Mexican soil. English filibusters occupied it two years after its founding in 1596, when the Spanish called it San Juan Bautista. The seventeenth and early eighteenth centuries saw incursions by booty-hungry pirates, and Frenchmen in the service of Napoleon III arrived in the 1860s.

The quest for and extraction of black gold has converted Villahermosa from a steamy, friendly provincial capital into a bustling city. Oil has detonated an explosion in the city's population from 33,000 inhabitants in 1950 to 99,000 in 1970 to 250,000 in 1979.[67] Many of the poor live in fetid shacks that cling precariously to the banks of the Grijalva River, until a few years ago alive with the splash of expatriated Mississippi steamboats.

With the possible exception of prostitutes who frequent tawdry hangouts in the *zona de tolerancia,* everything is in short supply—housing, schools, hotel rooms, navigable roads, transportation, potable water, and food. The upshot is the nation's sharpest inflation rate: It is estimated at 35 percent compared with a countrywide figure of 20 percent in 1979.[68] *Proceso* reported in 1979 that bananas selling for forty centavos 530 miles away in Mexico City commanded five times as much in Tabasco state, the country's second-largest banana-growing region. Meat, fish, and vegetables are similarly dear. Corrupt union officials exact seventy to eighty thousand pesos for permanent jobs with PEMEX; transitory positions, which last but a few weeks, command eight thousand pesos.[69]

Who can pay the city's exorbitant prices? The oil workers—a "privileged, separated class," according to the Most Reverend Rafael García González, Villahermosa's bishop, who has founded a small training and information center to help the poor secure just treatment from PEMEX and other governmental agencies. These people urgently need help. Despite the free-spending atmosphere, official figures show that two out of three "economically active" Tabasqueños work only part time or have no job at all. Many of the idle are uprooted farmers or fishermen who abandoned their traditional occupations to pursue a job with the national oil company whose lowest daily wage was ten times greater[70] than the thirty pesos ($1.30) paid *macheteros* who harvested bananas in 1979. Sometimes the position never materializes or lasts only a few months. On other occasions the prospective worker cannot pay the fee exacted by union bosses for almost all jobs in the industry not allotted to relatives of current employees (women frequently pay with sexual favors). In any case failure of the peasants to till their land further reduces output in a country whose outlays for basic food imports cut into the $12-billion annual oil revenues.

Their poverty often leads the peasants to borrow money from well-to-do cattle ranchers who by hook or crook manage to gain their land for pasturage. This drives *campesinos* desiring to return to farming into the inhospitable highlands to resume their ancient *"tumba-roza-quema-siembra"* ("cut-clear-burn-sow") cropping technique. Lack of technical assistance, hostility to the government, and resistance to collective agriculture exacerbate the production problem. Tabasco has become a major food importer.

PEMEX crews have spurred the flight from land- and water-related work by destroying large tracts of fertile terrain and contaminating productive rivers and estuaries. "With their dredges, they can make and remake rivers," says Francisco Iracheta, a biologist who heads a World Bank–funded project to stimulate employment in the fishing industry. "They have cut grooves across the entire state, creating an artificial water network almost as large as the natural one. This has turned the hydrological system upside down."[71]

The state enterprise has concentrated its drilling in waterways because this saves money, excites fewer peasant protests, and requires no expenditures for indemnities. While these bodies may eventually flush themselves clean, irreparable damage has been done to the marine spawning grounds, food chain, and habitats. Meanwhile, the government, of which the company is an integral part, has directed tax monies, World Bank support, and Japanese technical assistance to fostering seafood production.

In 1977 Alejandro Yáñez, director of the National University's Institute of Marine Science and Limnology, warned that drilling in the adjacent state of Campeche had so contaminated the Laguna de Términos that its shrimp population faced extinction.[72] The head of PEMEX's Department of Environmental Protection agreed that some pollution had occurred but insisted that "no estimate existed to show that the situation is grave and that marine fauna has suffered damage." He also stated that his firm would spend 2,300 million pesos in the next half-dozen years on ecological concerns. "Pollution exists," he said, "but it is not impossible to keep it to an innocuous and acceptable level."[73] Such honeyed words must have soured in his mouth in the wake of the Ixtoc 1 catastrophe.

The monopoly spends less than 1 percent of its development budget on environmental projects. Moreover, the country's laws in this area are nonexistent, anemic, or simply ignored. Protecting natural resources is often treated as a luxury affordable only by developed nations. As an envious Oklahoma oilman expressed it: "PEMEX doesn't have all those damn environmental impact statements to fill out. They just do it."[74]

What they "just do" with little planning or even consultation with local officials is use their dredges and drills to imperil the life of renewable resources—human, animal, and vegetable—to obtain a nonrenewable one.

Compared to the PEMEX invaders, the English freebooters, avaricious corsairs, and French soldiers may seem like tender-hearted missionaries.

The situation has led Tabasqueños to quote with bitterness López Portillo's words of January 4, 1976: "During my administration, I will not allow areas rich in petroleum to be exploited as if they were colonies in order to transfer the wealth to other parts of the country and leave the production zone beset by disease, despoliation, misery, isolation, and depleted wells."[75]

4

The Oil Workers' Union

Formation

Labor associations sprang up soon after the commercial exploitation of petroleum began in Mexico. The first oil workers' union in Tampico—and probably the first in the country because of this city's pivotal role in the industry—was the Gremio Unido of the Waters-Pierce Oil Corporation.[1] Others, also composed of refinery workers, soon emerged, and with them bitter struggles against the foreign companies.

The rapid inflation that beset the country during the Mexican Revolution intensified this combat. Initial demands often focused on payment of a minimum wage and the institution of an eight-hour work day. A rash of strikes occurred in 1917, when only the months of August and September were free of work stoppages. On June 15, 1917, the strike committee of the Huasteca Petroleum Company, a firm which had suffered its first general strike the year before, specified the recognition of a union as one of several preconditions for returning to work. Waters-Pierce broke a May 1919 strike by using military force and jailing the instigators. In this period a workers' association rarely lasted more than two weeks.[2] The firms simply fired or imprisoned "troublemakers" while using the army or their own white guards to intimidate potential union members. When these tactics failed to halt organizing efforts, they created company unions, ushering in a period of bloody strife between genuine independent unions and those manipulated by the trusts.[3] The companies, aided by civil and military authorities, also threw up every conceivable roadblock to cooperation between the oil workers and other proletarian groups in Tampico.[4]

The 1917 Constitution conferred notable benefits on Mexico's working class. Among the most important from the *petroleros'* point of view were the eight-hour day, payments for employees incapacitated by job-related illnesses or accidents, and compensation equal to three months' wages for workers unjustly dismissed or suspended for having belonged to a union or participated in a strike. Individuals generally appealed on their own behalves, rather than as members of an organization, to the appropriate authorities to secure these rights. But the government lacked resources and personnel to investigate complaints, and the companies—though willing to implement the shorter work day—found devious ways to avoid complying with other provisions.[5]

An uneasy truce characterized labor-management relations between 1920 and 1923. The relative calm was broken in mid-1921 when the companies suspended shipments of oil to protest the imposition of a new export tax. This action left twenty thousand workers throughout the country without employment for two months. The ensuing unrest prompted the United States to dispatch warships to the Tampico area as a "means of precaution." At this time the Grupo Hermanos Rojos and the Sindicato Unico de la Región Petrolera sought to raise the consciousness of the workers: The former presented dramatic works of Emile Zola and Ricardo Flores Magón in local theaters; the latter sponsored conferences with union themes presided over by the "most advanced" members of the proletariat.[6]

As previously mentioned, Mexico achieved a record volume of production in 1921, after which output declined and with it the number of workers. The industry's work force soon diminished from fifty thousand to sixteen thousand.[7] In the face of rising unemployment, many workers sought means to protect their positions. Yet the skilled workers at the Villa Cecilia refinery near Tampico felt secure in their jobs. After all, they had responsibility for manipulating valves, regulating pressure, maintaining precise boiler temperatures, and handling pumps. The peculiar design of the facility—pipes and related equipment were located underground—enhanced the value of the men, who had mastered the subterranean system. However, a strike by electrical workers, who had organized their union in August 1923, left these El Aguila installations without power, prompting the company to shut down and put its employees on leave without pay.

The experienced workers discovered that they were not indispensable after all. Thus they proved receptive to the appeals of Serapio Venegas, a stoker with an unusual talent for organization and publicity who set about to form a union. His efforts bore fruit on December 13, 1923, with the creation of the Sindicato de Obreros y Empleados de la Compañía de Petróleo "El Aguila." To strengthen its position, the union soon affiliated with the Confederación Regional Obrera Mexicana (CROM), a national movement led by Luis N. Morones that first aligned itself with President Alvaro Obregón and later with the revolutionary party created by Gen. Plutarco Elias Calles.[8]

After devoting seven weeks to the clandestine recruitment of new members, the union on February 2, 1924, publicly condemned the maltreatment of Mexicans by foreign refinery managers and sought the intervention of governmental authorities. Two days later the new *sindicato* notified the municipal president of Tampico that it had delivered a list of demands to the management of El Aguila, which had undertaken reprisals against workers whose only action was to organize a union. The firm responded by firing forty labor organizers.[9]

The list submitted to El Aguila made no mention of higher salaries. Instead it addressed recognition and preservation of the union; work safety, health, accident, death, and unemployment benefits; and the need to establish two job categories: permanent and transitory. The drafters included additional paragraphs fashioned to appeal to masons, boiler makers, and other work groups that the union hoped to attract.[10] Although two thousand signatures appeared on the document, only five hundred men participated in a strike that erupted in late March. Nonetheless, the bold action succeeded. Serapio Venegas and his fellow organizers, who immobilized the refinery by seizing important facilities, enjoyed the support of other unions.

Most important were the electrical workers, who had just emerged from a successful strike and were eager to share their experience and ideas. In addition, truck drivers, longshoremen, masons, carpenters, firefighters, and more than a dozen other labor organizations answered Venegas's call for help. These unions contributed 11,665.52 pesos to the strike committee, while approximately 1,150 of their members served as guards to help the oil workers retain control of the refinery and a large adjoining area. The strike committee reported daily to the municipal president of Tampico on security activities. The Longshoremen's Union even provided the strikers with occasional work loading and unloading vessels. Gen. Lorenzo Muñoz, Tampico's military chief, sympathized with the strikers and helped prevent strikebreakers from entering the plant.[11]

The strike lasted four months as Venegas and his followers spurned a CROM initiative for an earlier compromise settlement. El Aguila finally agreed to sign a collective contract after Lic. Emilio Portes Gil, senator from Tamaulipas and a candidate for governor, interceded as an arbiter. For the first time a petroleum company recognized a workers' association as the only representative of its employees in a given facility. A combination of factors—support of key authorities, labor solidarity, and the prime role of highly skilled employees—contributed to the outcome.[12] This victory stimulated the organization of petroleum workers throughout the country and laid the groundwork for a consolidated national union. One exception was Mexican Gulf which, during the strike, moved its Mexican employees out of Tampico to the firm's installation at Prieto. There it housed them comfortably with the result that Gulf workers neither joined a union nor took part in general labor disputes.

By 1935 the industry counted thirty-five independent oil workers' unions. Yet not a single petroleum firm was completely organized by union labor until that year as different contracts existed at each refinery and even within the same facility. There were contracts with fifteen of the trusts and three shipping companies, but provisions as to wages and conditions of work varied greatly.[13] Widespread dissatisfaction of labor and the tutelage of the Cárdenas administration led to the first Grand Congress of Petroleum Labor Organizations, which convened in Mexico City on August 15, 1935.[14] This congress led to the creation, four months later, of the Oil Workers' Union (Sindicato de Trabajadores Petroleros de la República Mexicana—STPRM). After resolving its most pressing organizational problems, the STPRM called a special convention for July 1936 to fashion a collective-bargaining agreement for all workers in the industry. As discussed in chapter 1, the demands made in this document gave rise to a protracted conflict with the foreign firms and the expropriation order of March 18, 1938.

Organization

During the Bermúdez years the thirty-five locals—the unions that existed before nationalization—had gradually ceded their cherished autonomy for the sake of unity and efficiency. The STPRM's current statutes reveal a commingling of centralized power, regional balance, and local autonomy that dates back to the organization's early years.[15]

The convention stands at the apex of the union's organization. Composed of elected representatives—three from each local, one from each delegation (a work center with fewer than twenty active members), and one from each of three zones on behalf of the professions—this body enjoys full executive, legislative, and judicial authority in union matters. "Once convened a Convention is automatically the highest authority of the Union and all of the other labor entities under its jurisdiction."[16] Ordinary conventions, called by either the General Executive Committee or the General Council of Vigilance, must be convened in early December every three years. Extraordinary conventions may be convoked either by the General Executive Committee or by locals that represent at least two thirds of those belonging to the union. In both cases sessions are held in Mexico City, although an extraordinary convention can take place at a site judged more appropriate to resolve the problem at hand. Proportional voting, based on the number of members represented by a local, obtains for "matters of great importance," which must win approval of delegates representing 66 percent of the national membership. Other questions are subject to majority rule. The leaders of locals, who receive the program before the convention begins, are expected to hold grassroots assemblies to discuss key issues and elect delegates from the active membership.

One of the most important functions of the convention is the selection of a General Executive Committee, which is charged with the "defense, direction, orientation, and general administration" of union affairs on a day-to-day basis. The STPRM's statutes vest the committee with twenty-nine separate functions, including the signing of the collective work contract, which establishes wages, benefits, and other conditions of employment. Approval of the contract by locals embracing at least 66 percent of the membership must precede final action on this accord. A secretary general heads the Executive Committee. He is empowered to chair committee meetings and those of any other union body when deemed necessary, "intervene in all union matters," resolve whatever problems may arise, monitor the activities of his fellow secretaries, review the treasurer's books, authorize expenditures, issue appropriate reports, convene conventions, visit each local at least twice a year, and preside over the General Strike Committee. He may also secure credits needed to finance consumer stores, savings and loan associations, funeral parlors, and similar social projects.

In theory, any member in good standing can aspire to be secretary general. In practice, the post rotates every three years among the most powerful locals in the three zones: northern zone (Ciudad Madero—No. 1), central zone (Poza Rica—No. 30), and southern zone (Minatitlán—No. 10). An alternate from the same local replaces a secretary general unable to discharge his duties. Incumbents of this office tend to share a number of traits. They are men around forty years of age or slightly older who have logged many years of union service, starting from the bottom. They will have served both as secretary general of one of the three large locals and as head of that organization's dominant political group, a kind of political party that competes for elected positions. In recent years the friendship of Joaquín Hernández Galicia, the redoubtable "La Quina," has also been an important consideration.[17] Table 7 identifies the men who have led the union.

Other members of the General Executive Committee include a secretary of interior, agendas, and agreements, who keeps and disseminates the minutes of meetings and "develops and maintains fraternal relations" among the locals; a secretary of exterior and information, who coordinates the union's affairs with kindred organizations both at home and abroad, runs the union newspapers, provides information to members, and handles publicity; a secretary of labor, who supervises compliance with the work register in the assignment of jobs, assures that local secretaries of labor meticulously observe this roster and the rights of workers, helps apportion men and women to new work centers where no local has jurisdiction, and renders opinions to the General Executive Committee regarding jurisdictional conflicts among locals; a treasurer, who manages the STPRM's finances and has the right to veto any expenditure believed by chief union officials to "endanger the union's economic stability"; and a secretary of organization and statistics, who collects, analyzes, and issues reports on data related to membership, accidents, deaths, and other matters of concern to locals and the headquarters in Mexico City.

A Social Security and Education Office (Cuerpo de Educación y Previsión Social) and an Office of Adjustments (Cuerpo de Ajustes) are also part of the General Executive Committee. Eight executive secretaries constitute the former: One comes from each zone; two are chosen by the zones on a rotating basis; and three are retirees elected by their peers in each zone. This body looks after the *petroleros'* educational advancement, distributes PEMEX-funded scholarships, encourages the establishment and maintenance of libraries, promotes improved industrial hygiene and medical care, fosters organized cultural and athletic activities, and champions consumer rights. The eleven secretaries of the Office of Adjustments are chosen on the basis of regional balance and occupational diversity. They specialize in the resolution of legal conflicts

TABLE 7
Secretaries General of the STPRM, 1936–1980

Name	Local	Period
Eduardo Soto Innes		1936/1937
Juan Gray		1938[1]
Rafael López T.		1938/1939
Aurelio Martínez		1939[2]
Rafael Suárez R.	30	1940/1941
Antonio Salmon	1	1942/1943
Isidoro Gutiérrez	10	1944/1945
Jorge R. Ortega	30	1946
Antonio H. Abrego	1	1947
Eulalio N. Ibáñez	1	1948/1949
Demetrio Martínez	10	1950/1951
Enrique López Naranja	30	1952/1953
Ignacio Pacheco Leon	1	1954/1955
Felipe Mortera Prieto	10	1956/1957/1958[3]
Pedro Vivanco García	30	1959/1960/1961
Joaquín Hernández Galicia	1	1962/1963/1964
Rafael Cárdenas Lomelí	10	1965/1966/1967
Samuel Terrazas Zozaya	30	1968/1969/1970
Salvador Barragán Camacho	1	1971/1972/1973
Sergio Martínez Mendoza	10	1974/1975/1976
Heriberto Kehoe Vincent	30	1977[4]
Oscar Torres Pancardo	30	1977/1978/1979
Salvador Barragán Camacho	1	1980/1981/1982/1983/1984[5]

Source: Information supplied by Brigido Piñeyro de los Santos, secretary of organization and statistics, STPRM.
1. January to May 1938.
2. End of 1939.
3. Conversion from a two- to a three-year term.
4. Killed on February 28, 1977.
5. Term extended to five years on a one-time basis.

involving the union, guaranteeing "the defense of the Organization's Interests in general and those of its members in particular."[18]

The union's ombudsman is the General Council of Vigilance (Consejo General de Vigilancia), which is "obligated to monitor the strict fulfillment of provisions found in the present Statutes and the faithful observation of legal accords as directed by the Organization's governing bodies." Three members—a president, secretary, and a director—constitute this body. These posts, which carry three-year terms, rotate among the union's three geographic zones. The council has broad authority to involve itself in union affairs and secure the cooperation of all bodies (except the convention) and members in carrying out its decisions. Its investigative powers extend to the actions of locals, officers, and individual members. For example, it may use the services of an auditor in rendering a monthly report on the treasurer's books, documents, and files. It may suspend members and take other disciplinary measures, veto certain actions of the General Executive Committee deemed inimical to union interests, and report to competent authorities any improper actions undertaken by members or officers.

According to a decree of July 20, 1939, the Oil Workers may name four union counselors to the board of directors of Petróleos Mexicanos. One is chosen from each of the three zones, while the fourth position rotates among the zones. Like most STPRM officials, the union counselors hold three-year terms. They are elected from the delegates accredited to ordinary conventions.

The statutes provide that each of the twenty-eight locals organize itself in a manner similar to the national administration. The most notable organizational difference is that each local's General Executive Committee includes a separate secretary of agenda (in addition to a secretary of interior and agreements) as well as a secretary of education and social security and a secretary of education and social security for retirees.

Although this tie does not appear in statutes, official documents, or organizational charts, union leaders have an intricate, longstanding relationship with the president of the Republic and, secondarily, the minister of labor. An authoritarian, hierarchical political system such as Mexico's often breeds reciprocal ties: Subordinates seek to ingratiate themselves with persons in more powerful positions in return for favors and advantages. At least since the Cárdenas epoch, union leaders have appealed to the president and minister of labor for assistance in resolving both internal union problems and difficulties arising between their organization and PEMEX. For instance, after Bermúdez fired fifty leaders involved in an illegal work stoppage and petitioned for a conflict-of-economic-interest order, the union held a special convention in January 1947. A delegation from the convention called upon President Alemán, who agreed to suspend the move for the order for two months and rehire those of the discharged leaders who had no role in the unlawful work stoppage.[19] One of the presidents most receptive to labor entreaties was López Mateos. He and Gutiérrez Roldán even devised a bargaining minuet for salary negotiations. In discussions with the union, the PEMEX chief would stop short of assenting to the maximum raise he knew the company could afford. He then relayed the acceptable figure to the National Palace. When the labor leaders begged support for their cause, López Mateos could then demonstrate his magnanimity in the paternalistic style that characterizes superior-subordinate relations in Mexico.[20]

Echeverría was the greatest friend of the STPRM to occupy Los Piños, the presidential residence. La Quina, stung by attacks on union practices by Reyes Heroles, cultivated close ties with Echeverría, who served as minister of the interior from 1964 to 1970. It is widely believed that the labor leader tendered Echeverría support in terms of manpower and financial contributions during the 1970 presidential campaign. Once in office Echeverría remembered his friends in the Oil Workers' Union. Most important, he threw his weight behind the movement, begun in the last year of Reyes Heroles's tenure as director general, to unionize *empleados de confianza*. Echeverría overrode opposition from Dovalí Jaime, Francisco Inguanzo, and other PEMEX executives to secure the desired syndicalization on March 18, 1976.

Their deference notwithstanding, union leaders have at times conspired with PEMEX officials to deceive the chief executive. During the Alemán period, they connected pipes from functioning wells to those yet to be completed at the site of a presidential visit. This conveyed the impression that work was on schedule and production was increasing as planned. Venezuelan president Carlos Andrés Pérez was reportedly shown these "political wells" at the time of his March 1975 visit to Mexico. López Portillo was also taken in on March 18, 1978, when he inaugurated a new airport and adjoining facilities in Poza Rica that, while apparently completed, required another nine months of work before commencing operations.[21]

Sources of Union Power

The STPRM is one of the most powerful unions in Mexico if not in all of Latin America. Its strength derives in part from the sector in which it operates. The ever-growing importance of petroleum to Mexico gives the *petroleros* a pivotal role in the nation's economic life. Although there has not been a major strike since the Bermúdez years, this threat is so forbidding a weapon that Dovalí Jaime readily acceded to union demands for significantly higher wages and benefits lest it be employed. The role of the oil workers in the collapse of the shah's regime in Iran has further underscored labor's key position. This position is enhanced by the leaders' skill in portraying their organization, which kept oil flowing in the difficult postexpropriation period, as a major contributor to the fulfillment of the Mexican Revolution.

The existence of locals throughout the country also nourishes the STPRM's power. These units are widely distributed, and most are in towns or small cities where few if any other organized groups can match their resources. This enables each local to play a dominant—if not *the* dominant—role in its area's social, economic, and political life. The locals invariably boast close ties with municipal, state, and federal officials, many of whom are union members. In recent years secretaries general have held seats in the federal Senate or House of Deputies. An exception is La Quina, who turned down López Mateos's offer of a Senate post from Tamaulipas.[22] Sen. Sergio Martínez Mendoza, a leader from Minatitlán who headed the union from 1974 to 1976, has been the most prominent union official in elective office in recent years. In the elections of July 1, 1979, the STPRM increased from three to five the number of its members who hold seats in Congress. Two of the deputies come from Veracruz, two from Tamaulipas, and one from Mexico City.

The union enjoys impressive financial holdings. These derive from dues that total 2.5 percent of the salaries or pensions of all members (active and retired), PEMEX contributions to the STPRM's "social fund," payments made to locals by private firms in lieu of hiring union personnel, and earnings of farms, supermarkets, credit unions, and other union-owned facilities. It is widely alleged that leaders augment their incomes through the sale of jobs, stealing of equipment, and manipulation of contracts.

The "closed shop" enjoyed by the Oil Workers and other Mexican labor organizations also contributes to the union's power. This means that the union hires all workers except *empleados de confianza.* As discussed below, this authority gives leaders a crucial role in suppressing dissidents. Other unions, which may envy the economic status of *petroleros,* nonetheless work with the STPRM within the Confederation of Mexican Workers. In 1979 a member of the Oil Workers' Union was secretary general of the labor council of the CTM in Veracruz. Meanwhile, Oscar Torres Pancardo, STPRM secretary general from 1977 to 1979, served nine months as president of the Congreso del Trabajo, a loosely knit organization that attempts to define and coordinate common policies for all labor groups. And Fidel Velázquez, the strong-willed head of the CTM and a man sometimes called "Mexico's George Meany," has extolled Hernández Galicia for "the advances he has achieved in social matters as well as in economic benefits."[23]

As important as any factor in the union's power is the leadership provided by Joaquín Hernández Galicia. Son of a union member and himself a welder by trade, La Quina fought his way through the ranks of Local 1, serving as secretary of labor, secretary general, and president of the Majoritarian Unification Group, a faction within the local.[24] In 1962 he won election as national secretary general of the STPRM. In recent years he has neither held an official post nor occupied an office in the syndicate's drab Mexico City headquarters. Rather, he conducts his affairs from a comfortable but unpretentious home in a quiet suburb known as Colonia Unidad Nacional, several miles from Ciudad Madero.

While waiting there on May 31, 1978, for an interview, I witnessed scenes reminiscent of an episode from *The Godfather* unfold outside his gate. Men gathered to plead for work or a handout; women came to have unfaithful husbands disciplined; children delivered shiny, cellophane-wrapped gifts; and politicians sought to ingratiate themselves with the man who can make or break their careers. He enjoys close ties with elected officials, including the governor of his adopted home state, Tamaulipas. That President Adolfo López Mateos dined in his home on March 18, 1963, symbolized this relationship between government and major labor groups. La Quina and his supporters regularly influence the appointment of government officials, especially in oil-producing regions.

During the six hours spent waiting to be admitted, I had ample opportunity to talk with his bodyguards, other retainers, and neighbors while observing the behavior of those who sought his time, advice, and largesse. Although it was known that La Quina was at a union ranch outside town and would not receive visitors that evening, at least sixty people came by to inquire about his schedule, chat with associates, or leave a message. He did arrive that evening, around 7:30, in a three-vehicle caravan, apparently for protection. As his truck appeared the twenty-five or so people on hand scurried to see and be seen. They obsequiously opened the gate to his yard, held doors, and cleared a path for him among the faithful, who formed into a semicircle at a respectful distance from their hero. One by one those fortunate enough to be summoned approached La Quina, who intently listened to the plea before giving an order to a subordinate or jotting a note to the appropriate person who could satisfy the request.

Born in a petroleum worker's family in Veracruz in 1922, the slightly built, ramrod-straight leader describes himself simply as an unofficial "administrator of social works" for Local 1.[25] In this capacity he shrewdly manages a multimillion-dollar operation, for the local boasts a credit union, a large food and department store, ranches embracing 19,760 acres, and commercial buildings such as the one in which the local PRI headquarters is located. He and the union also obtain money from the sources described above. Yet his influence derives from much more than his widely praised managerial prowess. *First,* he exhibits a magnetic, arresting personality. Whether talking to a humble day laborer or a well-heeled businessman, he appears absorbed in the conversation, as if no one else existed and the topic at hand was the most important in the world. Observation suggests that he evokes in many followers the combination of awe, respect, adulation, loyalty, and fear that one associates with mobster leaders, to whom detractors invariably compare him. A loyal, devoted *petrolero* official told me: "He is considered as a god and a king."[26] Mexicans often follow men more than ideas, and La Quina's magnetism has attracted a *camarilla,* an informal brotherhood of fiercely loyal followers whose own advancement, prosperity, and influence depend on Joaquín Hernández Galicia, a name that they utter with reverence.

Second, he has placed his confidants in key posts throughout the country as well as in the central headquarters. There his prime contact has been Ricardo Camero Cardiel, the shrewd, often disagreeable secretary of interior who is a committed *quinista* from Local 1, where he used to run the union's retail consumer stores. La Quina's supporters are frequently active in the Majoritarian Unification Group, one of several factions that competes in STPRM elections. It provides an unrivaled source of information about the organization's activities. In the words of one retired leader, "Not a leaf falls without his knowing about it." Knowledge is power, as is the ability to place adherents in highly remunerative jobs. *Third,* in addition to jobs, the "administrator of social works" has patronage in the form of credit, housing, medical care, and other items which form part of the social benefits enjoyed by the union. For a family whose breadwinner is approaching retirement, the prospect of a lifetime job for a son or a snug home in a *colonia petrolera* is a powerful inducement to back the leadership. Retirees, whose pensions are often modest, are particularly vulnerable to manipulation and sometimes become retainers for union chiefs.

Finally, Hernández Galicia knows how to wield the stick as well as the carrot. He and his cohorts readily employ blackmail, intimidation, and violence to maintain their hegemony, according to Samuel Terrazas Zozaya, a former senator and union head. Prospective workers have been required to sign blank sheets of paper as a prerequisite of employment, enabling the leadership to "accept their resignations" at an opportune time. The leader of an opposition movement within the union has charged that La Quina has a "paramilitary corps" composed of three thousand thugs to break up union assemblies when opponents mount serious challenges.[27]

Heriberto Kehoe Vincent, the union's extremely adroit secretary general from Local 30, was assassinated on February 28, 1977, in a killing that had some of the trappings of an execution in New York's "Little Italy." Kehoe, known by friends as "El Güero," was ostensibly in Poza Rica, his home base, to observe the progress of a school being built by Local 30. It was approximately 10:45 A.M., and he was leaving the Chalet restaurant, having just completed breakfast with fifty union leaders. On the sidewalk outside he gave a solicitor a contribution to the Red Cross. Then a bullet fired at close range from a thirty-eight-caliber pistol pierced his head, entering behind his right ear and leaving through his left ear. Someone in the crowd quickly gunned down the assassin, Antonio Madrigal Mendoza, a recently dismissed petroleum worker who had earlier told his wife: "I'm going to the union headquarters and if something happens to me, sell everything and go to Tampico to live. It's possible that they'll kill me."[28] A thorough search and the questioning of forty witnesses failed to turn up either the gun used on Kehoe or the revolver that killed Madrigal, whose mother died upon hearing of her son's demise. Kehoe was reportedly fashioning a base of support among younger workers and those in the Reforma area with the goal of supplanting La Quina and breaking up the *"mafia petrolera."*[29]

Even though La Quina and his chief confederates had been at odds with the victim, they dutifully hurried to Poza Rica, where twenty thousand people, including school children, watched the funeral cortege. For two hours virtually all activity ceased in the city. After Kehoe's burial in the La Santisima Trinidad cemetery two miles outside town and the personal conveying of condolences to the black-veiled widow and her children, La Quina and three leaders (Sen. Sergio Martínez Mendoza, Efraín Capitanachi, and Salvador Barragán Camacho) held a private meeting in a room next to the one in which Kehoe's remains had reposed several hours before. Later, when members of the family tried to contest the widow's right to her husband's property, La Quina reportedly came to the rescue, warning them to leave her alone. For this intercession the man whom many believe was responsible for the killing earned the gratitude of the bereaved woman.

Like a godfather, Hernández Galicia has cultivated the image of benefactor and protector of his flock. Among the most notable projects in Ciudad Madero are a general store, clothing store, appliance store, clothing factory, supermarket, cinema, recreation pavilion with an Olympic-sized swimming pool, credit union, kindergarten, and primary and secondary schools.[30] Conversations with an unscientifically selected sample of men, women, and children in the area reveal a genuine affection for La Quina, despite the myriad charges of wrongdoing. Moreover, prices charged in the various union-owned commercial establishments seem competitive with those obtaining for products sold in other stores, even though critics allege overreaching. Still, Ciudad Madero abounds with crumbling sidewalks, faulty drainage, deteriorating buildings, and pothole-filled streets, many of which are unpaved. Although other locals have copied some of its social projects, Local 1's hometown is hardly a monument to union enlightenment.

On the other hand, La Quina has helped secure generous wages and the best fringe benefits in Mexico for the 92 percent of PEMEX's approximately ninety-five thousand employees who belong to the union. In 1975 the average salary for all company employees was $4,177, compared to $2,442 for a cross-section of employees in 867 industrial establishments.[31] The year before Petróleos Mexicanos had spent the equivalent of $745 per worker on education, medical services, housing, recreation, and miscellaneous items. The union also looks after the "petroleum family." Over 50 percent of the individuals hired by PEMEX in 1976 were related to current employees.[32]

Corruption

We have seen in this and earlier chapters that corruption suffuses the Oil Workers' Union. How did it begin? What form does it take? What, if any, opposition has emerged to combat it?

Corrupt practices, no doubt present among the various unions that operated before expropriation, were observed during the earliest days of PEMEX. Antonio Vargas MacDonald suggested that the takeover of the industry represented a victory in which the workers had taken part. "And after the conquest comes the booty."[33] Yet the financial embarrassment of the state monopoly during World War II limited opportunities for enrichment. Such opportunities mushroomed under Bermúdez, especially in his second term, who believed it was better "to pay centavos to leaders than to lose millions of pesos in strikes." Beginning in 1946 the union received 2 percent of the value of contracts negotiated between PEMEX and private firms. These agreements loomed increasingly more important in the drilling field as the number of wells completed by outside contractors grew from one in 1947 to 44 in 1949 to 117 in 1951 to 149 in 1952. In theory, the union's percentage served as a means to finance educational and recreational activities; in practice, it provided an opportunity for enrichment.

The first secretaries general had come from the ranks of the workers. By the early 1950s, a new breed of professional leaders—intermediaries between the workers and the patron, Petróleos Mexicanos—emerged. PEMEX's willingness, first enshrined in a 1947 accord, to pay their expenses and allowances stimulated the development of professional syndicate officers. Many now enjoy salaries, lavish fringe benefits, and opportunities for graft without having worked for ten or fifteen years. Perhaps the first notorious leader was Enrique López Naranja, who served as secretary general in 1952 and 1953. The union boasted 139 full-time officials by 1971.[34]

An explosion in *vendeplazas* or job-selling, a practice traceable to the postexpropriation era, took place under Pedro Vivanco García, secretary general in the 1959–1961 period when Antonio Vargas MacDonald reported that a "river of gold" flowed into the leaders' pockets. Depending upon the region and job, permanent positions commanded between six thousand and fifteen thousand pesos.[35] Some leaders charged even more. On one occasion a sixteen-year-old boy elbowed his way through the crowd when Gutiérrez Roldán was making a visit to the South. The youngster complained that he could only afford ten thousand pesos for a job, half the amount demanded by union bosses. The director general introduced him to López Mateos, who happened to be in town to make a banquet speech, and the president granted him a permanent position in the industry.[36] Without presidential intercession less fortunate aspirants paid the going rate.

As in other fields of endeavor, inflation has beset *vendeplazas*. Transitory positions now bring two thousand pesos (twenty-eight days), four thousand pesos (sixty days), and six thousand pesos (ninety days). In mid-1978 unscrupulous leaders obtained up to forty thousand pesos for a "general worker's position (*obrero general*)," while a mechanical-engineering post went for one-hundred-fifty thousand pesos.[37] A critic of the union estimated that jobs related to the construction of the famous *gasoducto* from Cactus to the North would yield between fifty thousand and eighty thousand pesos.[38] In late 1979 refinery workers in Tula, Hidalgo, accused the secretary general of Local 35 of exacting eighty thousand to one hundred thousand pesos for permanent positions from workers who lacked even a day's experience in the industry.[39]

La Quina, who has called job-selling the most serious problem bedeviling the union, has urged support for legislation to increase from six months to six years the minimum penalty for conviction of this offense.[40] This proposal is of little use, since few workers are brave enough to prefer charges against those culpable. Soon after winning election as governor of Tabasco, Leandro Rovirosa Wade convened an open assembly in Villahermosa so that aggrieved persons could point the finger of guilt at bosses who had demanded money for employment. Despite the ubiquitous corruption, not a single charge was made by the men and women present, who feared for their safety and that of their families.

Union leaders often own or serve on the boards of directors of companies that have contracts with PEMEX. Such an apparent conflict of interest is legal under Mexican law. What is not legal is the practice of failing to do the work, completing only half the job, or using shoddy material. Also widespread is the leaders' embezzlement of funds earmarked by PEMEX for roads, kindergartens, clinics, or other public projects, according to Francisco Alonso González, a petroleum engineer, *Excelsior* columnist, and author of an important book on the history of Mexico's oil industry.[41]

Reports abound of abuse of personnel: Workers seeking transitory employment have been forced to perform free services for STPRM officials; retirees, who eke out a living on modest pensions, have been compelled to perform labor on union farms; and girls have been required to trade sexual favors for secretarial positions.

Union leaders often flout the law with impunity. For example, dissidents insist that beatings and murders are a way of life among refinery workers in Salamanca, where Ramón López Díaz, a member of La Quina's *camarilla*, owns ranches, subdivisions, and a newspaper. He also serves both as president of Local 24's Majoritarian Unification Group and as a representative of the Oil Workers on the PEMEX Administrative Council.[42]

López Díaz played the leading role in a January 12, 1978, scandal that demonstrated the union's venality and power. Two eyewitnesses identified the labor leader as having shot and killed Silvia María Priego Ferrer, a twenty-two-year-old transitory worker, during a predawn drunken orgy in the syndicate's Salamanca headquarters. A clumsily orchestrated effort to disguise the incident as a suicide fooled no one. Nevertheless, López Díaz—described by the governor of the state as "my best and closest friend"—not only escaped incarceration, conviction, and punishment, but the Salamanca police declined to investigate the accusations against him, serving instead as his protectors. Meanwhile, a funeral-home operator, who incidentally owed his position as a local legislator to López Díaz, assured the victim's parents that López Díaz couldn't possibly have fired the lethal shot because he was with the governor at the time the death occurred.[43]

The case of Pedro Vivanco García, generally recognized as one of the most unscrupulous leaders, demonstrates the role of PEMEX officials in the corruption process.[44] Vivanco García was a creature of the union movement in Poza Rica, where Jaime J. Merino had been the area superintendent for Petróleos Mexicanos since his appointment in 1942. As we have seen, Merino controlled not only the oil industry but the political and economic life of the city, in which he held the Ford agency, owned the Hotel Poza Rica, and profited from choice real estate holdings. He operated without supervision, central control, or auditing. He built a baseball park with PEMEX funds and required all *petroleros* to purchase season tickets, the proceeds of which were never reported. He constructed hundreds of homes with company funds and assigned them to political loyalists. He got a cut on many business activities, such as the three pesos a day he exacted from taxi drivers. He is also believed to have profited from the generous commissions doled out to favorite contractors such as a young man named Jorge Díaz Serrano.[45] So complete was his dominance that he manipulated the election of his city's municipal president and vetoed appointments made by the governor of Veracruz, the state in which Poza Rica is located. When a friend was rejected as head of the local transit system, Gov.

Antonio M. Quirasco said plaintively: "What do you want me to do; I run the rest of the state, but Merino is in charge of Poza Rica."[46]

Merino never hesitated to use company money to corrupt and manipulate labor leaders who in turn could help control the workers. He sometimes hired temporary workers to provide a majority for favorites in union assemblies; he even assigned permanent employees to inconvenient shifts to prevent them from attending meetings. He rewarded friends and bought off troublemakers by conferring jobs without duties, creating a corps of "aviators" who flew into the payroll office to collect their monthly checks, only to disappear for another thirty days. So widespread was the practice that checks totaling two million pesos lay unclaimed in the PEMEX office in both January and February 1959, after Merino left Poza Rica.[47] These paychecks bore the names of local notables, including the bishop of Papantla, who had never done a day's work for the oil company. Merino allowed others to supplement their PEMEX pay with government posts.

Vivanco became a disciple, prepared to employ violence to accomplish the *cacique*'s objectives. For example, he was accused of participating in the infamous "Poza Rica massacre" of October 6, 1958, when *pistoleros* fired at demonstrators in front of Local 30's headquarters. The marchers, members of the Poza Rica Democratic party (PDP), were protesting the fraud committed in the election for municipal president and proclaiming the victory of their candidate. Newspapers reported that at least five people were killed, but as many corpses were believed to have been buried secretly, the number of victims remains in doubt. Dr. Francisco Villa, head of the PDP and son of the famous revolutionary, named Merino as the intellectual author of the violence, to which he linked Vivanco. Neither was prosecuted.[48] In return for unswerving loyalty, Merino aided Vivanco financially, acquiesced in *vendeplazas*, and helped him secure a gasoline station as well as contracts with Petróleos Mexicanos.

Almost everyone in the industry knew of Merino's fiefdom. Francisco Inguanzo often urged Bermúdez to discharge the superintendent, and the director general even scheduled a trip to Poza Rica for that purpose. But he failed to follow through on the dismissal, possibly because President Ruiz Cortines intervened on Merino's behalf. Merino ran afoul of López Mateos over an appointment when the latter was minister of labor. Upon assuming the presidency, López Mateos had the arrogant official shifted from Poza Rica to the PEMEX office in Los Angeles on January 11, 1959. The next year Mexico sought to prosecute Merino for misappropriating two million pesos while superintendent. However, the United States government refused extradition.

Bereft of his patron, Vivanco soon fell on hard times. After completing his term as the union's national secretary general, he sought the same post in Local

30. He apparently won by a narrow margin, but the outcome was contested. At that point the minister of labor, Salomón González Blanco, warned that he might be certified the winner, but that any problems that developed in Poza Rica would be laid at his doorstep. In view of this not-so-veiled warning, Vivanco renounced union politics, sold his property and gas station to PEMEX, and moved to Mexico City. Gutiérrez Roldán subsequently withheld contracts from him, and his income sagged.[49] Later he and La Quina—once compared to John Dillinger and Baby Face Nelson—clashed,[50] and the former's star in the union movement declined as the latter's rose.

In the last several years, corruption has flourished as never before because of the volume of oil produced, the higher prices which it commands, and the rush to accelerate output. "Politicians and union leaders are destroying not only Petróleos Mexicanos, but the entire country," claimed Julián Amado Saade Atille, general superintendent of the petrochemical complex in Poza Rica.[51] The flagrancy of the fraud has precipitated attacks against union and company practices from three sources. To begin with, many Mexico City newspapers and magazines—*Excelsior, Uno Más Uno, Proceso, Impacto*—regularly excoriate the tawdry activities of the STPRM, exposing the difference in lifestyles between its affluent leaders and the penurious masses in oil-rich areas. *"Selva Petrolera,"* an editorial carried in the January 16, 1978, issue of *Excelsior,* typifies this aggressive journalism. The piece described union malfeasance and berated López Díaz as a "powerful and stupid drunk." One publication has blamed the epidemic of accidents, including one in Tabasco that took at least fifty-two lives, on PEMEX's "history of errors, fraud, concessions and favoritism" that has found unqualified personnel in critical positions.[52] In addition, militancy is growing within the ranks of the grossly abused transitory workers. Evidence of this appeared in a bitter and protracted sitdown strike against the selling of jobs that began in October 1977 outside the union hall in Agua Dulce, Veracruz.[53]

Moreover, seventy-two suspended temporary workers at the Tula refinery formed the "Movimiento Hidalgo" to protest job-selling and other corrupt labor practices, in which they alleged labor and management worked hand in glove.[54] Also vocal has been the Civil Association of Transitory Petroleum Workers, which has asked López Portillo to intercede to require the STPRM to recognize seventy thousand workers as *"miembros supernumerarios."* This status would give them preeminent right, according to seniority, to newly created permanent positions.[55] Furthermore, the leaders of the National Petroleum Movement, founded in the early 1970s, have become more and more skilled at publicizing union wrongdoing and attacking PEMEX for compromising national interests. Their demonstrations and those of other dissidents have attracted increasingly larger crowds.

The union and PEMEX have met this threat by revoking the work contract of Hebraicaz Vázquez Gutiérrez, the movement's president, who had not only alienated La Quina and his entourage but publicly urged an investigation into the sources of Díaz Serrano's "enormous fortune" as part of a sweeping inquiry into the ownership of drilling companies holding contracts with the state firm.[56] It was alleged that Vázquez, known as an incorruptible secretary of labor for Local 15 where he suppressed *vendeplazas,* had accepted payment for a day when he did not work. Thus after he had spent nineteen years with PEMEX, his contract was terminated.[57] Moreover, he found it nearly impossible to find another job, reportedly because the union movement had placed him on a "blacklist." In the past this kind of action exacerbated alienation, quiescence, and passivity—traits common among Mexico's working class. Men such as Vázquez have stressed that they will not tuck their tails and run but will resort to violence if no other means can be found to attain their rights.[58] Widespread violence could diminish the union's ability to "deliver" on deals, threaten the alliance between the STPRM and the company, and limit the amount of oil lifted in Mexico. Yet Vázquez, who finally found a minor post in a state-owned sugar plant in Morelos, has apparently renounced political activity. His failure to prepare others for leadership positions leaves the National Petroleum Movement but a feeble opponent of the ever-stronger Oil Workers.

La Quina, his *camarilla,* and the STPRM could not maintain their grip on the *petroleros* without the active support of authorities in Mexico City. It is widely believed that Jesús Reyes Heroles—a former director general of PEMEX, López Portillo's first minister of the interior, and architect of a controversial political reform designed to magnify the voice of opposition parties—advocated an attack on union corruption. PEMEX has shown no enthusiasm for such an effort, and recent actions have strengthened, not weakened, the STPRM's position. As a kind of "farewell present," President Echeverría granted the union the right to organize professionals and technicians who had heretofore served at management's pleasure. We have previously noted that in the view of at least one official, this step has diminished the efficiency of these skilled employees.[59] As these individuals advance into key management posts, the union can look forward to an even stronger hand in running PEMEX. One former union chief insists that the STPRM already has important responsibility in planning and that labor and management roles have switched: "[PEMEX] officials have acted as if they had union responsibilities while [STPRM leaders] discharged executive functions."[60] Díaz Serrano has recognized Echeverría's action, and as also noted, agreed that 2 to 2.5 percent of the value of every contract let to an outside firm would be contributed to the union's social fund—monies for which the government requires no systematic accounting. The gas pipeline from the Southeast to northern Mexico will generate over 1,250 million pesos, while all projects should produce 200 million pesos—ostensibly for the construction of schools, technological institutes, kindergartens, and streets.[61] In addition, PEMEX pledged to award 40 percent of third-party drilling contracts through the union's Committee on Contracts as a quid pro quo for the annulment of Clause 36 of the collective agreement with the STPRM, which required all work to be done by union labor. The union may farm out work that it lacks the capability to perform. Herein lies an opportunity to earn generous commissions on this second contract or to enter into arrangements with firms in which syndicate members have a financial interest.

Petróleos Mexicanos has helped retard the development of an anti-STPRM movement by disciplining or transferring workers who run afoul of union leaders.[62] When Salvador Barragán Camacho, La Quina's foremost ally, first headed the union (1971–1973), a candidate unacceptable to him and Hernández Galicia won election as secretary general of a local. A telephone call to the subdirector of PEMEX sufficed to have the man reassigned to another part of the country, where he could not make waves for the "Quinista empire." "Afterward came persecution, threats and even implication in some scandal to justify imprisonment."[63] STPRM leaders enjoy the protection of bodyguards whose names appear on the PEMEX payroll but who perform no productive functions. López Díaz's *guardias blancas* alone cost the firm an estimated two million pesos per month—without taking into account the materials they pilfer.[64]

Despite all this neither the Ministry of Labor, responsible for monitoring labor organizations, nor the attorney general's office has moved to require secret ballots in STPRM elections, expel López Díaz from PEMEX's Administrative Council, protect union assemblies from gangster-like assaults, insist upon an accounting of union funds, or demand that the seniority register be scrupulously respected when jobs are filled. In fact, Porfirio Muñoz Ledo, minister of labor under Echeverría, allegedly summoned a dissident leader to sign a memorandum that denied the existence of union corruption and handled him roughly when he refused to do so.[65] Under Echeverría the Ministry of Labor is believed to have required workers from the large Cadereyta refinery to join La Quina's Local 1 instead of Reynosa's Local 36, which is much nearer and apparently had jurisdiction. The affiliation of these *petroleros* would have enabled Local 36 to surpass Local 1 as the largest entity in the northern zone, thereby allowing it to select the union's national secretary general in rotation with locals in Poza Rica and Minatitlán.

Why do PEMEX and other agencies acquiesce in, if not nourish, such flagrant corruption? The most cynical explanation is that malfeasance permeates the monopoly itself, and that a "clean-up" drive might not be limited to the union's Augean stable. Financial and personal connections between companies and PEMEX executives have for years influenced the dispensing of equipment and service contracts as well as concessions to operate gas stations.

PEMEX's Subdirection of Projects and Construction plays the principal role in awarding contracts. The Production Department often paid less than the Projects and Construction Department (forerunner of the subdirection) for comparable ventures during the 1970–1976 period. In addition, Projects and Construction frequently hired contractors who failed to adhere to professional standards and evinced careless workmanship as they cut corners to increase profit margins. Well-connected politicians, who have used the firm as a source of patronage, often own both the trucking companies that transport PEMEX products and the country's handful of private petrochemical plants.[66] It used to be the practice to give the wives of politicians a brace of company filling stations.[67] A former STPRM spokesman has charged that La Quina and PEMEX officials jointly own a drilling firm that has a contract for work in the Chicontepec area and takes jobs from union members in Poza Rica.[68] In addition, PEMEX employees and concessionaires have colluded to steal over two hundred million pesos worth of diesel fuel from refineries, possibly for sale in the United States, where the price is five or six times greater than in Mexico.[69] During the final year of the presidential term, known as the Year of Hidalgo, corruption reaches a crescendo, giving rise to the doggerel: "*Este es el año de Hidalgo: Buey el que no roba algo.*" [This is the year of Hidalgo; he is a fool who doesn't steal something.][70]

The government often masks concern over the loyalty and durability of its political base by emphasizing PRI's long tenure in office and the "stability" of Mexican democracy. In recent years peasants, students, intellectuals, bureaucrats, labor dissidents, and unemployed urban masses have demonstrated their disenchantment with the system. The army is not popular and can be deployed only in an emergency. But the petroleum, electrical, railroad, and telephone workers and other powerful unions within the Mexican Confederation of Workers, an official component of the ruling party, represent sturdy pillars of support for the regime. Undermining one of these pillars could shake the entire structure. Moreover, the omnipresence of peculation in the oil sector indicates that La Quina and his entourage might be replaced by men just as wily and unscrupulous but without the ability to manipulate the membership and fulfill agreements with the government. Thus Mexican politicians "emphasize political control and prefer to limit behind-the-scenes decision-making to as few participants as possible."[71]

When confronted with documented charges of widespread theft from refineries, López Portillo expressed the regime's "live-and-let-live" philosophy by remarking: "Our search is for solutions, not people to blame."[72] He has stated that Mexico would be in an "abyss" if the oil workers had not reacted as they did to the challenges facing the country in 1976, "but now we are heading for the summit, thanks to this response." Referring to the doubling of production in three years, he told Secretary General Torres Pancardo on November 23, 1979: "If other sectors had performed like you, Mexico would now be even better off."[73] Meanwhile, union leaders boast that La Quina, who fulsomely compares López Portillo to Cárdenas, enjoys the right of *picaporte* or immediate access at Los Piños. He also has a strong voice in personnel matters, as evidenced by his having blocked the appointment in 1977 of a former PEMEX employee who had been hired to help develop a new wage scale for the personnel office.

Díaz Serrano, with whom La Quina claims to "get along well," and the union have apparently struck a bargain.[74] Petróleos Mexicanos will not contest the STPRM's many social, political, and economic prerogatives, there will be no full-scale attack on corruption, and the unionization of former confidential employees will not be challenged. In return, the union has backed the new director general to the hilt, agreed to the contracting out of work with private firms, and committed itself—rhetorically, at least—to curbing illicit practices.[75] It has apparently made a good-faith effort to spur productivity without pursuing a major expansion of the work force or sharply higher wages (fringe benefits have materially increased). Thus in 1979 Díaz Serrano could boast that output had doubled in recent years while the number of employees had grown only 5 percent. In response to charges of inefficiency, he argued that daily worker yields had climbed from 6.8 barrels (1938) to 9.2 (1958) to 15.6 (1978).[76] "[How] is it possible," he asked, "that a group of corrupt people has raised the petroleum industry to the level where it is now found?"[77] The director general failed to point out that bounteous reservoirs of oil, not increased efficiency, have assured greater output. Nor did he discuss the dangers posed by the hiring and promotion of unskilled or poorly trained employees.

The union has defended the government from nationalist attacks in proclamations, publications, and mass rallies. Torres Pancardo rebuffed charges that PEMEX policy—such as the proposed construction of the *gasoducto* to the U.S. border—constituted a "sell-out" to imperialism. The facility does not compromise the nation's sovereignty, for "with less debt we will be more solvent and independent," he said. He also backed PEMEX against charges made by the governor of Tabasco that his state was being exploited. "Subsoil resources belong to the entire nation, not to a single jurisdiction," the labor leader insisted.[78] The secretary general also discounted the loss of hydrocarbons caused by the Ixtoc 1 blowout, erroneously stating that PEMEX managed to recover 90 percent of the oil and gas that spewed from the runaway well.[79]

The growth of production in Reforma has shifted workers to the Villahermosa area, and with them has come the prospect that a new group of leaders will emerge to contest La Quina's control. Thus the shrewd *jefe* has moved to obviate this possibility. In late 1978 Barragán Camacho, his chief lieutenant and handpicked choice to become the next national secretary general, visited Local 11 in Nanchital, Veracruz, in the Southeast. School children, the military-zone commander, local politicians, and STPRM officials were on hand to give him an enthusiastic welcome. Ceremonies in the Lázaro Cárdenas cinema included a rousing speech by the guest, the unveiling of a huge picture of La Quina, and a presentation of a check for one million pesos from Local 1 to Local 11. This gesture of generosity marked the first step in a plan to replace Local 10 of Minatitlán with Local 11 as the most important in the southern zone. After all, La Quina resented the unwillingness of Cárdenas Lomelí and Sen. Martínez Mendoza to take a strong stand against the efforts of Terrazas and Reyes Heroles to purify the union a decade before. Therefore he was casting about for a new alliance to consolidate his grip on the *petroleros* from the Río Grande to the Guatemalan border. La Quina, who enjoys strong support within Local 31 (Coatzacoalcos), can look forward to opposition from Locals 22 (Agua Dulce) and 26 (Las Choapas) as he attempts to secure control of the South. Ever attentive to the wishes of Hernández Galicia, the STPRM's 1979 convention extended the secretary general's term from three to five years, to apply only to the period of the next union head (1979–1984). This move gives La Quina and "Chava" Barragán, his lieutenant, an additional two years to help Local 11 surpass Local 10 as the South's largest section—and thus the right to choose Barragán's successor as secretary general. A requirement that workers in La Cangrejera, a new refining center in Veracruz, join Local 11 rather than Local 10 will bolster the size and strength of Nanchital. As a pro–La Quina leader in a Mexico City local told me: "The [Soviet] Politburo is innocent compared to us." The success of this strategy would extinguish the last hope that the evolution of the industry will bring with it the impetus for reform.

Oil, Jobs, and Economic Development

Unemployment in Mexico

Mexico City's cartoonists have found in oil an irresistible theme for their work. Some have depicted *el petróleo* as a guardian angel, fluttering down to save the nation; others have shown it as a new Virgin of Guadalupe, also playing a redemptive role; finally, it has been sketched as black gold, flowing profusely from a cornucopia-like vessel resembling Mexico itself.

These artists have captured the hopes and ambitions of politicians, who see in petroleum the answer to their country's pressing economic and social problems. On the eve of President Carter's February 1979 visit, President López Portillo proclaimed that Mexicans "are bent on solving our own problems with our own resources based on our unity and always seeking political and economic independence." Even more effusive was Díaz Serrano, who said in a speech to his nation's Chamber of Deputies: "For the first time in its history, Mexico enjoys sufficient wealth to make possible not only the resolution of economic problems facing the country, but also the creation of a new permanently prosperous country, a rich country where the right to work will be a reality."[1]

López Portillo has correctly identified unemployment as "the source of all injustice" in a country where over half of the 18.5-million-member labor force lacks employment or works only a few weeks each year. Policies pursued over the last three decades have exacerbated unemployment and underemployment, long a trait of the Mexican economy. Beginning in the 1940s, the nation's elites aggressively sponsored import substitution to spur industrialization. It was "taken as an article of faith that as soon as the domestic demand for a product was large enough to offer some hope of domestic production on a scale appropriate to the technology of the industry, every effort should be made to stimulate the necessary domestic investment and to eliminate imports."[2] The emphasis of the 1950s was on such consumer goods as clothing, footwear, textiles, canned foods, and similar items of daily household use. During the 1960s more sophisticated manufactures, including some capital equipment, were produced. The government fostered the process through fiscal and credit incentives, cheap energy, and expenditures on communications, roads, warehouses, ports, and other public works.

Equally important was protectionism accomplished through import licenses and other quantitative controls. For example, pressure from businessmen and industrialists led to the creation of one thousand new categories of protected articles each year between 1964 and 1970. The protection often took the form of a permit which had to be acquired before a foreign-made item could be brought into the country. Such quantitative controls, first introduced in 1948, covered two thirds of all imports by the time Echeverría assumed office.[3] Meanwhile, the government subsidized the export of some Mexican goods. Economist Gustav Ranis argued that in terms of employment and income distribution, encouraging the export of manufactured articles makes more sense than indiscriminately extending import substitution to goods with high technical and capital requirements.[4]

The replacement strategy worked—to a point. Factories sprang up, the line of *"Hecho en México"* products increased, a new generation of managers received training, and profits soared, even though these earnings generally sprang from a high mark-up on relatively few transactions rather than mass sales. Still, members of a surging middle class again spoke rhapsodically of the "Mexican miracle."

Success exacted a formidable price. Bribes to helpful officials became an obligatory step in acquiring fiscal benefits, financial considerations, and import licenses. The industrialists used their growing political power to delay reform of a regressive tax structure, and the government manipulated the corrupt labor unions to keep wage increases below the rate of worker productivity. As a consequence the country's already lopsided income distribution grew more skewed in favor of these captains of industry and their economic allies. In 1950 the average income of the top 5 percent of Mexico's households was 22 times greater than that of the poorest 40 percent (that of the next highest 5 percent was 5.2 times greater). Two decades later the multiples had increased to 34 and 10.7 respectively.[5]

The goods fabricated under "hothouse" conditions tended to be extremely dear and of uneven quality. This militated against their sale abroad, and the percentage of manufactured exports—which had steadily climbed in the previous decade—started an inexorable decline in the 1960s.[6] The antiagriculture bias of the industrialization program distorted the economy and led to a decline in earnings from food and fiber exports as the country purchased more food abroad. Three factors were at play. The federal government lavished monies on industrialization at the expense of agriculture during the 1950s and 1960s. The population rose sharply. And to pacify union members who accepted modest salary increases, support prices for foods were fixed at low levels, especially between 1965 and 1970, thereby discouraging investment and plantings by the peasants. Imports shot up as exports flagged. The industrialists developed a voracious appetite for imported goods needed in their operations. A similar demand sprang from a proliferating number of quasi-public enterprises, many of which furnished the infrastructure demanded by the economic elites. At the same time, families at the apex of the social pyramid acquired a taste for Mercedes automobiles, Scotch whiskey, and Mediterranean vacations—all of which added to the foreign exchange bill. The current account deficit rose from $908.8 million in 1970 to $2,558.1 million in 1975. The government, increasingly responsible for economic activity, negotiated foreign loans to finance both the yawning trade gap and its own budgetary shortfalls. The pressures on prices proved irresistible, and they started to rise.

Mexico's economy began to spin wildly out of control. Echeverría attempted to take charge by reducing the current account deficit and diminishing inflationary currents. To accomplish this he sharply restricted government spending in 1971. In response the trade deficit fell by 15 percent, and prices rose at a more moderate pace. However, the rate of economic growth also plummeted—from 6.9 percent in 1970 to 3.4 percent in 1971. Fearful of the political repercussions that protracted stagnation would cause, the chief executive reversed his course. Thereafter he ordered increases in public expenditures combined with an expansionary monetary policy.

He began to rail against the business community with a new-found populist rhetoric designed to curry favor in union halls and on the *ejidos,* communally owned parcels of land distributed under the agrarian reform. He used the "antipatriotic" propensities of investors to justify massive government spending. His administration proceeded to construct airports and railways while building seventy thousand miles of paved and dirt roads. It boosted public expenditures on education sixfold, on industry fivefold, and on agriculture fourfold. As a result GDP growth doubled over its 1971 rate and continued to climb.[7]

The situation was roughly analogous to President Lyndon B. Johnson's simultaneous prosecution of wars against poverty and communism in the 1960s. Although taxes were increased in Mexico, government expenditures outstripped revenues, spurring aggregate demand. Prices then spiraled upward at a rate higher than that in the United States, Mexico's leading trading partner. As exports became uncompetitive, the trade deficit widened only to be financed by prodigious external borrowing. Economic elites, outraged by Echeverría's populism, perceived that the twenty-two-year-old link between the dollar and the now overvalued peso might be severed. Thus they converted their holdings into somewhere between $4 and $5 billion for deposit in U.S. and European banks. The president's only significant act of restraint before he left office was to devalue the peso in August 1976. Because it "floated like a stone" at first, a second devaluation was required.

The failure to generate jobs for Mexico's masses was the heaviest social cost of both import substitution and Echeverría's eccentric stewardship of the economy. Central to the substitution program were incentives for domestic producers, public and private, to use capital-intensive rather than labor-intensive machinery and production methods. These producers enjoyed cheap loans and investment subsidies tied to the purchase of capital goods, inexpensive credit, and access to capital imports that were tariff-free or carried a nominal levy. State agencies, ubiquitous and important in Mexico's economy, pay no import duties. Moreover, an increasingly overvalued peso in the late 1960s and the 1970s enhanced the attraction of capital goods from the United States and Europe. In contrast to their counterparts in Korea, Hong Kong, and Japan, Mexican industrialists regarded exporting as a marginal activity. Too, items shipped abroad tended to be heavier industrial products—chemicals, metal goods, transportation equipment, electrical and other machinery—produced through capital-intensive methods. Consequently, exports valued at $1 million generated, on the average, about one hundred jobs in Mexican manufacturing in 1970, while an equal amount of Korean manufactured exports created employment for nearly five hundred workers in 1969. Mexico's fastest-growing export industries were—and still are—the least labor-intensive, partly because they depend heavily on imported parts.[8] The Echeverría administration promulgated laws to monitor and control the "technology transfer" of foreign corporations operating in Mexico to assure introduction of only the most advanced techniques. These are invariably capital-intensive.

Further, a number of disincentives—a minimum wage, high industrial salaries, low educational levels, politically potent unions such as the STPRM, and fringe benefits which include a Christmas bonus of up to a month's pay—attach to employing workers. As noted, the "closed shop" exists in Mexico, and once hired, a union member has the job as long as he or she can report for duty, irrespective of the quality of work. Echeverría contributed to this situation by conferring higher wages, approving generous social benefits, and deciding most industrial conflicts in favor of the workers, making them even more expensive to employ.[9] A study published in 1974 showed that Mexican manufacturers spent, on the average, about three times more per worker than Korean entrepreneurs when wages, salaries, social security expenditures, and other welfare benefits were taken into consideration, while the productivity of Mexican labor—in terms of monetary gross output for each worker—was about double.[10]

Psychological and political factors reinforce this preference for machines over men. Mexican elites, bitter over a legacy of U.S. discrimination, see in shiny new equipment a sign of the country's technological *machismo,* while labor-intensive techniques betoken backwardness in a society that disdains manual labor. At a mid-1979 public discussion, a prominent Mexican congressman categorically rejected his country's use of appropriate technology instead of the latest capital-intensive methods. "You can no longer expect to treat us like greasy little brown people," he said.[11]

Government politics also encouraged the mechanization of agriculture, an important and socially desirable source of jobs in view of the country's hyperurbanization. A sharp increase in wheat production followed the introduction of new, high-yielding plant varieties combined with improved soil preparation, planting practices, land management, and insect and pest control. Publicly financed irrigation played an essential role in this "green revolution," and the resulting profits enabled large farmers in the North to invest heavily in labor-saving machinery. They expanded their earnings even more by buying or leasing land from small farmers. The upshot was that the proportion of adult rural landowners declined from 42 percent in 1940 to 33 percent in 1970. And by the early 1970s, there were between 3.2 and 3.6 million landless peasants despite the commitment of the revolutionary system to agrarian reform.[12]

It is difficult to obtain reliable data on Mexican employment. In an analysis of 1970 census data, Saul Trejo Reyes found that 3.8 percent of Mexico's eligible workers are openly unemployed and 44.8 percent suffer serious underemployment. A more recent study concludes that for 1977, the relative figures should be 6 percent unemployment and 49.3 percent underemployment.[13]

A report published by the Mexican census bureau strongly suggests that the situation will get worse in view of the economy's changing structure. Although average annual output in the three sectors analyzed—industry, commerce, and services—grew faster during 1970–1975 (4 percent) than in the 1965–1970 period (3.1 percent), the pace of job creation slowed. An additional 585,000 workers raised production by 179,257 million pesos in the first period compared with the 544,000 workers added during the second period to raise output by 446,385 million pesos. The sharpest differential took place in industry, which hired 258,000 workers in 1965–1970 but only 104,000 in 1970–1975.[14] Care must be used in evaluating these figures, which have been adjusted neither for inflation nor the fact that employment embraces only wage labor and plant personnel.

Nonetheless, the study indicates that a continuing shift from labor-intensive to capital-intensive production explains the declining demand for workers. During the period under scrutiny, the major growth occurred in heavily capital-intensive industries: petrochemicals, steel, and oil. If the creation of employment continues to lag behind economic growth, Mexico will have to find jobs for about thirty-eight million people by the year 2000. This projection, which is based on the assumption that most women will not attempt to join the labor force, may prove conservative. Employment would have to grow at an average annual rate of 4.1 percent simply to maintain its 1977 level.[15]

These cold numbers, which roll so easily off the tongues of journalists, economists, and urban planners, take human form when one's car becomes snarled in one of Mexico City's frequent traffic jams. A leathery-skinned grandfather darts between automobiles, hawking a fistful of lottery tickets. A snaggle-toothed woman of thirty, whose wrinkles and burned-out eyes add a score of years to her age, implores you to buy a yellow plastic brush as a pair of tatterdemalions tug at her skirts. Meanwhile, two youths have thrust calloused hands through your open window seeking a few pesos for the buff job just accomplished on the front fenders. Just then horns blare, fumes rise—and you race to the next obstacle, where a new cast of characters takes over. In the absence of unemployment insurance or welfare, these street denizens support several people besides themselves. They often live in the one-room, mud-floored huts that compose the hundreds of shantytowns or *barracas periféricos* that gird Mexico City and other urban centers. Many abandoned the countryside in search of work and a better life. More than two thousand men, women, and children flock to the capital every day from the poverty-blighted *campo,* where two thirds of Mexico's unemployed reside.[16]

Population Growth

Not only is half the economically active population without year-round work, but an additional eight hundred thousand men and women enter the labor force each year. The economy seldom generates enough new positions to meet half that need, and in 1976 no jobs were created at all. The prospects are exceedingly gloomy because the rapid increase in population will continue this alarming trend. Especially ominous is the fact that 45 percent of all Mexicans are under fifteen years of age and, in most cases, have yet to begin raising families.[17]

In 1980 a baby was born in Mexico every sixteen seconds—224 per hour, 5,370 per day, and 1.96 million in the entire year.[18] The eradication of yellow fever, cholera, and the bubonic plague fifty years ago and subsequent advances in nutrition, drugs, and public health have cut mortalities to a half-million annually. The difference between these life and death figures results in a population that is growing faster than that of any major country. A Mexican born in 1900 had a predicted lifespan of thirty years; his grandson born sixty years later can look forward to living twice as long; and life expectancy is still increasing.[19] This situation has profound implications for governments on both sides of the Río Grande.

When López Portillo, who has two daughters and a son, took the presidential oath on December 1, 1976, he became the political father of a nation of sixty-three million. Even though all population statistics in Mexico are at best tentative, conservative official projections indicate that when the chief executive leaves office after six years, his family will exceed seventy-eight million,[20] a figure that the National Population Council (CONAPO) hopes to limit to one hundred million by the end of the century if the annual growth rate can be reduced to 1 percent. By then the population of Mexico City, which grew at 4.5 percent a year in the late 1970s, will have more than doubled.

Until recently few Mexican politicians considered their country's population explosion a problem. Luís Echeverría, father of nine, campaigned in 1970 with the slogan: *"Gobernar es Popular"* [To Govern is to Populate], and the ruling PRI talked earnestly of constructing more plants, creating more jobs, building more schools, and producing more food for the country's growing numbers. Leftists condemned family planning as either a plot by the Central Intelligence Agency to restrain Mexico's rising power or a scheme by multinational firms to sell birth-control pills to *campesinas* in Chihuahua and IUDs to well-to-do housewives in the capital. Women who sought contraceptives were treated as delinquents, according to Dr. Manuel Mateos Cardano, a leading Mexico City physician who strongly supports family planning. The Roman Catholic Church, far less influential in Mexico than in the rest of Latin America, also opposed artificial birth control. Family planning was left completely to private organizations.

Confronted by the sharply growing cost of social programs, rising unemployment, and serious economic problems, Echeverría endorsed the General Population Law, which took effect on January 7, 1974. This statute provides for free birth-control instruction and the distribution of contraceptives in three thousand centers, and the theme *"La Familia Pequeña Vive Mejor"* [The Small Family Lives Better] has been disseminated through multifloral posters, media advertisements, radio soap operas, and theatrical performances. The number of "users" doubled during the first four years of the program, to embrace 1.5 to 2 million of the 13.8 million women of childbearing age (fifteen to forty-nine). In late 1979 Dr. Jorge Martínez Manautou, national coordinator of the family planning program, estimated that 44 percent of the country's fertile women (whom he numbered at 11 million) practiced some form of contraception.[21] Profam, a private group founded in 1978, has made contraceptives readily available by setting up highly visible displays in pharmacies throughout the country.

Abortion is an increasingly popular form of birth control. Even though Mexican law prohibits abortion except under special circumstances—depending upon the state, the seldom-imposed prison term varies from three months to six years, with poor women having large families accorded leniency in Chiapas—five hundred thousand to eight hundred thousand are induced each year, according to a highly regarded study co-authored by Dr. Alan Keller, an expert in Mexican demographics.[22] CONAPO estimates the number of abortions to be eight hundred thousand, but women's organizations believe 1.2 million is a more realistic figure.

Although private hospitals perform this procedure openly without legal sanctions, over 75 percent of all abortions are either self-induced or carried out by unqualified practitioners, often under unhygienic conditions. This situation led the Communist party to introduce legislation to legalize abortions, which would be free in government-run hospitals. The measure, under consideration by Congress in mid-1980, aroused scathing criticism from the Roman Catholic Church, which acquiesced in the widespread promotion of contraception on the tacit understanding that abortion would remain illegal. Monseñor Genaro Alamilla, secretary general of the Conference of Mexican Bishops, asserted that powerful industrial nations might be behind the birth-control effort in order to keep Mexico in a dependent position. The National Action party, with traditional ties to the Church, has proposed right-to-life legislation as an alternative to the Communist initiative.[23]

Mexico's overall birth rate is believed to have fallen to 36 per 1,000, which would mean that the rate of population growth declined to approximately 2.9 percent in 1979, with 2.5 percent as the target for 1982.[24]

While expenditures for *planeación familiar* have increased along with earnings from Mexico's vast petroleum reserves, the government seems to work at cross-purposes. Women with exceptionally large families are acclaimed and rewarded in mother's day celebrations every May 10; public employees covered by social security for at least three years are eligible for a bonus equal to one or two months pay upon marrying; and the legal marriage age for women and men is fourteen and sixteen respectively—and twelve and fourteen in Puebla. While socially desirable, various material benefits—exemption from heavy or unhealthy work during pregnancy, six weeks of paid leave both before and after giving birth, two half-hour rest periods daily for nursing, required day-care centers for youngsters through age four—ease the difficulties of having children. Authorities have taken steps to halt Profam's distribution of contraceptive pills and other products in rural areas where medical supervision was deemed inadequate. Rather than curb Profam's campaign, the government might have mobilized recent medical school graduates, many of whom cannot find work, to participate in the program.[25]

Despite the bureaucratic problems that spring from the participation of a dozen governmental, quasi-governmental, and private agencies in Mexican family planning, the results in major cities are encouraging, especially among older, somewhat better-educated women who have had four children and five pregnancies. But these are the easiest people to reach. The program has barely touched many small cities, much less the countryside, where 38 percent of the population lives; where there are ninety-seven thousand towns with fewer than twenty-five hundred inhabitants (eighty-three thousand have populations under five hundred); where the average number of children per family is 5.6 as compared with 4.4 in urban areas; and where less than 6 percent of the population is believed to practice scientific birth control.

The dearth of doctors and clinics in remote areas is complemented by ignorance of effective birth-control methods, superstition, and cultural barriers. "*Machismo* is the greatest obstacle in the *campo*," stated Lucia Mier y Terán de Muñoz, a social anthropologist employed by the National Population Council who has done extensive field work. "Men demonstrate their masculinity by having children and also keep their wives 'faithful' when they are away from home—such as when they spend the harvest season in the U.S.," she added.[26]

Whether in cities or the countryside, Mexican males often oppose their wives' recourse to contraceptives and fear their own sexual prowess will be attentuated by a vasectomy, of which only several thousand have been performed in Mexico. No more than 4 percent of the men are believed to use condoms at home, as this birth-control method is associated with prostitutes.

Dr. Keller describes another dimension of *machismo*. This is the woman's desire to satisfy with children the emotional and physical needs poorly met by her he-man husband, to whom she is submissive—a peasant woman recently confessed doubt of her husband's love because "he has stopped beating me"—and with whom she simply does not communicate about sexual matters. Bold black letters on green posters sprouting up in working-class sections of Mexico City implore couples: "You Know About Family Planning—Talk It Over." But even well-educated middle-class women admit that their husbands resist talking about sex and procreation with them, and the subject is considered taboo in public discussions, according to sociologist Alfonso López Juárez, coordinator of CONAPO's national sex-education program.[27]

Because one or two of her children died as infants or at an early age, the *campesina* views a large family as necessary to assure the survival of a sufficient number of children to an economically productive age; after all, she has no social security, and her husband is an erratic provider. Such feelings persist even though the mortality rate of minors was cut in half between 1960 and 1970 and the population per square mile has leaped from 3.2 (1930) to 12 (1976).

Peer pressure for large families is almost irresistible in the countryside, where old wives' tales abound. "Diaphragms cause cancer," "an IUD will shock my husband's penis," "the pill will make me sterile"—these are but a few of the grassroots reactions recounted to me by health and social workers.

A confluence of factors—more clinics, a saturation media campaign, less living space, greater anonymity, new female roles—augurs well for birth control in center cities. But the greatest challenge facing family planning is to find means to tumble the barriers of ignorance, superstition, and custom in the *campo* and in the peasant-laden shantytowns that surround the major cities. Midwives, army recruits, local merchants, and practitioners of folk medicine known as *curanderos* might offer traditional and legitimate channels for penetrating these areas if an incentive system can be designed to elicit their cooperation. Ultimately, however, jobs are the answer. Until employment opportunities are created, the marginal contribution of another child to a family's security and wealth will exceed the marginal cost of the youngster. In the meantime population problems will persist as the countryside continues to export large numbers of its people to search elsewhere for jobs and a better life.

Illegal Immigration

Jobless Mexicans not only pour into major Mexican cities but also flock to the United States, for "nowhere else in the free world is there a boundary between two nations of such remarkably different levels of wealth."[28] Early each weekday morning Mexicans of all ages form a queue in front of the Río Danubio entrance to the fortress-like American embassy to fill out the forms, answer the questions, and submit the documentation that may eventually help them secure entry visas. Many gather but few are selected, as only about ninety thousand Mexicans may enter the United States legally each year. Even the fortunate have to wait several years before the paperwork is completed and an opening becomes available.

For the great majority who wish to go north, clandestine entry offers more promise of success. While there is no such thing as a "typical" illegal alien, most share certain characteristics.[29] They are young males, usually twenty-two to thirty years old, with only a few years of schooling at best. About half are married at the time of their first journey to the United States, though fewer than 1 percent who go as temporary residents take their families because of the additional expense involved. While less than 40 percent of the population lives in rural areas, the countryside contributes approximately 60 percent of the flow—with five north central states generating the largest number. Landless farm workers and residents of *ejidos* are most prone to migrate. They may have first learned of opportunities north of the border from their fathers who participated in the *bracero* program, which began during World War II and lasted until 1964 as a government-sponsored effort to provide Mexican workers to help harvest American crops. Wayne A. Cornelius, the leading American authority on the subject, found that the illegal aliens are highly motivated, seek opportunities to achieve, and exhibit less fatalism and submissiveness than do their peers.[30]

While an increasing number of aliens want to become permanent residents in the United States, the pattern of emigration remains "heavily temporary." As Cornelius points out, that permanency may vary inversely with proximity to the border: "The farther he is from the border, the less likely the migrant is to remain in the U.S. permanently."[31] Of course, some of the illegal aliens are middle-class students, tourists, or businessmen who lawfully gained access to the country only to melt into the masses in Chicago, New York, or Los Angeles.

For the great majority the act of migration, whether temporary or permanent, involves borrowing money from family, friends, or the local moneylender and traveling by bus or truck to a fetid border town. Tijuana, across from Chula Vista, California, and Juárez, the sister city of El Paso, Texas, are the most popular entry points. An increasing number of migrants rely on a *coyote* to facilitate the crossing. This is a cruel, cutthroat business, and the smugglers exacted between $250 and $300 per person for their services in 1980. Many clients are cheated: They pay their money to a contact who simply disappears. Others are told to rendezvous at dusk for a nighttime crossing that may be attempted by any one of a dozen imaginative methods—hiding in the recesses of a *coyote*'s vehicle, walking and wading through a labyrinth that includes traversing the Río Grande, or crouching in the hull of a boat that anchors at some isolated point off the California coast. Determined migrants have even achieved their mission through meandering sewage pipes that join some Mexican cities to their American counterparts. The migrants must be ever vigilant for U.S. authorities. In fiscal year 1979 the Immigration and Naturalization Service (INS) apprehended 998,761 illegal migrants, over 90 percent of whom were Mexicans.[32] Although an illegal entry is punishable by a jail term of up to six months, the volume of offenders would so strain federal prison facilities that most are returned to the border—from where they frequently initiate their next entry attempt.

Success may follow multiple failures, and once over the border the undocumented worker looks for a job. He may have to pay a broker to procure work; however, friends and family already in the United States may help him get started. A growing number of illegals reportedly hold skilled positions, but a first job will probably be picking fruit, gathering vegetables, washing dishes, cleaning tables, scrubbing floors, or performing basic construction chores.

Illegal aliens do not advertise their presence in the United States, as was evident to census takers in 1980. No one knows approximately, let alone precisely, how many there are. Gen. Leonard Chapman, former commissioner of the INS, spoke of an estimated range of four to twelve million.[33] At least five groups have voiced alarm at the size and vagueness of these numbers. Trade unions contend that the aliens are seriously "undermining and undercutting . . . wage and working conditions for U.S citizen workers and permanent residents."[34] The National Association for the Advancement of Colored People has expressed "grave concerns" about the illegals,[35] and black leaders have argued that the unemployment afflicting their people will continue at an unacceptably high rate as long as foreigners stream across the frontier. Some Chicanos (U.S. citizens of Mexican origin) express similar beliefs. Mexican-American labor leaders in El Paso complain that "wetbacks" from Mexico "are eating our lunch." In an impressionistic survey of Texans with Spanish surnames, Sen. Lloyd Bentsen found that 80.3 percent of the respondents favored legislation to deal with the "illegal alien problem" and 40 percent identified such workers as threats to their jobs.[36] In contrast, many Hispanic-American organizations welcome the newcomers who, by dint of sheer numbers, may enhance the political power of the Spanish-speaking population.

Ecologists take a different tack. They argue that millions of new people make impossible the stabilization of the country's population that is essential in order to keep up the environment without severe economic dislocation. The influx of Mexicans necessitates more factories, more houses, more automobiles—the very things that must be restricted to save natural resources.[37] Finally, a group of congressmen has warned that the migrants are involved in break-ins and drug-smuggling. Rep. Lester L. Wolff (D-New York) led members of his House Select Committee on Narcotics Abuse on a tour of the U.S.–Mexican border. Upon returning he compared it to a pre–World War II fortification: "We really have a Maginot Line. It is outflanked, overflown and infiltrated. And you know what happened to the French."[38] Meanwhile, Rep. James H. Scheuer (D-New York), chairman of the House Select Committee on Population, has advocated "a firm, hard sealing of the border."[39]

President Carter responded to these constituencies, most notably organized labor, with legislation in August 1977. His bill granted amnesty to aliens who entered the country illegally before January 1, 1970, and have resided in the United States continuously since then. They could apply for permanent residence. It also conferred the status of "temporary residents" on aliens who arrived between 1970 and January 1, 1977. After registering they could work and travel abroad but not receive welfare, food stamps, Medicaid, or other social benefits for five years. The government would reevaluate their position at the end of the period. Undocumented aliens who entered the United States after January 1, 1977, would be ineligible for temporary resident alien status. Other provisions would have increased the size of the U.S. border patrol 40 percent by adding two thousand agents and imposed civil penalties of up to $1,000 per illegal worker for employers who "knowingly" hired illegal aliens and criminal penalties on those who serve as brokers for such employers. The attorney general could seek both a civil penalty and injunctive relief where "cause exists to believe that an employer has engaged in a pattern or practice of employing aliens" in violation of the law.

The Illegal Alien Control Bill failed because of opposition in the Senate Judiciary Committee and because of mounting apprehension on both ends of Pennsylvania Avenue that a resolute border program would offend Hispanic-American spokesmen in a half-dozen populous states whose electoral votes are crucial to the outcome of presidential elections. Sen. Edward M. Kennedy's twenty-hour, highly publicized trip to Mexico, which included a meeting with López Portillo, came on the eve of the May 3, 1980, Texas Democratic primary and clearly was an attempt to use contacts with Mexico for domestic political purposes. The emergence of Mexico as a substantial oil-possessing nation has further diminished enthusiasm—in the executive branch as well as in Congress—for action at the border. This sentiment hardened into a negative reaction toward a proposed security fence, twelve feet high and six and one-half miles long, between Juárez and El Paso, where the Border Patrol arrests eleven thousand illegal aliens each month. Authorities estimate that five times that number evade detection. Critics viewed the "Tortilla Curtain" as a powerfully offensive symbol of tension between the two countries. "I don't believe that we and Mexico should have any sort of Berlin [Wall] on our borders," said William P. Clements, the successful 1978 Republican candidate for governor of Texas. Others decried the structure as useless because highly motivated migrants would surely walk around it. The Immigration and Naturalization Service, which had conceived the $2-million project, admitted that this would happen, but insisted that the wall would narrow the number of entry points and make patrols easier and more effective.

Because of the uproar in Texas and in California, where a similar fence was planned for San Ysidro, INS officials said in late 1978 that the undertaking had been "put on hold" and was being "re-evaluated" by the Carter administration.[40] The idea was subsequently interred as the president named Father Theodore M. Hesburgh, president of Notre Dame University, to head the Select Commission on Immigration and Refugee Policy, which will recommend a new, comprehensive policy to Congress by the spring of 1981. The commission has voiced a belief that the number of illegal Mexican immigrants living in the United States does not exceed three million. The appointment of this body effectively shelved the issue until after the 1980 presidential campaign. By that time an economic recession and rising unemployment may sharpen concern over the entry of undocumented workers.

Washington's swift retreat on illegal immigration also reflects an awareness of Mexican sensitivity toward the matter. The press and other opinion leaders in Mexico City invariably treat the subject in terms of benefits to the American economy, police brutality against workers, the failure of employers to honor contracts, and the paucity of social benefits conferred. Glossed over are the considerable advantages to Mexico's economy and the human-rights issue raised when a country cannot—or will not—provide opportunities for its own people. Also overlooked are the difficulties that Americans face in attempting to obtain employment or purchase property in Mexico. Announcement by the Ku Klux Klan that it would organize vigilante patrols along the 1,948-mile-long border precipitated a flurry of articles decrying the danger posed to Mexican citizens. The president of the minuscule Revolutionary Socialist party even alleged that the action had the blessing of the government of the United States, where "respect for human rights is nonexistent."[41]

Mexican spokesmen argue that the illegal workers contribute to United States prosperity by performing vital, "dead-end" jobs that American citizens will not accept and by keeping prices down because of the low wages they receive. While there is a strong element of truth in this contention, it neglects the fact that illegals are enjoying job mobility in the construction and automobile industries as well as in other fields. Moreover, what may be a menial, "dead-end" job for a foreigner may offer the possibility of mobility to citizens with roots in the community. For example, the language barrier, dread of deportation, and rootlessness often consign an illegal alien to a restaurant's steamy kitchen as a dishwasher for the duration of his employment. Yet a black or Chicano who makes good as a dishwasher may become a busboy and then a waiter, earning several hundred dollars per week in a good establishment. While all of these jobs may be considered menial, the waiter's higher pay materially affects his family's quality of life.

Though seldom openly expressed, an element of irredentism informs some comments about migration. The view may be that since the present American Southwest belonged to Mexico before the 1848 Treaty of Guadalupe Hidalgo, signed by Mexico only under duress, Mexican citizens have every right to come and go there as they please. After all, "movements of persons across America's southern border are really movements between two areas which were, for centuries, governed by Mexico City."[42] A declaration by Foreign Minister Jorge Castañeda carried a hint of this argument: "Mexico cannot constitutionally, politically, juridically, or morally take restrictive, repressive steps to impede the movement of Mexicans into or out of the country, or the entry or departure of Mexicans from United States territory."[43] But Mexican authorities made news in late June 1979 when, for the first time in ten years, they prevented workers from crossing illegally into California. At least one hundred persons—undocumented aliens and accomplices—were jailed at the military barracks in Aguaje de la Tuna.[44]

President López Portillo has sought to minimize the problem, comparing it to the flow of Europeans to the United States at the end of the nineteenth century. Like the European phenomenon, the Mexican migration embraces large numbers of poor Catholics, most of whom are young, unattached males who return home after working in America for limited periods. Similarly, the Mexicans speak a different language and are products of a different culture. But the geographic distance involved in the earlier movement and the arrival of all aliens by boat enhanced the element of control by limiting the speed and size of the flow. The European phenomenon also consisted largely of legal immigrants who entered during a strongly expansionist economic period when the country was predominantly rural and the demand was great for miners and factory workers. Recessions would then generate a return of Italians, thrown out of work, to Italy. In contrast, a *bracero* encountering hard times in the United States can anticipate worse conditions back home because of the linkage of the two economies. Finally, López Portillo should realize that despite the more favorable environment for the European immigration, it aroused such opposition that Congress enacted highly restrictionist legislation in the 1920s.

In the final analysis, the illegal immigration question may be so sensitive to López Portillo and other elites because it provides dramatic evidence of Mexico's inability to provide work for its people and its concomitant dependence on a neighboring country. Mexican policy makers insist that oil-based progress will overcome this situation, an extremely questionable assumption in view of the industrial development plan discussed later in this chapter.

Venezuela and Ecuador

Have other countries succeeded in channeling oil money into the pockets of the poor? What has been the record of Venezuela and Ecuador, two other Latin American nations that have attempted petroleum-fueled development?

The Spanish, hungry for gold and silver, took scant interest in the black sulfurous seepages they found in what is present-day Venezuela. They did send bottles of this heavy liquid, known derisively as "the devil's excrement," back to Iberia as a cure for the gout. But the substance appeared to have little other use since it could neither be drunk nor ploughed back into the soil.[45]

Although the commercial exploitation of oil did not begin until 1912, by 1930 Venezuela was the world's second-largest producer. North American and European companies rapidly gained control of the petroleum industry thanks to concessions granted by Juan Vicente Gómez, a cattle rancher, military man, and venal dictator who ruled the country as his own fiefdom for twenty-seven years. Between 1919 and 1936, the government of Venezuela obtained only 7 percent of the value of the oil lifted in its country, while some of the seventy foreign companies operating there occasionally recovered annual profits of 1,000 percent on their investments.[46]

Beginning in the 1940s, surging nationalism led to a more equitable distribution of earnings. A "fifty-fifty" split between the government and companies achieved in 1948 rose to seventy-eight–twenty-two by 1970. On January 1, 1976, President Carlos Andrés Pérez presided over the nationalization of the assets of the multinational firms. Earnings derived from the increase in crude prices since October 1973 financed this takeover. As a founding member of OPEC, Venezuela consistently urged higher prices for the product that generates 90 percent of its foreign exchange.

To long-time observers, Venezuela was again heading into the kind of upswing that had alternated with downturns for thirty years. The last such cycle began during the Suez crisis of the 1950s, when Venezuela was the world's largest oil exporter and dictator Marcos Pérez Jiménez justified a spending binge with modernizing, developmentalist rhetoric. Colonial buildings toppled before bulldozers that paved the way for skyscrapers and freeways. The armed forces were bought off with tanks, submarines, jet fighters, and a lavish Círculo Militar. The English-language *Caracas Daily Journal* wrote of streets paved with gold, and building proceeded at such a feverish pace that some readers wished for zippers in the streets so that they could be torn up and restored more easily. After reaching a peak in 1957, oil prices began to decline and with them the fortunes of Pérez Jiménez, who in the face of a military and popular insurrection fled the country in 1958, his luggage jammed with stolen funds.

Venezuela was again riding high in the 1970s. The initial impact of the cartel's 1973 action was to convert the country's current account deficit of $29 million in 1972 to a surplus of $787 million in a period of several years. The Gross National Product climbed from $18 billion in 1972 to an estimated $35.8 billion in 1978 when the Andean republic boasted the highest per capita income in Latin America ($2,652).[47] Especially favored were members of powerful unions such as the Oil Workers. To reduce inflationary pressures and control development, the government placed half the total oil income in the Venezuelan Investment Fund. It would help finance a new petrochemical industry, shipbuilding, completion of a massive steel-making complex in the Guayana region, and long-range agricultural programs designed to obtain self-sufficiency in food within three years. This fund obviated the need for foreign aid, as Venezuela made available larger sums than before to international financial agencies while providing economic assistance to Central American and Caribbean neighbors.

A significant portion of the revenues did not contribute to development. Venezuela deserves credit for developing and nurturing one of the hemisphere's few democratically elected governments. It has championed land reform, the organization of workers, educational improvements, and a host of other social programs. But its civilian leaders must constantly ingratiate themselves with the military, which demands a share of the accelerating flow of dollars. Even in the absence of a foreign threat, expenditures on the armed forces rose from $494 million in 1975 to $615 million in 1978, an amount equal to 5.9 percent of public expenditures. (The percentage in Mexico was 2.8 for the same year.)[48]

The military bureaucracy is eclipsed by its civilian counterpart. Oil revenues enabled the government's income to jump from $3.8 billion in 1973 to almost $11.8 billion in 1978.[49] A swelling army of public servants, billeted in shiny new office buildings, administers programs financed by these monies. Unfortunately, corruption and lassitude often attend the discharge of official duties.

The monies also flowed into monument-type projects—modern, showy, and expensive, but of dubious value to the country's economic progress. Undistinguished new buildings replaced some of the handsome old ones that Pérez Jiménez had left standing in Caracas, where the sidewalks have been torn up to make room for the ever-growing volume of automobiles that create some of the worst traffic jams in the world. Exemplary of the new architecture are the steel and glass towers that soar above the Parque Central.

Well-to-do Venezuelans have become conspicuous spenders in a nation which imports 85 percent of its consumer goods. Like their counterparts in other countries, they have acquired a keen taste for French fashions, Japanese electronic gadgets, and American automobiles, which have contributed to an 11-percent annual rise in fuel consumption in a country where "gasoline is cheaper than drinking water."[50] They also enjoy horseback-riding, servant-filled villas, swimming pools, and vacations abroad. In 1978 Venezuelans spent almost $1 billion more on travel and tourism than did visitors to their nation.[51]

Juxtaposed to this affluence is the "other Venezuela." "We are a poor rich nation," wrote Romulo Betancourt, father of Venezuelan democracy. "The spectacular volume of money is distributed in a repugnantly unequal fashion among Venezuelans." According to the former president, the top 20 percent of the population received 65 percent of the national income, the 40 percent in the middle earned 27.1 percent, and the lowest 40 percent got but 7.9 percent.[52] Carlos Andrés Pérez fulfilled one campaign promise soon after taking office in 1974 when he temporarily froze prices, doubled the monthly minimum wage for domestics to $70 (plus food and social security), and granted a 5 to 25 percent across-the-board pay raise to all persons earning under $1,200 per year.[53]

Still, the "have nots" at the bottom of the social pyramid are buffeted by inflation, the inevitable concomitant of public and private spending which outstrips the country's productive capacity. According to the Central Bank index, which tends to understate the situation, the rise in the cost of living in Caracas increased from 8.1 percent in 1977 to 12.3 percent in 1979 as the government cut back oil output from 2.36 million bpd in 1979 to 1.95 million in early 1980, exercised some budgetary restraint, and reduced the growth of the money supply.[54] But inflationary pressures propelled by escalating demand continue to pose a threat, although increased oil earnings that sprang from higher OPEC prices totaled less than one half the $6 billion originally projected for 1980 because of reduced demand for heavy fuel oils in the U.S. market.[55] Politicians have long since raided the Venezuelan Investment Fund, designed in part to control the influx of new monies into the economy.

The country's poor have steadily left rural areas for the perceived opportunities of urban life. Between 1950 and 1976 the percentage of the population residing in the countryside declined from 46 to 17.4. The poor live clustered together in the *ranchos* or shantytowns that cling like swallows' nests to the steep hillsides around Caracas and other cities. Many fail to find full-time jobs and must eke out a living shining shoes, washing clothes, or selling assorted wares along the car-clogged main streets of El Silencio, one of the capital's most congested areas. Some "urban peasants," free of the social constraint imposed by rural communities, have become thieves—a factor in Caracas's spiraling crime rate. In the mid-1970s the government put the force of law behind "make-work." It commanded enterprises employing more than ten people to increase their staffs by 5 percent; it also decreed that all elevators must be manned and washrooms attended. Such efforts have had but limited success, and industry remains highly capital-intensive. One reason for the attractiveness of machines may be the high wages and generous fringe benefits enjoyed by trade-union members. Moreover, the law entitles a person leaving his work voluntarily to one month's pay for every year employed. If he is discharged, the entitlement increases to two months salary for every year with the firm. While the official unemployment rate in 1978 was only 6 percent, around 20 percent of the population is underemployed in the sense of earning less than the legally mandated minimum wage.[56] The need for technically trained personnel has led the government both to send thousands of Venezuelans abroad to study and to launch a massive recruitment campaign directed at foreign professionals.

The migration of small farmers to the cities in search of jobs has weakened the agricultural sector, long the least dynamic component of the economy. At the same time, the demand for food has risen in response to a high birth rate and expanded purchasing power made possible by increased petroleum revenues after 1973–1974. These factors, combined with crop failures in 1975, have produced a rapid rise in agricultural imports. For example, in 1976 Venezuela bought $814 million worth of processed food from other countries, an amount equal to 9 percent of total export earnings.[57] Food imports contributed to a $2,053-million balance-of-payments shortfall in 1978, up from $156 million the year before. The financing of such deficits as well as huge investment projects has given rise to a $7,740-million foreign debt, as large—on a per capita basis—as that of Mexico and greater than Venezuela's foreign reserves ($5,878 million on January 31, 1979). The proportion of the federal budget devoted to serving this burden grew from 15 percent in 1978 to 20 percent in 1979.[58] The nation's external position was expected to improve as the completion of heavy industrial ventures in steel, aluminum, and hydroelectricity reduced the need for capital improvements, and oil revenues rose sharply.

President Luis Herrera Campíns, who took office on March 12, 1979, claimed to have inherited a "mortgaged Venezuela" beset by undisciplined public spending, a corrupt bureaucracy, deteriorating public services, profligate consumerism, swelling prices, housing shortages, a growing crime rate, and a balance-of-payments deficit.[59] He reiterated his commitment to retrenchment in government spending and talked about enforcing a modest income-tax reform reluctantly passed by Congress in 1978. He promised to stimulate agricultural production through fiscal and price incentives. Initial indications point to an active but less flamboyant international role complemented by emphasis at home on stabilization, tariff relaxation, greater price freedom, fiscal austerity, a reallocation of spending priorities, and reduced growth. Attention has also been devoted to the alleged involvement of Carlos Andrés Pérez during his presidency in an $8-million excess payment for the refrigerated storage vessel *Sierra Nevada.*

The threat of recession unexpectedly gave way to the danger of an overheated economy as higher prices increased oil earnings 53 percent in 1979, generating an additional $5.5 billion. The only major project announced by the Herrera administration in its first year was a pledge to spend more on low-income housing. Meanwhile, the government relied heavily on monetary policy to restrain the inflation rate that reached double-digit proportions in 1979. Prices increased by approximately 20 percent in 1980; as a result,

Herrera, whose popularity declined precipitously in public opinion polls, shelved much of his austerity program while drawing on petrodollars to raise government spending.[60] The additional revenues, which have postponed both a far-reaching tax reform and reorganization of inefficient public companies, may persuade Venezuela to embark upon another spending binge when the market for its oil improves.

Ecuador has also faced the challenge of prudently spending increased oil revenues. Its economy—long dependent on bananas, coffee, sugar, and cacao—changed in August 1972 when oil began flowing through a 313-mile pipeline from the Lago Agrio wells in the Amazon Basin across the Andes Mountains to the port of Esmeraldas. As production climbed to 210,000 bpd, the GNP rose from $2.2 billion in 1971 to $5.9 billion in 1977.[61] The country became the second Latin American member of OPEC the year after exports began. The government's income quadrupled overnight, and populist rhetoric filled the air as the military regime announced plans to provide roads, schools, health clinics, electricity, potable water, and sewage facilities to at least half the country's seven million residents by 1975. A quarter of the 1975 budget was earmarked for education, allocated to the National Development Bank for rural loans, and channeled to the National Development Fund for roads and a massive refinery.[62]

Divisions beset Ecuador's generals-turned-politicians, and their reformist ardor cooled. They did agree on higher military expenditures, which grew from $42 million in 1971 to $98 million in 1976.[63] Apart from military outlays per se, the armed forces received half of the government's 16-percent oil royalties or approximately $250,000 per day. These monies enabled the army, navy, and air force to apportion responsibility among themselves for such economic activities as cement production, metal works, petroleum tankers, fishing fleets, and a new state refinery.[64]

Quito, described by Simón Bolívar as a "convent" because of its conservatism and quaintness, has become a boom town. Cars crowd its picturesque squares, the English spoken by equipment salesmen competes with Spanish and Quechua along its narrow streets, and steel and concrete buildings fill a skyline that was once dominated by colonial-style churches. Many of these new buildings house employees of a bureaucracy that doubled to 150,000 in the five years after oil shipments began.[65]

Production could not match newly generated demand in this predominantly rural nation, and the rate of inflation increased from 9 percent at the beginning of the decade to 12 percent in 1979, with a 20- to 30-percent level forecast for 1980.[66] This meant that the poor, unable to protect their purchasing power, suffered a diminution in real income amid overall prosperity. A study produced by the University of Guayaquil revealed that an increasing concentration of wealth accompanied Ecuador's expanding oil revenues. In 1968 the poorest 20 percent of the population earned 4 percent of the national income, and the richest 5 percent received 26 percent. Seven years later the share of the poorest 20 percent had declined to 2.5 percent, while the richest 5 percent boasted 33 percent.[67]

A confidential 1979 World Bank report, made available to me by an outside source, proved more optimistic with respect to the distribution of wealth. It found that the 20 percent at the base of the economic pyramid saw their portion of National Urban Income (essentially that income generated in the market economy) fall from 3.4 percent in 1968 to 3 percent in 1975. During the same period, the most affluent 25 percent of Ecuadoreans experienced a decline from 63.3 percent to 57.5 percent. However, a growing middle class, estimated to embrace just over half the population, consistently registered gains.

Despite repeated promises the Ecuadorean regime has failed to devise an effective rural-development strategy. Imports of wheat, barley, and milk rose 38 percent, 184 percent, and 218 percent respectively between 1975 and 1978 as population growth outstripped agricultural production.[68] Unemployment and underemployment afflict nearly one half of the *campesinos*. A heralded five-year plan failed for lack of political support, financial commitment, adequate planning, and sound management. Another such plan, for the 1980–1984 period, projects large increases in the production and sales abroad of bananas, coffee, and cacao, although an overall decline from 18.6 percent (1979) to 17.5 (1984) is forecast in agriculture's contribution to GDP.[69]

One possibility for increasing rural employment and raising output lies in the Foderuma program, created by the Banco Central. It enables communities with an average annual income of under $300 to apply for loans for integrated rural development. The financing of such a project is contingent upon local involvement in the diagnosis of the problem at hand and a determination of how best to resolve it. So extreme is rural poverty that two million Ecuadoreans qualify for participation in Foderuma.[70] The greatest promise for the countryside rests with the Daule-Peripa dam system, a $450-million project financed by the Inter-American Development Bank and designed to provide hydroelectricity, irrigate 247,000 acres north of Guayaquil, handle flood control, and generate rural employment.

In the fall of 1978, the country's finance minister warned that the oil boom was nearing an end. But the upheaval in Iran and sharply higher crude prices extended its life as the country derived an additional $40,000 a day in revenues. Production in 1979 (76.7 million barrels) exceeded that of 1978 (73.7 million barrels) and 1977 (66.3 million), and the 1980 output—announced at approximately 225,000 bpd early in the year—promised to be even higher.[71] The civilian government of Jaime Roldós, elected in August 1979, has attempted to reduce expenditures on military and capital projects while fulfilling campaign promises to spend more on education, health, welfare, and the salaries of civil servants. On January 1, 1980, the center-left regime reduced the work week to forty hours and raised the minimum wage 100 percent for agricultural and industrial workers. The new president's political path is strewn with obstacles, not the least of which are opposition within his own movement, an obstructionist Congress, surging inflation, and the likelihood of having to import oil by 1985 unless new reserves are discovered.

What conclusions spring from the Venezuelan and Ecuadorean experiences? In both countries oil earnings, especially after the OPEC price increases of 1973–1974, ballooned the Gross National Product and nourished the national treasury. These funds permitted the leaders to rule by promise and pronouncement. They often employed either a developmentalist or populist idiom in pledging that the new earnings would bridge the chasm between the rich and poor in housing, food, employment, education, and medical services. The politicians used this goal to justify a greatly expanded government role in the economy, and henceforth private investment generated a smaller proportion of national income. Moreover, while the Ecuadorean military kept a tight rein on unions, Venezuela's petroleum industry spawned "a small, elite, labor group that possesses great coercive power over the entire economy, as well as an administrative bureaucracy that controls so much revenue that its decisions tend to become largely autonomous and to draw engineering resources of the country almost exclusively to the exploration for and extraction of oil."[72]

Some social progress resulted, and services, notably to urban residents, were expanded. However, both regimes raised more expectations than they satisfied because of mismanagement, corruption, and the inability of their economies to absorb major injections of spending without inflation, waste, and a sharp rise in imports. Meanwhile, ostentatious consumption took the form of sophisticated jet fighter planes, high-rise office buildings, and luxury goods purchased from abroad. Amid plenty, the governments, which have frequently adopted a paternalistic attitude toward the people, proved unable or unwilling to raise taxes substantially or impose other austerity measures to redistribute income. "The domestic tax systems, relying as they do on the easy tapping of the stream of oil revenue, require few sacrifices of the upper income classes who benefit most by the new wealth and the inflation associated with it."[73] Thus the gap between the very rich and the abjectly poor widened.

The new prosperity benefited the middle class, which secured jobs in proliferating state agencies, positions in private industry, and a plethora of expensive consumer imports. At the same time, inflation diminished the purchasing power of the poor, a large percentage of whom lack full-time work. Many live in the shadows of the machines which deprive them of a livelihood or in shantytowns—halfway houses between rural poverty and urban promise. Mass migrations to the cities uproot peasants from their traditional communities, increase the concentration of people in already crowded areas, and reveal the inability of governments to fashion coherent programs for expanding food production and creating jobs in the *campo*.

Oil-rich regimes invariably nationalize or increase control over foreign firms operating on their soil. Despite such shrilly publicized blows against an "exploitative economic presence," they become ever-more dependent on developed countries, foreign banks, and international lending agencies because of the need to finance expanded imports of food, machinery, and consumer goods. Externally, dependence on outsiders remains high; internally, dependence of the "have nots" on the "haves" increases. In both Ecuador and Venezuela, oil seems to have produced as many problems as it solved.

Oil and Mexican Development

When it comes to using oil to put people to work, Mexico enjoys advantages over Venezuela, Ecuador, and other hydrocarbon-rich countries. *First,* thanks to OPEC actions in recent years, Mexican oil commands a higher price per barrel than did that of other countries when they spurred their economic development. This means that López Portillo has more leeway to manipulate production in accordance with revenue requirements and the economy's capacity to absorb the mounting flow of petrodollars. *Second,* Mexico borders one of the world's greatest military powers. This location makes a large military establishment unnecessary. Small specialized units most effectively combat the guerrilla bands which periodically spring up in Guerrero, Oaxaca, Nayarit, and elsewhere. The United States would respond to any attack on its neighbor from a third nation, and in the extremely unlikely event that the United States launched an invasion, Mexico would scarcely be able to defend itself. Consequently, Mexico devoted only 2.8 percent of its budget to the armed forces in 1978, down from 8.1 percent in 1967 and 6.2 percent in 1972.[74] At the beginning of López Portillo's term, Mexico's generals showed strong interest in purchasing twenty-six F-5 fighters from the Northrop Corporation for approximately $150 million. However, a combination of unanticipated press coverage and the reluctance of the U.S. government to permit the spread of supersonic aircraft chilled the negotiations.[75] Mexico's defense minister has made trips to Brazil, France, and the Soviet Union, but no major purchases had been announced as of mid-1980. A relatively modest military budget frees pesos to be used for improving the lot of Mexico's masses.

Third, despite all of the rhetoric about being so far from God and so near the United States, contiguity to the world's largest market offers unparalleled commercial as well as military benefits. No other developing nation has so many economic opportunities literally in its backyard. *Fourth,* in contrast to every OPEC nation with the possible exception of Venezuela, Mexico boasts an impressively large middle class which embraces tens of thousands of technical experts. Such expertise, though attenuated by positions owed to corruption and political influence, abounds in PEMEX and the Mexican Petroleum Institute—organizations unequaled in OPEC countries. Thus Mexico requires not legions of foreign advisers to guide its economy's development to benefit all citizens but a determination by elites to use its growing wealth prudently.

Fifth, Mexico has gone beyond lip service to evince a strong interest in comprehensive rural development. PIDER, a World Bank–initiated rural-development–public-investment program operated jointly between the World Bank and the Mexican government, offers the greatest promise of raising productivity, constructing needed facilities, and generating jobs in the *campo*. Whether the interest turns into a major commitment remains to be seen. *Finally,* as an emerging hydrocarbon exporter, Mexico can profit from the mistakes made in other petroleum-endowed nations. López Portillo has reiterated a determination to avoid the traps that have ensnared similarly situated countries.

These advantages notwithstanding, López Portillo's first set of economic priorities did not include the massive creation of jobs. Instead, his goals were to (1) make government operate in a rational and efficient manner; (2) convince the business community of his enthusiasm for a mixed economy in which the state acts primarily as an "inducer, not a producer"; and (3) curb inflation.

He wasted no time pursuing his objectives. Within hours of being sworn in on December 1, 1976, López Portillo, "a sophisticated student of administrative theory and practice" who had held a number of high-level administrative posts, sent to Congress an organic law on federal public administration.[76] Three weeks later the measure passed unanimously in a testament to executive power and his party's domination of an often comatose legislative body. "López Portillo believed that the reforms were an integrated whole, that they had to be implemented immediately and simultaneously to break the inertia of tradition and convince both the public and government of his seriousness of purpose."[77]

The reform sought to gain control over and reshape a public administration that had grown profusely during nearly a half-century of one-party dominance. In 1930 a national government of 48,730 employees, working in 21 agencies, served a population of 17 million; by 1976 there were approximately 1.3 million federal employees in 1,018 agencies, serving 63 million Mexicans. Echeverría had contributed to this proliferation by enlarging the federal work force by 60 percent, increasing state spending from 120 billion to 520 billion pesos, and expanding the public sector share of GDP from 26.8 to 39.6 percent during his *sexenio*. He apparently viewed the creation of a new agency as the most appropriate means to resolve a problem. As a consequence over two hundred economically specialized federal agencies called *fideicomisos* and approximately thirty major government institutions sprang up—often in competition with existing entities.[78] The spread of agencies led to administrative incoherence, rivalries, redundancy, ill-defined lines of responsibility, and a lack of accountability. Average citizens seeking services often encountered a "come-back-tomorrow" bureaucracy. According to a member of the business community, "had he [Echeverría] been a good administrator, Mexico would now be socialist. It isn't."[79]

Particularly serious problems afflicted agencies responsible for such economic functions as planning, finance, trade, credit, agriculture, and industrial production. For instance, uncoordinated efforts in credit, infrastructure works, and industrial and administrative decentralization hobbled Echeverría's much-publicized regional-development policy. Similarly, no working connection existed between those official entities responsible for promoting and regulating industry on the one hand and those charged with directing state participation in production on the other.[80]

The imperative of economic recovery dictated a rational administrative structure. Thus the goals of the reform were (1) the reallocation of economic functions to four "super ministries" with clearly defined duties; (2) the "sectorization" of the *fideicomisos*—that is, their placement under eleven "sector heads" for purposes of coordination and accountability; and (3) the implementation of program or goal-based budgeting.[81] The newly created Ministry of Planning and Budget emerged as the *primus inter pares* with authority to draft national- and regional-development plans, draw up the federal budget, and monitor public outlays.

While Planning and Budget was to exercise central control over planning, budgeting, and evaluation of programs—a role similar to that discharged by the Office of Management and Budget in the United States—the president insisted that sector heads should have autonomy in setting goals, allocating resources, and evaluating results.[82] The organic law stripped the Ministry of Finance and Public Credit, also called Treasury, of control over the expenditure of tax monies and left it with the obligation to raise revenue, manage the public debt, and fashion credit and monetary policies. Because of these duties, its traditional preeminence in planning, its capable leadership, the high caliber and experience of its personnel, and its responsibility for controlling credit-financed spending, Treasury remained a formidable player on the bureaucratic stage.

The legislation also established the Ministry of Patrimony and Industrial

Development to monitor, conserve, and administer public lands and non-renewable resources, including the nation's prodigious oil and gas holdings. Additionally, the new ministry promotes and participates in, when necessary, manufacturing, mining, and electricity-generating industries. In this capacity it grants and supervises concessions and permits. Patrimony also oversees foreign investment and technology transfers. Finally, the reform bestowed on the Ministry of Commerce responsibility for trade, export promotion, pricing policies, and consumer protection. CONASUPO, the state retailing firm that buys basic food and necessities from farmers at support prices and sells them to consumers, also falls within this jurisdiction.[83]

At the head of each ministry, López Portillo placed a man in his thirties or forties with a background in economics, accounting, or finance. Thus a team of technicians, loyal to the president and conversant with modern administrative methods, replaced *echeverristas,* some of whom indulged in the Third World rhetoric and political opportunism of their leader.

The reform, personnel changes, and new style kindled the hope that future economic policy would be conceived, executed, and evaluated in a systematic and coordinated fashion. Admittedly the reform looked good—and appeared logical—on paper, but logic does not necessarily prevail in a political system known for webs of personal relationships, secrecy, corruption, and horizons that seldom extend beyond the six-year presidential mandate. The intended role of the Ministry of Planning and Budget as a super agency aroused suspicion and hostility, especially in view of its leading responsibility for evaluation. "The notion of collegial, professional evaluation of program performance proved difficult to introduce in a system in which thousands of jobs are at stake in a presidential transition."[84] If he proved effective, the country's chief planner might become a candidate for president—another consideration that engenders misgivings among fellow cabinet members.

The reform threw the participating ministries into turmoil for six months without producing tidy jurisdictional lines according to economic functions, something which appears impossible to accomplish. Twilight zones and over-lapping responsibilities persist, and, for example, no fewer than three ministries—Patrimony, Treasury, and Commerce—are concerned with tax incentives to promote trade. Cooperation has been difficult to achieve as demonstrated by the difficulty of formulating a five-year "global plan" to guide annual programming and budgeting.

Once the new structure has been designed, the decision makers appointed, and the style unveiled, what would be the substance of the new policy? The president announced its outline in his inaugural address. It was essential, he said, "to create jobs, spur agrarian reform, adopt a healthy price policy, provide efficient public services, and ensure judicious management of public enterprises." The redistribution of income required to accomplish these goals must be preceded by sustained economic growth. But real wealth could not be created unless the country overcame the "bureaucratic inertia, the coupon-clipping mentality, fraud, and inefficiency which vitiate our best intentions and convictions." He proposed to convert the suspicion, mistrust, and conflict among Mexico's key economic actors in recent years[85] into a "popular, national and democratic alliance for production" that would involve all productive forces in society.[86]

López Portillo's speech and subsequent pronouncements disclosed three components of the "Alliance for Production": government, industry, and labor. Each would make short-term sacrifices for the sake of protracted economic growth and for its own long-term interests. A reorganized and more efficient government would tighten its budgetary belt, cut wasteful outlays, and exercise restraint when managing the money supply. Such austerity comported with Mexico's agreement with the International Monetary Fund to cut public-sector investment from 8.7 to 8 percent of GDP and the budgetary deficit from 1.2 to 0.5 percent of GDP over the 1976–1979 period. The accord specified that the rise in the public-sector debt would be reduced to 1 percent by 1979. To assure economic growth amid a retrenchment in state activity, it was anticipated that private-sector investment would rise as a percentage of GDP from 14.3 percent to 18 percent, assuring growth rates of 5 percent (1977), 6 percent (1978), and 8 percent (1979). Increased domestic savings—from 0.5 to 5.5 percent of GDP—would finance the private-sector activity.[87] The government would fashion a financial policy to generate funds needed to stimulate capital formation and production, resorting to foreign financing only to the degree required to import necessary capital goods. It would seek to channel investment into certain priority areas: agriculture and fisheries, energy and fuels, petrochemicals, mining, steel, capital goods, transport, and basic consumer goods.

In return for the government's putting its house in order, recognizing the essential role of the private sector in a mixed economy, and providing economic incentives, the business community pledged to operate more efficiently and expand investment. Labor, for its part, would benefit from both the new jobs generated by investment and the vigorous attack on inflation. As a quid pro quo, Fidel Velázquez, the chief of the Confederation of Mexican Workers, agreed that unions should hold wage demands to 10 percent (later 13 percent) and foster harmony in work places.

The president launched the alliance on December 10, 1976, when two hundred businessmen crowded into the ornate reception hall of the National Palace cheek by jowl with cabinet members and labor leaders. All told, López Portillo signed ten agreements that day with 140 private and mixed industrial firms. These letters of intent contemplated one hundred billion pesos ($5 billion) in investment projects designed to create three hundred thousand jobs.[88]

After six years of mutual antagonism, a Mexican president had apparently recaptured the hearts and minds of the business community. Bernardo Garza Sada, a key member of the highly renowned "Monterrey Group," a conservative and powerful coalition of northern industrialists, avowed to López Portillo: "Your presence here consolidates the tranquillity the country now enjoys and which makes this meeting possible." What we offer "here and now," he added, "is an effort of investment, productivity, efficiency, organization and justice. . . . the confirmation of our positive attitude in dealing with national problems, our renewed commitment to Mexico."[89]

Several months elapsed before the industrialists translated this stirring rhetoric into investment. In fact, the president became so vexed at their foot-dragging that he offered to relieve labor from the 10-percent ceiling on salary increases, declaring that "the only responsible, all-embracing and fully united move to control the economic crisis has been made by the labor sector. I am quite aware, and fully convinced, that the Mexican worker cannot live on gratitude or praise for his sacrifices."[90]

The president's stern language had the desired effect. Just before his September 1977 state of the nation address, associations of chambers of commerce and industry rushed forward with a ten-point program to show their good faith and commitment to the Alliance for Production. Specifically, they pledged to pay a bonus to workers, increase employment by 2 percent of the work force of the participating organizations, restrain prices, encourage investment, provide loans to small- and medium-sized companies, and stimulate import substitution and exports.[91]

The alliance helped the government regain the confidence of once alienated businessmen. But the employment picture remained bleak. As early as May 1977, Fidel Velázquez complained that 10,000 workers had lost jobs in the automobile industry and 2,500 in textiles. The National Chamber of Construction Industries reported 400,000 layoffs in the construction business, and Labor Minister Pedro Ojeda Paullada placed unemployment at 2 million and underemployment at between 5 and 7 million among 17.5 million able-bodied workers.[92]

Not only did the employment situation remain critical, but prices continued to rise faster than the salaries of workers. During the early part of his administration, López Portillo maintained the support of Velázquez by granting the labor confederation formal sway over INFONAVIT, the association established under Echeverría to help workers acquire housing. Even though dues from nonunion as well as union members helped to fund the agency, some 90 percent of the 130,000 houses financed thus far had gone to union members. The CTM sought legal recognition of organized labor's de facto dominance as one more weapon to use against the constantly expanding number of independent unions whose success in recruitment improved with the rise in inflation and unemployment.[93]

This deal did not mollify many workers, who saw their 12-percent average wage increase outstripped by a 17-percent official rate of inflation in 1978 and a 16.8-percent average increase exceeded by a 20-percent rise in prices in 1979. The figures are not available for the latter year, but in 1978 the number of work stoppages rose by 558, 40 percent above the 1977 level.[94] Most of these strikes were unsuccessful because of the government's ability to manipulate a large component of organized labor. Still, many workers at the grassroots, convinced that the Alliance for Production constituted an alliance against their interests, began to pressure the official labor movement to stand up for their well being in the face of declining living standards or lose control to independent unions that have emerged in recent years. As a result the old-line *sindicatos*, traditionally a responsive arm of the state, succeeded in late 1979 in negotiating an increase of 21.5 percent in the minimum wage, halfway between the private sector's offer of 18 percent and the Labor Congress's demand of 25 percent.[95] By 1980 the CTM was instructing affiliates to ask for wage rises of 30 percent, and the railroad workers were demanding 50 percent.

By 1979 there were few references to the alliance, which appeared to have outlived its usefulness. López Portillo's assurances combined with the oil boom to instill confidence in the business community, even if private investment did not rise as fast as anticipated—in part because of continued inflation. Foreign investors again looked with favor on Mexico, and international bankers swarmed over Mexico City to seek participation in financing PEMEX's development scheme. Yet except for the surging petroleum sector, there had been no fundamental change in the structure of the economy, as presidential advisers were divided over the proper development strategy. One group, associated with the ministries of Planning and Budget and Patrimony and Industrial Development, advocated *desarrollo compartido*; namely, increasing government outlays on education, health, and other social needs while at the same time channeling resources to quasi-public agencies involved in heavy industry and the manufacture of capital goods. Tariff protection and restrictions on foreign investors would continue as import-substitution was extended throughout the industrial sector. In contrast, the Treasury and the Bank of Mexico favored *desarrollo estabilizador*, a variant of which was found in the austerity program followed during the first two years of López Portillo's term. This strategy relies on tax incentives and ample credit to stimulate private investment. Mexican manufacturing would be encouraged to become more competitive by restraining wages, lowering tariffs, and welcoming foreign investment. Where *desarrollo compartido* anticipates redistributing income from capital to labor to expand aggregate demand and stimulate economic activity, *desarrollo estabilizador*, at least until sustained growth is achieved, directs more resources to capital with a view to spurring private investment. Amid such competing theories, would López Portillo unveil a bold new development plan? After all, a Mexican president has only about four-and-one-half years to institute new programs if they are to bear fruit before the selection and election of a successor consumes his last eighteen months in office.

Industrial Development Plan

For this reason keen interest attended the March 12, 1979, announcement of a national plan of industrial development that would use oil earnings, expected to exceed $12 billion from sales abroad in 1980 alone, to propel national growth.[96] José Andrés Oteyza, secretary of patrimony and industrial development, stressed that this "is not simply one more plan . . . it is the first time that the government has formally and thoroughly come to grips with planning in this field." Petroleum would save the Mexican economy. As Oteyza expressed it: "We are going to depend on this resource in the short run in order to free ourselves of it in a few years."[97] The document identified unemployment as the major long-term challenge facing the country, but it cited inflation as the most pressing short-term problem and emphasized the need to combat higher prices.

The plan divides preferred industrial activities into two categories. The first, to enjoy the greatest incentives, includes capital goods, engines, trucks, buses, railroad cars, cement, iron and steel, and processed foods for human and animal consumption. The second embraces textiles, shoes, clothing, furniture, cardboard boxes, automobile parts, electronics equipment, petrochemicals, plate glass, and metal works. No criteria are presented to explain how these items, which account for 60 percent of gross industrial production, were selected. They are believed to be industries with an actual or potential export capability that are not dependent on the export earnings of other industries to finance their necessary imports.

It is anticipated that GDP will increase from 7.1 percent in 1979 to 9.5 percent in 1981 to an annual rate of between 10.2 and 10.6 percent throughout the rest of the decade. Industry (12 percent) and capital goods (20 percent) will attain even higher rates. Production targets appear for thirty-three industries for 1979, 1982, and 1990.

A number of incentives will encourage investment along desirable paths. For example, private investors can look forward to a 5-percent tax credit when they purchase nationally produced capital goods. State enterprises, expected to generate about one third of gross fixed capital investment between 1979 and 1982, are required to buy capital goods from domestic producers provided the price, including freight costs, does not exceed that of a comparable imported item by more than 15 percent. PEMEX, the Federal Electricity Commission, and other governmental agencies must prepare detailed acquisition programs for coming years to inform Mexican companies interested in supplying their capital goods.

Incentives apply to regions as well as products. Mexican planners are committed to decentralizing industry because of the congested, polluted character of Mexico City, Guadalajara, and Monterrey, beset by rapid and disorderly growth, and because of the need to provide opportunities in other parts of the country to stem the migration to major cities. The government proposes to reduce the Valley of Mexico's share in generating gross industrial output from 50 (1979) to 40 percent (1982). In pursuit of this goal, it will grant credit supports, preferential rates for public services, tax credits of up to 25 percent, and a 30-percent reduction on the price of electricity, natural gas, fuel oil, and petrochemicals to firms that invest in eleven priority areas. These include the northern border, various points along the coast, cities served by the new gas line, and the industrial ports of Tampico, Coatzacoalcos, Las Truchas, and Salina Cruz.

For the first time in history, Mexico has offered stimuli specifically for the creation of jobs. Employers who generate new jobs through investment or who put people to work through additional shifts (outside Mexico City and its environs) will receive for each new position a credit on their federal taxes equivalent to 20 percent of the annual minimum wage for the economic zone in question. The plan stipulates an additional 5-percent tax credit to smaller firms, defined as those with total fixed assets less than two hundred times the annual minimum wage in Mexico City, or approximately $4.4 million in 1979. This preference is justified on the grounds that "small industry plays a nationalistic and democratic role in our economy by creating proportionally more jobs, generating foreign exchange income, widely distributing the ownership of capital and working in association with the basic economy of the region in which it operates."[98]

Also important is an interest in halting the emergence of oligarchies, increasingly visible in the Mexican economy. The plan encourages large corporations to subcontract with medium- and small-sized companies to encourage the endeavors of the latter. A 3.6-percent increase in employment is anticipated between 1979 and 1982, reaching 6.7 percent by the last half of the 1980s. These figures are substantially above the 2.8-percent rate recorded between 1970 and 1975. Between 1978 and 1982, an average of 600,000 jobs are to be created annually, rising to 810,000 in 1982. By 1990, 12.6 million more positions will have been generated than there were twelve years before, with 4.6 million arising directly from policies advanced by the plan. The largest rise in new jobs is in commercial and service activities, expanding from 37.4 percent (1975) to 42.2 percent (1982) to 52.9 percent (1990). Meanwhile, the proportion of workers in agriculture is expected to diminish from 35 percent (1975) to 29.2 (1982) to 19 (1990). All told, the plan contemplates a 30-percent expansion of employment in industries that enjoy incentives. Total employment, 14.6 million in 1979, is expected to reach 26.9 million in 1990.

The plan also reiterates the imperative to break bottlenecks that frustrate development. Consequently, it promises early attention to constructing ports, warehouses, and transportation systems. This proposal dovetails with an earlier announcement that Mexico would spend $1.75 billion over the next two years to rejuvenate the dilapidated state-owned railways notorious for poor management and flagrant corruption.[99]

The impact of the program on the balance of payments will be substantial. It was previously estimated that mounting oil revenues would enable Mexico to accumulate a current account surplus of $3.2 billion between 1979 and 1982. As a result of the imports required to accomplish the industrial plan, a deficit of $4.2 billion occurred in 1979, with a $3.5-billion shortfall anticipated for 1980 as Mexico's non-oil exports become less competitive due to rampant inflation. Oteyza made clear that industrial goals would be pursued jointly by government, industry, and labor—components of the Alliance for Production. "This action entails development programs that must be carried out by the private and government sectors," he said.[100]

The government has vested responsibility for coordinating, promoting, and evaluating development in the hands of an interagency body called the National Commission of Industrial Development. This commission, presided over by the minister of national patrimony and industrial development, also consists of the ministers of finance and public credit, programming and budget, commerce, agriculture and water resources, communications and transportation, and human settlements and public works as well as the directors general of the Bank SOMEX and Nacional Financiera, a federal development agency.

In the apparent absence of an underlying econometric model, the plan relies on an input-output table in advancing a macroeconomic and sectoral framework for an eleven-year period. Many of its goals—spurring productivity, increasing efficiency, encouraging decentralization, combating oligopolies, and creating jobs—are laudable. Yet the original version of the plan, which will no doubt appear in other incarnations, raises as many questions as it answers.

First, in what products other than oil and natural gas does Mexico enjoy a comparative advantage to assure an expansion of exports? Quantitative restrictions have shielded producers from the bracing winds of the marketplace, and, although boasting the world's fourteenth-largest GNP, Mexico ranks only about fortieth among international traders. The ambitious document emphasizes a wide array of processed foods, capital items, and durable and nondurable consumer goods. Government economists assert that cheap energy will encourage exports, but little evidence exists that Mexico can compete successfully in the world market. López Portillo's announcement on March 18, 1980, that his country would keep out of the General Agreement on Tariffs and Trade (GATT), an organization designed to encourage freer trade among nations, confirms the uncompetitiveness of Mexican products, whose prices rise with domestic inflation.[101] The plan's preferred industries, apparently chosen using loose criteria, may prove to be as inefficient, politically protected, uncompetitive, and capital-focused as existing ones. For instance, the plan stresses textiles and shoes, which are precisely the items that encounter the most intractable protection. Reference is made to reducing trade barriers, but there is no mention of either the products to be affected or the timetable to be followed. Protracted inflation will lessen the ability to sell Mexican goods abroad, and López Portillo seems opposed to devaluation that would make his country's products more competitive. Without a change of policy, Mexico can be expected to rely more and more on oil exports to cover its current account deficit, a move that will inhibit diversification.

Second, the program implies that concentration on capital goods such as iron and steel, railroad cars, trucks, and earth-moving equipment will give rise to supporting industries. However, it is silent on the kinds of linkages that will occur. While the idea is theoretically attractive, it is exceedingly difficult to forge chains of development, as demonstrated in the Las Truchas steel works. Originally designed as a pole of development for the city of Lázaro Cárdenas, the project, located on the border of Michoacán and Guerrero states, became a colossal and expensive boondoggle because of shoddy management, inadequate infrastructure, inferior quality control, soaring prices, and inattention to social services for workers.[102]

Third, developing nations, even those with oil holdings, have mixed records in adhering to long-term, comprehensive plans. The Mexican situation is complicated by the ubiquitous presence of disorganization in national life and the disjunctive character of government policies, which at best endure only the length of a presidential *sexenio.* Chief executives have had little luck in binding their successors, and the prospect of massive oil revenues—with all of the possibilities for programs thus presented—may even enhance the tendency toward discontinuity. In 1975 the Revolutionary party's Instituto de Estudios Políticos, Económicos y Sociales fashioned a development program to guide Mexico's economic destinies for the years ahead. A few weeks later López

Portillo was tapped to succeed Echeverría, and the scheme was never heard of again.

Fourth, the plan minimizes the impact of inflation, stressing only the avoidance of production bottlenecks that might lead to upward pressure on price levels. Specific figures are avoided, although it is assumed that prices will rise at a moderate rate, possibly 8 or 9 percent. Yet massive deficit spending and monetary expansion (33 percent) caused prices to shoot up 20 percent in 1979 alone, and the prospect for the early 1980s is even more dismal because of budgetary extravagance, the expansion of the money supply, imported infla-tion, wage and salary hikes resulting from revised collective agreements such as that in the petroleum sector, the persistence of bottlenecks in transportation and other sectors, and insufficient production capacity in some industries. Accelerating prices illuminate both the lack of governmental restraint and the economy's difficulty in absorbing billions of fresh petrodollars. Continuing inflation was a prime factor in frustrating Venezuela's hope of becoming an exporter of manufactured goods.

Fifth, problems inhere in the plan's methodology. While used in the United States, economic models have major limitations—especially when it comes to forecasting investment, the most volatile element in GNP. The model assumes that the past is prologue to the future. Planners may have a reasonable grasp of the relations among factors at the outset, but it is extremely difficult to predict what relative changes will occur, particularly over a long period. The Mexican plan even contains incentives to enhance the use of labor relative to capital, although, as will be discussed below, the opposite result may take place.

Sixth, a mechanism has yet to be fashioned for capturing and allocating the petroleum earnings that the industrial plan proposes to channel into productive investment. Naturally Díaz Serrano wishes to retain a major portion for PEMEX's ever-growing—some say insatiable—needs. The 1980 budget ear-marked almost one third of total government expenditures for energy, with the lion's share lavished on the state oil company and the Federal Electricity Commission. The devotion of constantly increasing sums to PEMEX multi-plied pressures to breach the 2.25-million bpd production target for the early 1980s, and, as discussed in chapter 10, a new plateau was established in March 1980.[103] After the issue of the industry's portion is settled, there is still the question of whether remaining funds should be included in the government's general budget, permitting flexibility in allotment, or segregated into one or more special-purpose funds, allowing greater accountability of expenditures. Attempts to control the vast amounts have excited a high-level debate among key officials who realize the importance of the stakes for their particular bureaucracy.[104]

Seventh, if this is not simply "one more plan," and Mexico is serious about coherent development, who is going to assume responsibility for planning—not just of the industrial sector, but of the entire, extremely complex, economy? Ricardo García Sainz, then minister of budget and planning, the office with the most persuasive claim to this function, made a bid for the assignment a few days after Oteyza released his plan. On March 15, 1979, he publicized the outline of a comprehensive development plan (*plan global de desarrollo*) to establish short-, medium-, and long-range goals as well as integrate the sectoral programs of key ministries with those of states and the private sector. According to García Sainz, his blueprint would also (1) provide targets for socioeconomic growth and the political administration in order to (2) raise the standard of living of all groups, particularly the downtrodden, in a manner designed to (3) strengthen personal and social freedoms as well as enhance the nation's capacity to determine its own future.[105]

The removal of García Sainz, widely adjudged incompetent, in May 1979 suggested that his peculiarly timed proposal (shouldn't a "comprehensive" plan have preceded a sectoral one?) had not been coordinated with Oteyza's and that it represented only the most recent example of rivalries that have plagued Mexican planning. The Ministry of Patrimony and Industrial Development got a head start on other agencies because it boasts a team of planners, including Vladimiro Brailovsky, who was trained at Cambridge University, and Ernesto Marcos, who studied at Notre Dame. Rather than work in concert with other key ministries, Patrimony seized the initiative, much to the dismay of bureaucratic competitors, to promote the idea that Mexico should be fully industrialized. Miguel de la Madrid, García Sainz's highly respected successor, may take exception to Oteyza's approach, which assures continued protectionism as the country pursues an industrial growth strategy. Because of his position, the new minister "is ideally placed to cripple the industrial plan, even though it is not official policy."

A revised version of the "global plan," released in April 1980, covers only the remainder of the López Portillo *sexenio* as it offers a twenty-two-point strategy to promote the general goals of economic development, increased employment, improved social services, and income redistribution.[106] The president's full backing of a strong person to direct planning will enhance the likelihood of success. Such an individual must apply political pressures to effect needed changes in the bureaucracy and state agencies, which constitute some of the country's most inefficient enterprises, while enticing businessmen to invest in key areas. Otherwise the economy will become even more distorted as the army of jobless mushrooms. The decision over the 1980 budget to postpone for still another year the adoption of program budgeting, thereby leaving public expenditure decisions in the hands of ministries, autonomous agencies, and state governors, may symbolize the "abandoning [of] the commitment to planning in general and administrative reform in particular."[107]

Eighth, in the area of job creation, the plan—though a symbolic first step—contains apparent inadequacies, not the least of which is the absence of a rationale as to why the preferred industries will generate employment. Furthermore, while there are fiscal and other incentives to stimulate new jobs, they may turn out to have the opposite effect. For example, the 20-percent (25 percent for small firms) investment credit reduces the actual investment costs and constitutes a subsidy for the life of the project. In contrast, the tax credit for new jobs and additional shifts has only a two-year duration. The upshot is a reduction in capital costs relative to those of labor. Moreover, this last credit, as a fraction (20 percent) of the minimum wage, represents less of an inducement when skilled workers, earning well above the minimum, are involved. The government estimates that the plan will help create an average of 600,000 new jobs annually between 1979 and 1982. This is still 200,000 short of satisfying the demand for about 800,000 new jobs a year. Also, a significant portion of future employment springs neither from industry nor commerce but from government itself, expected to furnish 4.3 million new positions in the next decade. This target, yet to be broken down by agency, numbs the mind in view of the deplorable reputation enjoyed by the bureaucracy. In any case Oteyza, who believes that the development of vibrant, integrated industrial structure must precede massive job formation, has stated that the problem's resolution will not occur until the "last decade of the century." By then, cynics insist, the current generation of politicians will have retired in affluence. A national employment plan, announced in late 1979 and designed to generate 2.2 million jobs before López Portillo leaves office and 37.6 million by the end of the century, could give greater coherence to work-creating activities.[108]

Ninth, and especially worrisome, is the projected employment picture in the countryside. The plan contemplates the creation of only 400,000 new jobs between 1978 and 1990 in rural areas, or less than 35,000 per year. The unmistakable focus is on highly mechanized, highly capitalized commercial farming. Such an approach may further impoverish the traditional *campesino* economy while accelerating the flow of peasants to urban areas both inside and outside of Mexico.[109] In the twenty years after 1950, for example, over 4.5 million *campesinos* flocked to cities in search of work. For every 100 babies born in the countryside in 1974, 57 rural inhabitants migrated to cities. Should mechanization fail to increase annual output 3 percent or so, food imports will absorb 21 percent of oil earnings in 1982 and 54 percent in 1990.[110] López Portillo demonstrated greater sensitivity than his patrimony minister to the plight of small farmers when in March 1980 he announced a plan known as the

Mexican Food System (Sistema Alimentario Mexicano—SAM), intended to accomplish self-sufficiency in basic foodstuffs through greater governmental support of peasants. Like so many of the president's proposals, the idea sounds promising but lacks a coherent blueprint for implementation.[111]

Finally, apart from the issue of new employment, the plan fails to set forth a mechanism for redistributing income to improve the quality of life for workers. It mentions neither direct (new wage and fringe-benefit levels) nor indirect (progressive taxation) means for achieving this goal. As *Comercio exterior de México* pointed out, "while the NPID proposes an incomes policy for the private industrial sector, it makes no mention of a policy regarding payment to the labor factor."[112] Oil earnings are already flowing toward upper-income groups who prefer consumption over investment. Labor receives a slowly increasing return from growth, but this income is maldistributed and disproportionately winds up in the hands of members of strong unions. The priority should be to increase the percentage of GDP that flows to labor overall, not to enhance the privileged position of labor elites. However, because the capital-biased character of Mexican development is continued in this plan, relatively affluent workers with jobs in preferred sectors of the economy and with salaries and benefits rivaling those of their American counterparts can anticipate higher incomes. At the same time, workers who are poorly employed, partially employed, or unemployed lack organized, effective advocacy of their interests and will see little change in their lives.

Oteyza himself has pointed out that "the opportunity that oil offers us will come only once. This places on us the obligation to make rational use of it; to transform its nonrenewable character into something permanent. If this does not happen, the benefits will be ephemeral and will gradually disappear as the years go by, without any radical modification of the country's production having taken place."[113] Sadly, from Mexico's perspective Oteyza's scheme offers little promise of converting oil into jobs. Upon studying it, one is left with the feeling that the plan implicitly assumes that the flow of workers to the United States will not only continue but will also accelerate. Such a strategy, which depends on Washington's willingness to maintain a permeable border, may meet with only short-term success, postponing rather than resolving the employment problem. Angels, virgins, and streams of black gold may disappear almost as fast as they appear. It may be impossible to convince future generations of "have nots" that the social benefits of the 1910 revolution, withheld from their parents and grandparents for lack of resources, were legitimately denied them by a government that boasted highly publicized oil reserves.

6

Mexico and the Organization
of Petroleum-Exporting Countries

Development of OPEC

Two decades ago oil-endowed Third World nations banded together into the Organization of Petroleum-Exporting Countries to counter the power of the multinational oil firms that enjoyed concessions within their territories. OPEC evolved from a tentative forum for the exchange of views and technical information to an exceedingly powerful collectivity of states whose actions have raised crude oil prices tenfold between 1973 and 1980. Thus a public cartel of producing governments supplanted a private cartel of corporations as the dominant force in international oil.

The "Seven Sisters," the giants of international oil, dominated the petroleum industries of the countries in which they operated. Individually or through consortia, these powerful firms controlled exploration, production, pricing, transportation, and marketing. They often worked hand in glove with Western governments. When Dr. Mohammed Mossadeq-Salteneh, an Iranian nationalist and determined reformer, came to power in 1951, Iran seized the properties of British Petroleum (formerly the Anglo-Iranian Oil Company) in the first nationalization since President Lázaro Cárdenas took over Mexico's oil industry thirteen years before.

The companies, whose networks include shipping and refining, retaliated by refusing to move or process Persian crude. To compensate for the oil lost from world trade, they expanded output in Saudi Arabia, Iraq, and, especially, Kuwait, where British Petroleum held 50 percent of the concession. After Mossadeq's ascent to power, the firms also acted to increase their proven reserves in other Persian Gulf nations.[1] Such manipulation was only surpassed by that of the Central Intelligence Agency, which may have feared that the Russians and their client Tudeh party would take advantage of the instability that beset Iran at this time to enhance their influence. Therefore in 1953 the CIA helped to overthrow Mossadeq's regime, which had been enfeebled by a sharp decline in foreign exchange earnings and inept economic policies. Kermit Roosevelt, grandson of Theodore Roosevelt, is alleged to have slipped secretly into Tehran to mastermind the coup.[2] Shah Mohammed Reza Pahlavi, then thirty-three years old, triumphantly returned to his throne. The Compagnie

Française de Pétrole, the French national oil firm, joined the seven multinational corporations in a consortium to buy and develop Iran's petroleum.

By the late 1950s, however, a number of factors converged to weaken the position of the once omnipotent sisters. The Russians sold oil to Italy at sixty cents below the Mideast price. Independent companies managed to secure footholds in oil-rich nations, as evidenced by the concession that Occidental Petroleum obtained in Libya. And oil was being marketed at huge discounts in Japan.[3]

The companies moved to lower their selling prices below the formally listed or "posted price" in order to preserve their shares of energy sales. Therein lay the difficulty. The fifty-fifty tax payments which provided an equal division of profits between companies and governments were pegged to posted prices. Selling below these prices meant that corporate tax obligations shot above 50 percent. As a consequence Standard Oil of New Jersey (Exxon), a giant among giants, announced on August 8, 1960, its intention to cut average posted prices from $1.92 to $1.81 per barrel. The other sisters subsequently made their own reductions.

This step came less than a year and a half after an eighteen-cent cut in posted prices had deprived Saudi Arabia, Iran, Iraq, and Kuwait of $132 million a year in income. The unexpected move not only hit the producing states in their respective pocketbooks but once again revealed the vulnerability of individual countries when confronted by an orchestrated action of the tightly interlocked firms.

Could the producers overcome religious, economic, geographic, and ideological cleavages to unite to defend common interests? As early as 1953, Saudi Arabia and Iraq agreed to exchange information and consult each other on petroleum matters. But the first concerted effort to organize the producing countries did not come until six years later, when Juan Pérez Alfonso, Venezuela's minister of mines and hydrocarbons, led a delegation of his country's oil experts to the First Arab Petroleum Congress in Cairo. Pérez Alfonso, who had strongly denounced the companies' price cuts, urged the formation of an international producers' organization. He received enthusiastic support from Sheikh Abdullah Tariki, his Saudi counterpart. But other delegates balked at the idea, and Pérez Alfonso and Tariki had to content themselves with a "gentlemen's agreement" which bound the delegates to convey three recommendations to their governments: (1) That each producing nation would examine the creation of an agency to coordinate policies; (2) that none of the countries would take advantage of another's difficulties with the oil firms to enhance its own position; and (3) that consultation with other producers would precede any authorization to permit the companies to change prices.[4]

The sharp reduction in posted prices of August 1960 supplied the catalyst for more formal action. Persian Gulf states, which faced the loss of nearly $100 million per year in export earnings, wanted to protect their positions.[5] On September 10, 1960, representatives of Saudi Arabia, Iran, Iraq, Kuwait, and Venezuela, which collectively accounted for eight out of ten barrels of oil entering world trade, gathered in Baghdad. The result was the establishment of OPEC as the delegates pledged to work together to restore prices to their previous levels and to "study and formulate a system to ensure the stabilization of prices by, among other means, the regulation of production."[6] This agreement, the first in a series of assertive acts by petroleum-producing nations, was later adhered to by Qatar (1961), Libya (1962), Indonesia (1962), Abu Dhabi (1967), Algeria (1969), Nigeria (1971), Ecuador (1973), and Gabon (1975).

Pérez Alfonso hoped that the formation of OPEC would lead to the allocation of production among its members to control the supply of oil entering the world market.[7] Such a scheme aroused little enthusiasm among his new allies, who believed their self-interest lay not in rationing but in securing a larger share of the profits garnered by the Seven Sisters. They also questioned the motives of Venezuelan officials whose country, a high-cost producer with a diminishing portion of the world market, wanted a special relationship with the United States which might exclude Arab exporters. Rivalries among Middle Eastern nations—especially for leadership of the "Arab Revolution"—further threatened OPEC cohesion, and the nascent organization boasted only limited accomplishments during its first years of life. Still, its actions militated against further reductions in posted prices, set in each country a uniform rate of royalties that could not be deducted from income taxes, and initiated the exchange of technical data—a matter heretofore under exclusive jurisdiction of the oil companies, which found it increasingly difficult to play producing countries against each other.

A decade passed before the OPEC states began to comprehend their potential power. Two events sharpened this self-awareness. Libya's Col. Muammar el-Qadaffi, who had ousted a conservative monarch in 1969, applied unrelenting pressure on companies—beginning with Occidental, an independent firm whose Northern African concession was its major source of crude—to secure production cuts, a higher posted price, and an increased tax rate. In addition to acting in concert to prevent reprisals against Libya, the OPEC members followed Qadaffi's bold example and made greater demands on their concessionaires. The 1970–1971 period marked a watershed in relations between the governments and the companies. For the first time, the cartel nations challenged the firms' grip on the production process. Their tactics have been succinctly described as "more cash, more control, then more cash and yet more control, and then still more cash."[8]

The October 1973 Arab-Israeli war further enhanced OPEC's militancy. In the wake of the hostilities, the Arab members of the cartel, members of the Organization of the Arab Petroleum-Exporting Countries (OAPEC), temporarily reduced overall production and refused to sell oil to Israel's supporters, the Netherlands and the United States. They did not, however, apply effective destination controls to prevent transshipments to the offending nations. Within three months after the fighting began, OPEC's entire membership had taken steps to raise the price of crude fourfold. The cartel has continued to boost prices as additional members have entered into "participation agreements," enabling them to acquire total or substantial ownership of the drilling companies operating within their boundaries. The sisters still transport, refine, and market OPEC crude, but their role has been significantly altered. In 1973 they controlled 80 to 90 percent of OPEC production; in 1980 they obtained only 40 to 50 percent of the oil produced by cartel members.

Venezuela offered the most dramatic example of nationalization when, on January 1, 1976, it assumed ownership of its entire petroleum industry. Petróleos de Venezuela, the newly created state oil enterprise, became the parent of fourteen operating companies which had belonged to foreign corporations. These foreign firms now hold service contracts under which they provide technical assistance and are allowed to purchase specified quantities of the nation's output.

Mexico's Oil Industry Compared with OPEC

Mexico, though it shares the bond of petroleum, is quite different from the thirteen OPEC members. The majority of these countries export one principal item—oil—whereas Mexico, despite growing dependence on oil, also earns substantial foreign exchange from tourism, silver, coffee, and farm products. As table 8 reveals, oil accounted for only 43 percent of the value of Mexico's exports in 1979, compared with 84 percent of the dollars which it generates on the average for the cartel members.

Even more significant, the cartel sprang up in reaction to price manipulation by multinational corporations on which OPEC nations depended for the exploitation, transportation, refining, and marketing of their hydrocarbons. In contrast, Mexico freed itself over four decades ago from American, British, and Dutch firms which dominated its industry.

Unlike the OPEC countries, Mexico boasts a national oil monopoly that has gained invaluable experience and know-how since 1938. While all OPEC members have now created state companies, these tend to specialize in lifting crude as compared to the fully integrated structure of PEMEX. For several years the Arab members of OPEC have committed themselves to constructing refineries to obtain a "downstream" capability. Mexico already has nine refineries, including a 439,600-bpd facility at Minatitlán. The country also boasts a sophisticated petrochemical industry. The Mexican Petroleum Institute, a research center and think tank, has scores of patents to its credit, performs a high percentage of the industry's engineering and design work, and has trained thousands of Mexican and foreign technicians. As a result Mexico is the only Third World oil-exporting nation that can effectively run its petroleum industry without assistance from foreign companies.

Mexican Membership in OPEC

Many observers believed that Mexico would attempt to enter OPEC as soon as its exports reached a sufficiently high level. President Echeverría had championed his nation's leadership of the Third World, and a place in the cartel might have advanced his goal. It was not, as suggested by one writer, "a distinct possibility" that Mexico could aspire to a key role within the organization, inasmuch as the cartel was dominated by Saudi Arabia and Iran.[9] But membership certainly would have emphasized Mexican independence of the

TABLE 8
Oil as a Percentage of Exports of OPEC Members
(Compared with Mexico)

Country	Year	Value of Oil Exports (millions of dollars)	Value of All Exports (millions of dollars)	Oil as % of All Exports
Algeria	1977	$ 5,274	$ 5,809	91
Ecuador	1976	739	1,256	59
Gabon	1976	895	1,136	79
Indonesia	1977	6,826	10,853	63
Iran	1976	13,433	15,095	89
Iraq	1977	8,605	8,749	98
Kuwait	1977	7,268	9,754	75
Libya	1978	9,186	9,683	95
Nigeria	1977	11,039	11,891	93
Qatar	1978	2,294	2,368	97
Saudi Arabia	1977	41,232	43,458	95
United Arab Emirates	1977	8,993	9,400	96
Venezuela	1976	5,669	9,466	60
Average		9,342	10,686	84
Mexico	1979	3,789	8,913	43

Sources: United Nations Statistical Office, *Yearbook of International Trade Statistics, 1978* (New York: United Nations, 1979), I, passim. The U.S. Department of Commerce supplied the data on Mexico.

United States in international relations—an objective seriously pursued by Echeverría.

Such considerations no doubt weighed in the deliberations of Mexican policy makers. In October 1974 Horacio Flores de la Peña, a self-described "leftist militant" who served as minister of national patrimony, stated that his country would seek observer status in OPEC and join the cartel as soon as possible. At approximately the same time, however, Echeverría reportedly assured President Ford that Mexico entertained no such ambition. The subsequent dismissal of Flores de la Peña temporarily laid the matter to rest.[10]

The new patrimony minister, Francisco Javier Alejo, at first repudiated his predecessor's position. Yet in mid-1976, after a "confused melange of statements and counter-statements," Alejo announced Mexico's willingness to join the cartel.[11] The ensuing denials turned to a firm rejection of membership by the López Portillo administration, which assumed office on December 1, 1976.

The director general of PEMEX, the minister of patrimony, and other prominent officials rejected the idea of affiliating, but the president himself delivered the most emphatic statement of his country's position in response to a reporter's question on September 9, 1977: "No sir," he said, "we are not going to enter OPEC. I see nothing at this time that necessitates joining the organization in the short, medium, or long term."[12]

The chief executive and his cabinet members have cited four reasons for their country's policy. *First,* there is a traditional reluctance to affiliate with organizations that might compromise Mexican independence in international relations. Mexico has not even become a member of the "Group of 77," a loosely organized and extremely diverse collection of developing countries which have sought to advance their interests vis-à-vis the industrialized nations. Mexico does not want to compromise the autonomy that it now enjoys over operations and trade in the petroleum field. Its highly prized freedom of action "would have been curtailed if we had joined OPEC."[13]

Second, López Portillo has raised the question of whether Mexico satisfies the prerequisite, set forth in Article 7(c) of the OPEC statutes, of "exporting oil in a substantial quantity."[14] Although occasionally raised in discussions about Mexico's role with respect to the cartel, the "substantial quantity" concept is extremely nebulous and has not prevented small producers such as Gabon and Ecuador from becoming members.

Third, Mexicans argue that their industry is fundamentally different from those of OPEC members. As we have seen, the organization sprang to life because key producers that had given generous concessions to foreign firms wanted to set oil prices themselves and reduce their dependence. Even after the cartel's formation, these same companies played a major role in providing technical assistance, transportation, refining capacity, and market outlets. In contrast, Mexico has controlled its industry through PEMEX since the late 1930s, leading Sheikh Ahmed Zaki Yamani, Saudi Arabia's oil minister, to emphasize how much his country could learn from Mexico in terms of technology related to exploration, refining, and petrochemicals.[15]

Fourth, as a non-OPEC producer, Mexico obtains higher prices on the international market than many cartel members because of lower transportation costs to the U.S. market. Table 9 shows the differential between OPEC and Mexican prices.

Other reasons, never articulated by Mexican officials, may have contributed to their country's noninvolvement. In recent years the cartel has had difficulty arriving at a common price. The downfall of the shah in early 1979 accentuated this tendency as it became evident that Iran would no longer be an assured

TABLE 9
Recent OPEC and Mexican Oil Prices: A Comparison

Date	OPEC Price[1] (dollars per barrel)	Mexican Price[2] (dollars per barrel)	% Difference between OPEC and Mexican Prices
January 1, 1977	$11.50	$12.65	10.0
January 1, 1978	12.70	13.10	3.2
January 1, 1979	13.33	14.10	5.8
April 1, 1979	14.55	17.10	17.6
July 1, 1979	18.00–23.50[3]	22.60	
October 1, 1979	18.00–23.50	24.60	
January 1, 1980	21.43–34.72	32.00[4]	
May 15, 1980	21.43–34.72	33.15	
July 1, 1980	28.28–37.00	34.50	

Sources: Daily Report (Latin America), December 29, 1978, p. M-1; April 5, 1979, p. M-1; October 12, 1979, p. V-1; *Latin America Economic Report,* April 13, 1979, p. 119; *Wall Street Journal,* June 11, 1980, p. 2.

1. This price obtains for Saudi Light, the so-called "marker crude" of OPEC.

2. This price obtains for "Isthmus" crude, a blend of oils lifted from the Reforma fields that is roughly equivalent to Saudi Light. In early 1980, PEMEX charged $28.00 for "Maya," a heavy crude with a high sulfur content (similar to Venezuela's Tia Juana Medium) produced in the Campeche fields. To promote sales of the latter, PEMEX required customers during the first quarter of 1980 to purchase one barrel of Maya with every four barrels of Isthmus. This light-heavy ratio would change to 50-50 in mid-year, according to the firm's foreign commerce director. PEMEX, which does not sell in the volatile Rotterdam spot market, began monitoring the destination of some cargoes in 1979 to guard against resales that yielded "undeserved profits." See *Petroleum Intelligence Weekly,* October 15, 1979, p. 6, and February 11, 1980, p. 2.

3. While all thirteen cartel members charged different prices, these fell into three groups: (a) Saudi Arabia, the United Arab Emirates, and Qatar, $18.00; (b) Kuwait, Iraq, Iran, Venezuela, and Gabon, between $20.00 and slightly more than $22.00; and (c) Nigeria, Algeria, Libya, Ecuador, and probably Indonesia, $23.50. See the *Washington Post,* June 29, 1979, p. A-30.

4. After that of Libya ($34.72) and Nigeria ($34.48), Mexico's oil was the most expensive of any major producer selling under contract in early 1980. See *Times-Herald* (Newport News, Va.), January 3, 1980, p. 4.

supplier of 3.5 to 4 million bpd to the world market. Before 1979 Iran, which urged higher prices, and Saudi Arabia, an advocate of restraint, hammered out a price between them. The two Persian Gulf monarchies played such a dominant role because together they supplied nearly half of the cartel's 30.5-million-bpd production in 1978. "The Saudis and Iranians understood each other's limits and knew that they would not push a fellow monarch over the brink," according to a delegate to a recent OPEC meeting. "Other countries had a good sense of where the final compromise would be and could cast themselves on the high side or low side, depending on what their domestic or diplomatic needs were."[16]

The emergence of the Ayatollah Ruhollah Khomeini, who sided with extremists on pricing questions, shattered the two-power hegemony. At the June 1979 Geneva session, the representatives of the thirteen-member collectivity approved the sharpest price increase since the 1973 war. This move came in the wake of increased purchases by industrial nations, which viewed with alarm the situation in Iran and furiously began to stockpile supplies. However, failure of the OPEC countries to agree on a unified pricing system led to the emergence of a three-tier plan, with prices ranging from $18 to $23.50 per barrel.

Six months later differences prevented the delegates at the Caracas meeting from even agreeing on guidelines for prices. Before the gathering Saudi Arabia—joined by Kuwait, the United Arab Emirates, Iran, and Venezuela—urged an increase in prices to $24. Libya, Algeria, Iran, and other members rejected this figure, insisting they would charge what the market would pay. One journalist graphically noted the isolation of the Saudis: "The imagery of past OPEC meetings was that of two giants wrestling in friendly combat, with the smaller producers checking the gate receipts cheerfully. At Caracas, Saudi Arabia was a lonely Gulliver, hamstrung by self-imposed production limitations and by the hearty band of Lilliputians who insist that OPEC in times of high prices must become a real cartel."[17]

The higher prices in the aftermath of the December 1979 meeting precipitated greater efforts in conservation by industrial nations as well as a recession, which in turn produced a temporary glut of oil. This situation could encourage some OPEC members to advocate restraint on production, further diminishing the organization's attractiveness to a newcomer such as Mexico which intends to expand exports in the 1980s.

Other potential liabilities attach to membership. The U.S. Trade Act of 1974 awards preferential tariff treatment to selected exports of less-developed countries. In partial retaliation against the 1973–1974 oil embargo, Congress excluded OPEC members from this system. Consequently Ecuador and Venezuela, the two Latin American cartel members, lost valuable commercial preferences even though they continued to ship oil to the United States during and after the 1973 Mideast conflict. Mexico, whose products benefited in the amount of $458 million under U.S. legislation in 1978, could look forward to the same treatment until Congress repealed the "anti-OPEC" provision in late 1979. For this reason Gregory F. Treverton, staff member of the National Security Council, concluded that membership for Mexico would be "an economic catastrophe."[18]

There is also the question of Mexico's relationship to the Third World. While the industrialized countries can cushion the impact of higher energy costs by adjusting the price of manufactured goods upward, most LDCs enjoy no such hedge. The sharp rise in oil prices since 1973 has ravaged their economies. OPEC pledged an additional $800 million in aid to developing nations in mid-1979, but the contemporaneous price increase meant that the cartel would return only one dollar for each additional ten collected from Third World customers.[19] That such nations might follow the suggestion of Venezuela and other OPEC members and form their own cartels in other vital materials seems at best a remote possibility. A series of factors—pervasive use, strategic importance, ease of storage, production flexibility, lack of short-term alternatives, limited sources—distinguish oil from sugar, bananas, coffee, cotton, and most other raw materials. Equally important is the common political objective of OPEC's Arab states, which share a religious faith and whose most prolific supplier, Saudi Arabia, has shown a willingness to adjust output to advance the cartel's interests.

Thus Mexico may prefer to remain outside the body responsible for these price increases, especially if it wants to gain influence at the expense of Venezuela in Central America and the Caribbean, where OPEC prices have exacerbated economic problems. Mexico has not applied for membership, choosing instead to monitor carefully actions by the cartel. As such it enjoys the best of both worlds: the ability to charge more than many OPEC members for its oil while avoiding the negative aspects of membership.

Contacts Between Mexico and OPEC Countries

Mexico's decision to remain outside OPEC has not prevented it from developing close ties with individual members, eight of which have embassies in Mexico City. Most important have been its relations with Venezuela and Iran, the cartel's most influential non-Arab states.

Mexico and Venezuela have much in common. In addition to language, culture, geographic propinquity, and hydrocarbons, both have a strong state and an electoral form of government, permit widespread civil liberties, boast relatively high average incomes, exhibit an impressive degree of industrialization, seek a position of leadership among Third World nations, and aspire to reduce dependence on foreign countries and corporations. Cooperation was especially close during the administration of Echeverría, who along with President Carlos Andrés Pérez promoted the Latin American Economic System (Sistema Económico Latinoamericano—SELA) to reinvigorate regional economic integration and encourage concerted Latin American action within the international economy. Their joint efforts bore fruit on October 17, 1975, when twenty-five nations—including Cuba but excluding the United States—signed the SELA charter. The document called for regional cooperation and development through establishing Latin American multinational companies, promoting the formation of producers' associations, and improving the terms of trade for regional products vis-à-vis imported capital goods and technology. The Convention of Panama, which created SELA, also advocated exchanging technical and scientific information, ensuring that multinational companies serve the interests of host governments, and supporting regional efforts in economic integration.[20]

President Pérez reportedly favored Mexico's becoming the third Latin American member of OPEC. His successor, Luis Herrera Campíns, has described Mexico as an ally, not a competitor, that should affiliate with the cartel to be in the "avant-garde of the developing countries" which "defend the price of basic materials and propose a rational means of exploiting them.[21]

Mexico once showed almost as much interest in Iran as in Venezuela. The shah visited Mexico in May 1975 to examine investment possibilities, and two months later Echeverría visited Tehran as part of a world tour. The López Portillo administration continued to promote close ties with the Persian Gulf monarchy. Iran and Venezuela were the only two OPEC nations visited by Jorge Díaz Serrano during a November 1977 journey to three continents to promote his country's hydrocarbon interests. Even before Díaz Serrano left Mexico City, PEMEX invited Abdass Ghaffari, the North American representative of the Iranian National Oil Company, to visit the new deposits in Campeche and Tabasco and to explore technical exchanges between the two state oil companies. Ghaffari, who disclaimed any attempt to recruit Mexico into OPEC, stated that "Mexico has a great future in natural energy sources, not only in petroleum."[22]

Bahman Ahanin, Iran's ambassador to Mexico, reiterated this interest in energy cooperation when he presented his credentials in April 1978. The new envoy emphasized that both countries had important oil and gas deposits, that opportunities abounded for technical cooperation, and that Mexico's emergence as a major exporter would not affect the traditional producers because it has taken place at a time when OPEC has already achieved its objectives.[23] At this time the Mexican government disclosed that President López Portillo would accept an invitation of the shah to visit Iran. Political upheaval in the key

Persian Gulf state has indefinitely postponed the presidential trip. Instead, Reza Pahlavi, who owned an exquisite villa in Acapulco, came to Mexico as a political exile in June 1979. In granting him a six-month tourist visa, the government rejected a thinly veiled threat from the Islamic Republic of Iran that it would consider admitting the deposed monarch a "hostile act," after which the Iranian ambassador left Mexico City. The Foreign Ministry replied that Mexico, with a tradition of granting political asylum without attention to political philosophy, "does not accept warnings or suggestions by any government to deny or grant entry visas to Mexican territory."[24] While in Mexico Pahlavi divided his time between the Acapulco villa and a luxurious mansion and gardens in Cuernavaca. The latter, formally called La Serena, became known as "the bunker" because of its sophisticated radio equipment, closed-circuit television system, and portholes manned by machine-gun-carrying guards.

Following the shah's journey to a New York hospital for cancer and gall-bladder treatment and the seizure of the American embassy in Tehran by militant students, Mexico announced on November 12, 1979, that it was temporarily closing its embassy and withdrawing personnel from the Iranian capital. Meanwhile, it was said that the shah's agents spent $500,000 to remodel and redecorate the family home in Acapulco in preparation for his possible return. On several occasions Mexican officials stated that he would again be welcome.

Foreign Minister Jorge Castañeda pulled in the welcome mat on November 29 when he announced that renewal of the shah's visa, due to expire on December 9, "would be contrary to Mexico's interests." López Portillo made clear that vetoing the former Iranian leader's return was his "personal decision." Not only did Mexico reportedly come under intense pressure from Third World nations, but there was mounting concern that its citizens and property abroad might become the object of attacks. While deploring the occupation of the U.S. embassy, López Portillo stated that in the event of the shah's presence on their territory, "the Mexican people would have their hands tied if a Mexican were harmed anywhere in the world."[25] The shah's sojourn in the country and the attendant diplomatic maneuvering strained the once close relations between Mexico and Iran.

Algeria has claimed to have a keen interest in obtaining Mexican technology for building gas pipelines. Technicians from SONATRACH, the Algerian state oil company, arrived in Mexico on January 18, 1979, for conversations with their PEMEX counterparts. Their main purpose was to observe the gas pipeline and associated installations then under construction from the Reforma fields to the northern part of the country.[26]

Mexico has also had important contacts with Kuwait and Bahrain. In late

1977 the Kuwait Investment Company, a financial group that places large sums of Kuwaiti dinars in Europe and the United States, handled the issue of ten-year PEMEX bonds worth approximately $2 million.[27] Meanwhile, a leading financial institution in Bahrain, a non-OPEC emirate securely linked to Saudi Arabia, took prime responsibility for the issuance of $59.3 million in ten-year PEMEX bonds.

Relations with Israel

The first tanker to weigh anchor from the offshore buoy of Dos Bocas, a new port in Mexico's steamy, petroleum-rich Southeast, glided through the placid waters of Campeche Bay toward Israel. This departure once again confirmed the adage that oil and politics make strange bedfellows.

Mexico has long prided itself on having been the first country to supply oil to Israel after the founding of the Jewish homeland in 1948. Encouraged by the United States, the bonds of friendship grew stronger between the two nations. Relations, which began to change slowly after the 1973 Arab-Israeli war, took an abrupt turn on November 10, 1975, when Mexico's ambassador to the United Nations voted in favor of an Arab-inspired resolution defining Zionism as a "form of racism and racial discrimination." The measure overwhelmingly passed the General Assembly by a vote of seventy-five to thirty-five with thirty-two abstentions. Mexico's action probably sprang from a desire to demonstrate solidarity with oil-producing nations and the Third World rather than from animus toward Israel. After all, Echeverría hoped to become the UN's next secretary general. Nonetheless, American and Canadian Jews retaliated at once against this perceived act of hostility.

The American Jewish Congress suspended its tours to Mexico and excoriated the country for having "allied itself with a Soviet-Arab bloc engaged in a program of political anti-Semitism and anti-Americanism." Advertisements filled the pages of U.S. newspapers urging "good people of all faiths" not to go to Mexico as it had become a "less desirable place to visit or do business with."[28] B'nai B'rith and other Jewish groups cancelled charter tours and moved conventions from Mexico City to Caribbean or North American sites. Jewish families took similar actions. The boycott, which came just before the 1975 Christmas tourist season, sparked 128,000 cancellations at hotels in Acapulco and Mexico City according to the Mexican Travel Agents' Association.[29] The economic assault cost Mexico tens of millions of dollars at a time of a severe economic recession, featuring a $2.84 billion balance-of-trade deficit.

President Echeverría met with leaders of ten important North American Jewish organizations to reassure them that despite the vote, his government "in no way identified Zionism with racism." He insisted that Mexico viewed the

UN resolution as an instrument to promote peace talks in the Middle East.[30] This plea fell on deaf ears when, on December 15, his government backed two other resolutions which denounced Zionism. The next day Rabbi Arthur Hertzberg, then president of the American Jewish Congress, reaffirmed his organization's commitment to the boycott.

The Mexican regime wasted little time before trying to ingratiate itself with the Israelis, who—it was hoped—could reduce the economic pressures. Foreign Minister Emilio O. Rabasa flew to Tel Aviv in a gesture of amends arranged with Israeli officials. He laid a wreath on the tomb in Jerusalem of Dr. Theodore Herzl, founder of the Zionist movement; referred to the nation as "this land of Zion"; and said that it had been "created for a people that deserves our respect and admiration." He also stated publicly that "there is no discrimination in Zion and where there is no discrimination there is no racism."[31]

At the end of his five-day visit, he informed reporters that the misunderstandings had been "forgotten, forgiven and buried." This symbol of contrition provoked fierce press criticism. An *Excelsior* columnist called Rabasa's statement "intolerable" and averred that "these acts are without parallel and degrade our foreign policy to the lowest level in our history." Another publication took the government to task for "begging Israeli forgiveness on its knees."[32] Shortly thereafter the foreign minister stated that he was "painfully obliged" to resign, which he did on December 29, 1975. Echeverría then told the Mexican Congress that no one should confuse "a courteous explanation of his country's UN vote with an appeal for forgiveness, starting with the President of the Republic," as the legislators rose and applauded wildly.[33] This stentorian speech did not prevent his accepting Rabasa's resignation.

Nevertheless, the ex-foreign minister's diplomatic effort seemed to have borne fruit. The press in Jerusalem reported that Israeli diplomats in the United States were urging American Jews to terminate their boycott inasmuch as Mexico, the recent UN flap notwithstanding, was a valuable friend in an increasingly anti-Israel Third World. In January 1976 the Conference of Presidents of Major American Jewish Organizations spoke of "a decided reversal" of Mexican sentiment and called off the travel ban. "We are no longer confronting an adversary, but have regained a friend," said the conference leaders, who represent thirty-two religious and secular groups. No mention was made of Israeli pressures. Instead, a conference spokesman noted that Mexican diplomats had avoided meetings of UN agencies considering anti-Zionist resolutions and that statements by the new foreign minister had reassured the group.[34]

Several unresolved issues concerned the Jewish community. Despite its public reversal on the Zionism question, Mexico still expressed a belief that Israel should pull out of all territories occupied in the June 1967 Middle East war. Even more vexing was Echeverría's decision to allow the Palestine Liberation Organization to open an information office in Mexico City. Yigal Allon, then Israel's foreign minister, declared: "On the basis of experience we have gathered, that would be an encouragement of extremism around the world and not moderation."[35]

López Portillo's ascent to the presidency served as a balm for diplomatic wounds. Soon afterward El Al, the Israeli national airline, began flights to Mexico City. Rabbi Hertzberg said that López Portillo's election "marks the beginning of a new era not only in Mexico's relations with the United States, but also with Jews of this country and with the people of Israel."[36] The American Jewish Congress reinstituted its travel program to Mexico, and Jewish tourists returned to the sands of Acapulco and the silver-laden specialty shops of Mexico City's "pink zone." Earnings from tourism in the last quarter of 1976 and the first quarter of 1977 totaled $409 million, an increase of 60 percent over a similar period twelve months before.[37]

President Ephraim Katzir's November 1977 trip to Mexico City symbolized the improved atmosphere. Amid the tinny blare of mariachi music and the multifloral gyrations of a folkloric ballet, the Israeli head of state received the keys to Mexico City, home of most of the country's forty-five thousand Jews. He also heard López Portillo stress the central role played by Jerusalem in the life of the Jewish people—a declaration hailed in radio broadcasts in Israel as the most far-reaching support of the city's status ever made by a Mexican chief executive or the head of a Catholic nation.[38]

The visit spurred cultural exchanges which include promoting "sister-city" ties, bolstering the Hebrew-language program at UNAM, celebrating "Mexico days" and "Israel days" in each other's countries, undertaking joint archeological digs, and broadening the activities of the Mexican-Israeli Cultural Institute in Mexico City and the Central Institute of Cultural Relations with Latin America in Jerusalem.

The *Military Balance,* a publication of the London-based International Institute of Strategic Studies, has identified Israel as a prime arms supplier to Mexico.[39] And for several years talks have been held over the possible construction in Yucatán of a factory to assemble and service Israeli-designed ARAVAs, small turbo-prop transport planes that form part of the mixed collection of the five-thousand-man Mexican air force. The future of these discussions is clouded by Israel's interest in establishing an aircraft industry at home.

The Israeli delegates to the UN have not yet succeeded in persuading their Mexican counterparts to vote with them on sensitive Mideast issues. "But at least they are now more likely to abstain," one Israeli diplomat told me after his anonymity was assured. This assessment may have been overly optimistic because on February 27, 1980, Porfirio Muñoz Ledo, Mexico's UN ambassador, made a speech strongly backing the right of the Palestinians to self-determination and endorsing the creation of an independent homeland. He also criticized Israel's "colonial" policy of establishing settlements in Palestinian-occupied territory in the Gaza Strip and the West Bank of the Jordan River, while recommending international action "proportional to the seriousness" of such violations of UN resolutions. "The violation of such principles as the territorial integrity of States, the inadmissibility of acquiring territory by force and the right of free determination of people affect all nations, but particularly the developing countries which find in them [the principles] the best guaranty of their independence," Muñoz Ledo said.[40]

The increasing importance of Mexican oil makes it unthinkable that American and Canadian Jews would launch another boycott in response to Mexico's support for Palestinian rights. In fact, leaders of the Jewish community have urged closer ties to Mexico, in whose oil and gas they see an opportunity to diminish reliance on Persian Gulf suppliers. Most outspoken has been the *New Republic*, whose publisher Martin Peretz is a generous and articulate friend of Israel. The magazine advocates an "extra-special relationship" with Mexico to offset the "special relationship" with Saudi Arabia.

Meanwhile, Mexican exports to Israel climbed from $72.6 million in 1976 to over $100 million in 1978.[41] This figure will continue to grow sharply because of oil sales. Israel purchased Mexican crude even when alternative, cheaper supplies were available from Iran in order to keep open as many options as possible. This turned out to be an intelligent strategy when Ayatollah Khomeini struck Israel from Iran's export list soon after the shah's fall. Mexico—attentive to North American Jewry—immediately stepped to the forefront and promised to help fill Israel's needs, an obligation taken quite seriously by López Portillo. PEMEX pledged to supply 45,000 of the estimated 150,000 barrels per day consumed in Israel.[42] In March 1980 Mexico supplied 121,000 bpd, possibly because of shortfalls in previous months. At the $32 per barrel charged on January 1, 1980, such sales would generate over $525 million in that year alone, and more as prices continue to rise.

Expanding technical intercourse has complemented the oil trade as chemists, physicists, engineers, agronomists, nuclear scientists, and specialists in biomedicine shuttle between the two countries. Ironically, one of the programs, planned in early 1979, found twenty-five Mexicans flying to Israel to study the fine art of hotel administration.

Prospects for Mexico and OPEC

What are the prospects for Mexico's future relations with OPEC? What will be the country's position when it increases exports from an average of 672,000 bpd in 1979 to possibly 1.8 to 2.5 million bpd in the mid-1980s? Three policy alternatives exist for Mexico. It could (1) eventually join OPEC, (2) become a competitor to the cartel, either as an individual seller or in league with other non-OPEC exporters, or (3) remain an independent producer supportive of OPEC's efforts to keep prices high.

For reasons cited above, Mexico's affiliation with OPEC is exceedingly unlikely, at least during López Portillo's administration, which will conclude on November 30, 1982. The president has never closed the door completely to membership. In his strong statement of September 9, 1977, he said that ''I see nothing at this time that necessitates joining.'' However, a number of factors—his own moderate philosophy, a desire for good or at least correct relations with Washington, and the liabilities associated with membership— militate against a change in his position.

Sources in the Middle East have expressed concern that Mexico might try to undermine OPEC. In August 1978 the Abu Dhabi daily *Al Ittihad* described Mexican oil as ''a real threat and a dangerous competitor.''[43] A London-based publication favorable to Arab interests cited three reasons for such trepidation: (1) The large amount of Mexican crude that has been sold to the United States for its strategic stockpile; (2) the readiness of López Portillo to sell ''unlimited'' quantities of oil to Israel; and (3) the possibility that Mexico might attempt to recruit new petroleum producers such as England into a non-OPEC cartel.[44]

The ultimate concern of the Arabs is political.

In the case of Mexico and other oil producing countries outside OPEC, the worry therefore seems to centre not on pricing policies, but rather on the fact that these countries could shift the geographical centre of world oil production out of the Arab world. Since most oil outside the Arab world is far costlier to produce, OPEC sees such a move as primarily politically motivated. Like the build up of strategic stockpiles, it fears the excessive emphasis on developing (sic) non-OPEC oil as yet another attempt to undercut OPEC's bargaining power.[45]

Mana Said Al Otaiba, petroleum minister of the United Arab Emirates, urged Mexico and new producers to join the cartel instead of moving to form a second OPEC.[46] Indeed, Mexican officials have held discussions with their British counterparts over the shipment of PEMEX crude to Europe in tankers that leave U.S. ports empty after delivering cargoes from the North Sea. They have also met over energy questions with Canadian and Norwegian policy makers. No doubt such contacts will continue, but the formation of a second producers' association seems extremely improbable, especially since the Britains,

Canadas, Norways, and Mexicos of the world can take full advantage of OPEC actions.

Mexico, though unenthusiastic about joining either OPEC or a "little OPEC," has responded favorably to the idea of informal talks among a limited number of producers. Under a plan that originated in Caracas in mid-1978, such conversations would involve three OPEC nations (Saudi Arabia, Kuwait, and Venezuela) and three non-OPEC producers (Mexico, Great Britain, and Norway).[47] Algeria and Canada were later added to the list of participants. Tony Benn, minister of energy in Callahan's Labour government, effusively embraced the Venezuelan initiative and proposed that the meeting be held in London. It was to include only energy ministers (without staffs), take place in a country home outside the city, and be held on March 3–4, 1979. No formal agenda would guide the session, but among the most likely topics for discussion would be the operations of national oil companies, the economic situation of LDCs, the world oil picture, and—inevitably—the prospects for petroleum prices. The meeting was postponed because of a schedule conflict involving the Saudi minister and because of the May 1979 British elections captured by the Conservative party. Mexico expressed its willingness to attend at some future date, but no one has given recent impetus to the project.

Mexico, sponsor of the Tlaltelolco treaty to ban nuclear weapons from Latin America, proposed a dialogue between producing and consuming countries to avoid another international oil crisis that might lead to war. López Portillo broached the idea with Valéry Giscard d'Estaing during the French president's visit to Mexico City from February 28 to March 1, 1979. France, since 1979 an advocate of a similar plan, then presented it to the nine-member commission of the European Economic Community, where it received unanimous approval.

López Portillo did not specify the scope of his proposal. Conversations might begin between the EEC and OPEC. Later they could be expanded to embrace such large independent producers as England, Norway, and Mexico as well as such important consumers as the United States and Japan. It might also be possible to include representatives of the Third World such as Brazil, India, or Zaire.[48]

The Mexican chief executive advanced an even bolder idea during a September 27, 1979, speech to the United Nations. Again, his goal was to fashion a framework for the peaceful resolution of energy-related problems, thereby avoiding a possible war between consumers and producers. He advocated the formulation of a world energy plan "to assure a fair, integrated, progressive, and orderly transition from one stage to another in the history of humanity"; namely, until solar, nuclear, and geothermal energy are readily available. Among the nine points constituting his proposal were guarantees of each nation's sovereignty over its natural resources; assuring easier access to and transfer of energy technology; assisting developing and fuel-importing countries in meeting their long-term and emergency energy needs; and reducing speculation in the Rotterdam spot market, where prices fluctuate wildly. He also urged formation of a financial agency to help meet the needs of the poorest countries, the encouragement in these nations of industries complementary to the energy sector, and the establishment of an international energy institute. To explore such a comprehensive scheme, López Portillo suggested the creation of a working group composed of representatives from exporting countries, industrial states, and developing nations that import large quantities of petroleum.[49]

The speech, praised by President Carter as the "most profound and beautiful I have ever read," received a warm reception—except from OPEC leaders. From conservative Kuwait to radical Libya, government spokesmen decried the prospect of discussions that focused on energy to the exclusion of such other important issues to the Third World as trade, aid, investment, technology transfers, etc. "We don't like it [the plan]. It will die," was the brusque reaction of Kuwait's UN ambassador, who reiterated OPEC's opposition to "any separate treatment of energy as a theme." Libya favors the North-South dialogue on various matters, avowed its foreign minister, "but we will never permit a separate discussion of energy." Such a venture "is going to create an enormous quantity of problems and if Mexico is interested in this, it should enter OPEC as a producer," he added.[50] The bristling reaction of cartel members revealed less concern over Mexico as a competitor in production than fear that the important Latin American country, which boasts a tradition of social reform, would organize a new bloc of developing countries committed to a discussion of energy which could divide the ranks of the Third World. As Luis Herrera Campíns of Venezuela expressed it: "Some day Mexico may need the solidarity of the developing countries when it feels particularly vulnerable. This must never be forgotten."[51] Intense lobbying by OPEC nations had helped sidetrack the initiative.

Despite occasionally ambiguous statements on membership, two Mexican administrations have spurned the notion that they would participate in an attack on OPEC. All signs point to a continuance of this policy. President Echeverría stated that "Mexico will never sell a single barrel of oil below the price established in the international market." His ministers affirmed that their country would be neither a "Trojan Horse" nor a "strikebreaker."[52]

López Portillo has reiterated this sentiment in response to suggestions that industrialized nations wished to use Mexico to undermine OPEC.[53] For example, he stated on September 1, 1978: "Our potential and our geographic position is such that our stand on hydrocarbons could give us considerable leverage on a world-wide scale. For this reason, we again ratify Mexico's line: We have upheld our determined desire to give raw materials their just value.

We are not, nor will be . . . against those who, like us, struggle for this."[54] Lest the price of oil plummet, Mexico has stated its intention to follow a development policy that is "deliberately conservative."[55] In so doing it remains a supportive observer of OPEC activities without diminishing its jealously guarded freedom of action.

7

Oil and U.S.-Mexican Relations

Background

Mexicans sometimes exhibit either a Cuauhtemoc or Quetzalcoatl complex. The former connotes a dread of contact with foreigners, just as Aztec chieftain Cuauhtemoc feared Hernán Cortés, his sixteenth-century Spanish conqueror; the latter describes a frame of mind like that of the Aztecs who believed that the great white god Quetzalcoatl would bring salvation from across the sea.

Such an admixture of love and hate permeates relations with the United States. It springs from Mexico's dependence on its northern neighbor: a condition painfully obvious since the war over Texas and Mexico's eventual loss of nearly one half its territory in the 1848 Treaty of Guadalupe Hidalgo. Memories of this period, a nadir in the relationship between Mexico and the United States, are revived every time the Marine Band booms out "From the Halls of Montezuma."

American capitalists later secured generous concessions below the Río Grande, and American military forces frequently intervened to restore order in a country whose leaders were perceived as incapable of managing their own affairs. And American entrepreneurs played a deft and avaricious role in exploiting the commercial oil deposits discovered in Mexico after the turn of the century.

That local citizens often conspired with Americans did not justify Mexico's becoming a fiefdom for foreign interests in the eyes of the new generation of nationalist leaders that emerged following the 1910 revolution. As we have seen, President Lázaro Cárdenas struck the most resounding blow against this exploitative dependence when on March 18, 1938, he nationalized seventeen U.S. and European oil companies operating in his country.

More than two years of diplomatic pressure and official scoldings gave way to a "special relationship" as the Roosevelt administration declared war against the Axis powers, which had sought to curry favor with a country occupying a 1,948-mile border with a prospective adversary. The war years marked the period of closest economic, political, and military ties ever enjoyed between the two countries. However, the vitality of the special relationship diminished during the Cold War. The United States, its attention riveted on the Soviet Union, seemed to take Mexico for granted as a stable, quiescent nation along its southern flank. Mexico, compared to other countries of the hemisphere, enjoyed relative freedom from terrorism, guerrilla activity, and coups d'état. Even when the López Mateos government refused to condemn Castro, vote to suspend Cuba from the Organization of American States, or sever diplomatic ties with the Marxist regime, a U.S. official confidently stated that "Mexico is the best friend that the United States has."[1]

In 1964 President Lyndon B. Johnson responded to entreaties from organized labor and terminated the *bracero* program under which Mexican farm workers had entered the country during harvests since World War II. Five years later U.S. Customs agents began rigorous, time-consuming border searches to inspire Mexican participation in "Operation Intercept," a drive to interdict illegal drug traffic. And in 1971 Washington declined to exempt Mexico from a 10-percent import surchange imposed to impede the influx of foreign products and improve the country's balance of payments.

These American initiatives demonstrated that in fact the special characteristic of bilateral affairs was Mexico's extreme dependence on the United States. Sensitivity to this status combined with the bitter legacy of United States involvement in its territory has steeled Mexico's commitment to nonintervention, the absolute sovereignty of states, peaceful resolution of conflicts, detachment from power blocs, and adherence to international organizations as a buffer against American influence.[2] Since the presidency of dictator Porfiro Díaz (1876–1910), Mexico has endorsed the Calvo Doctrine, which holds that foreigners are entitled to only the same protection of life, liberty, and property enjoyed by nationals, and the Drago Doctrine, which stipulates that force must not be used to collect international debts. For a half-century it has also adhered to the Estrada Doctrine, whereby Mexico City automatically and immediately recognizes a new government, regardless of how it attained power.[3]

A disdain for U.S. intrusion into the affairs of Latin American nations led Mexico to oppose sanctions against Cuba. It also led it to eschew both military aid from its powerful neighbor under the Mutual Security Act of 1951 and participation in an Inter-American police force, proposed after the dispatch of American troops to the Dominican Republic in 1965. Further, Mexico chose not to receive volunteers under the Peace Corps program launched by President John F. Kennedy.

Echeverría attempted to reduce psychological dependence on the United States and give Mexico a twentieth-century identity by making his country a leader of the Third World. As discussed earlier, he drafted the Charter of Economic Rights and Duties of Nations, which sets forth the obligations of developed nations vis-à-vis poorer countries. Along with the president of

Venezuela, he formulated the Latin American Economic System as a vehicle for "economic consultation and cooperation" in Latin America. He visited thirty-six nations, held meetings with sixty-four heads of government, and forged diplomatic links with sixty-seven additional countries as evidence of Mexico's global role.[4]

Meanwhile, he also backed legislation both to control the use of technology from abroad and to limit the percentage of foreign ownership of Mexican firms. The second measure, the Law to Promote Mexican Investment and to Regulate Foreign Investment, provided that non-Mexicans can hold no more than 49 percent of the equity in new ventures. Despite mordant criticism of the United States in international bodies and highly publicized efforts to regulate American investment, relations with Washington remained open during the Echeverría years, thanks in large measure to the efforts of ambassadors Robert McBride and John J. Jova.

Rhetoric about sovereignty and independence notwithstanding, Mexico's reliance on the United States—measured by external debt, foreign investment, destination of exports, source of exports, and the role of American banks—increased under Echeverría.[5] U.S. assistance became crucial to Mexico's economy in late 1976, when two devaluations of the peso occurred. The dependence continues: The United States provides 90 percent of Mexico's tourism, 70 percent of its foreign investment, and two thirds of its trade. In 1979, for example, Mexico exported $6.2 billion worth of goods to its neighbor and imported $7.5 billion.

López Portillo views the massive oil holdings as an opportunity for his country to convert exploitative dependence into greater independence of its energy-hungry neighbor. In his first state of the nation address, he elicited prolonged applause from the Mexican Congress by stating: "Today, nations can be divided into those that have oil and those that do not. We have it." After a glowing tribute to Cárdenas for nationalizing the industry, he added that "oil has become the strongest pillar of our economic independence and a factor by means of which we can correct our deficiencies if we act moderately and skillfully."[6] Díaz Serrano reiterated these sentiments, insisting that "our energy supplies defend us, protect us from economic aggression, and make us increasingly less dependent."[7]

Bilateral Energy Policy

The United States and Mexico have vastly different means of formulating foreign policy. The U.S. Constitution confers authority in this area on both the executive and legislative branches. But the judiciary, regulatory bodies, intelligence agencies, state governments, mass media, and a bewildering array of special interests often play a key role. While any nation's political process confuses outsiders, the U.S. system is particularly bewildering because of the number of diverse participants. Generally speaking, the State Department and other agencies charged with fashioning policy prefer to deal with issues individually, reducing them to their technical aspects. Blurred lines of authority and bureaucratic rivalries may require creation of intra-agency task forces to coordinate policy-making. Energy questions, for example, can involve the departments of State, Energy, Defense, Treasury, Commerce, and Health and Human Resources as well as the Office of Management and Budget, the Central Intelligence Agency, the Export-Import Bank, and the Office of Special Trade Representative. The upshot may be discord, deadlock, and delay, preserving rather than upsetting the status quo while diffusing responsibility. It has been pointed out that "Mexico-related issues often get so mixed with domestic pressures and broader multilateral objectives, especially in trade and financial matters, that the State Department frequently is not the central determinant of U.S. policy toward Mexico."[8]

Mexico's political system is highly centralized despite formal separation of powers and a federal structure. This centralization pervades and influences the formulation of domestic and foreign policy. Naturally the president consults his advisers and appropriate leaders inside and outside the government, but once he or the high-level official to whom he has delegated authority reaches a decision, it is rarely criticized, much less reversed, by Congress, party officials, or interest-group spokesmen.

Centralized decision-making contributes to what has been called "closet diplomacy."[9] In the foreign-policy field, Mexicans prefer to establish personal contacts with high-level American officials. Their goals are to nurture a mutual confidence, engage in unpublicized, informal discussion, and seek solutions to a number of issues affecting bilateral relations. Santiago Roel García, foreign minister from December 1976 to May 1979, even retained a private Washington attorney, A. Lee Fentress of the firm of Dell, Craighill, Fentress & Benton, best known for its representation of professional tennis players and other athletes. Mr. Fentress, who operated independently of the Mexican embassy, collected information, prepared reports, and established contacts with key figures in the bureaucracy and on Capitol Hill.

Mexico's style stands in sharp contrast to the fragmented, bureaucratic approach of the Americans and lends itself to the linking of issues, a concept favored by López Portillo when he proposed examining the "total picture" and considering a "package" approach to resolving problems of mutual concern.[10]

Linkage poses enormous difficulties whether attempted in U.S.–Soviet or U.S.–Mexican relations. The most intractable problem is assigning weights to various issues to help determine the feasibility of trade-offs. For instance, can oil and gas sales by PEMEX compensate for American flexibility on border and trade questions? If so, how many barrels of crude must be placed on the scale to balance the annual entry of, say, two hundred thousand undocumented aliens? How many tons of tomatoes must be admitted at reduced tariffs to justify a twenty-year supply contract for natural gas?

Inextricably bound to the "weighting" problem is that of authority: namely, determining the individual or group assigned responsibility for bargaining once weights have been calculated. During the Nixon presidency Henry Kissinger played this role, for which he has been sharply criticized in recent years. No one currently involved in fashioning American foreign policy enjoys a similar position of power and confidence. National security adviser Zbigniew Brzezinski has often shown enthusiasm for linkage. Although receptive to linkages that clearly advance U.S. interests, former Secretary of State Cyrus R. Vance preferred careful, methodical negotiations on individual matters. Of course, the two presidents could meet to strike bargains. But this is a dangerous way to conduct diplomacy unless each man has a comprehensive command of the outstanding issues.

In addition, there is the problem of "delivery": the capability of either side to deliver what it promises after weights have been determined and trade-offs concluded. Issues between the United States and Mexico impinge upon domestic affairs more so than those outstanding between Washington and Moscow. Energy, jobs, trade, drugs, and law enforcement are only the most prominent of many bilateral concerns. For example, in return for lavish financial assistance or extremely high natural gas prices, could a Mexican chief executive stem the flow of illegal aliens even if he wanted to?

The executive branch's stress on linkage may whet Congress's appetite for this approach, thereby restricting the president's ability to maneuver in foreign affairs. The Nixon administration attempted to link increased trade with the Soviet Union to the Kremlin's support of its Vietnam policy. In contrast, Senator Henry M. Jackson (D-Wash.) and others in Congress moved to join trade to greater opportunities for Jewish emigration from Russia—an initiative that consistently complicated agreement on a Strategic Arms Limitation Treaty.

Another difficulty springs from United States participation in more international organizations and agreements than any other nation in the world. A case in point is membership in the General Agreement on Tariffs and Trade (GATT), under which trade concessions are negotiated on a multilateral rather than a bilateral basis. Such an accord militates against preferential arrangements with another country, although it by no means prevents linkages in the area of trade. Accomplishing linkage in relations with Mexico poses a formidable challenge to U.S. officials. Nonetheless, after Mexico rejected entry into

GATT in March 1980, a systematic exploration of possible trade-offs began in the unlikely but not impossible event that opportunities should arise.

López Portillo's rather vague notion of "package" negotiations may entail attempting to resolve one outstanding problem to improve the climate for addressing another—and, upon settling that one, still another. In any case he attempted to develop ties with his American counterpart that might foster personal diplomacy. Three months before donning the green, white, and red presidential sash, he visited Washington where he was courteously received by President Gerald R. Ford, both in a private family dinner and in a White House meeting. The Mexican leader skillfully presented the agenda of issues between the two countries. In reply President Ford made a pro forma suggestion that these matters be taken up with the appropriate cabinet members. After the president-elect and his party returned to Mexico, the Republican administration made no attempt to follow up on questions raised at the White House session.

Jimmy Carter took a different tack in dealing with Mexico. In addition to a twelve-member delegation headed by Secretary Kissinger, Mrs. Rosalyn Carter attended López Portillo's December 1, 1976, inauguration as a private guest of the new chief executive and his wife. She brought a message of "special friendship" for Mexico from her husband, who himself would take office on January 20, 1977. Once in the White House, President Carter invited López Portillo to become the first foreign head of state to visit Washington during his administration. Upon welcoming his guest on February 14, Carter pledged to work closely on a personal and official basis to bind the two countries in a "continual demonstration of common purpose, common hope, common confidence and common friendship."[11]

The two men, although extremely different, share a moderate political philosophy, a penchant for administrative reform, and election to the presidency with relatively few debts to traditional power brokers. López Portillo gained his party's support thanks to the endorsement of Echeverría, a friend since law-school years. Carter won his nomination by strong showings in primary contests held throughout the country. Mrs. Carter claims a long-standing interest in Latin America, and the American president has worked hard to learn Spanish.

The chief executives reportedly got along quite well. In a White House meeting with several staff and cabinet members, Carter acted quite differently than had Ford six months earlier. He took charge of the session, expressed a keen interest in Mexican affairs, and promised to follow up on matters of mutual concern. Energy, immigration, trade, narcotics, and tourism were among matters identified as of mutual interest; however, no in-depth discussions were held. Instead, President Carter proposed the creation of a consultative mechanism to focus on outstanding issues and lay the groundwork for future high-level meetings between representatives of the two countries. During a visit by ranking Mexican officials on May 26, 1977, this mechanism took shape in the establishment of three working groups to concentrate on politics, social matters, and finance.

The Political Group, chaired by the U.S. secretary of state and the Mexican minister of foreign relations, convened every six months. Sessions took place alternately in Washington and Mexico City. Because of the group's composition, its discussions covered the entire spectrum of U.S.–Mexican relations.

The Social Group was the most active of the three. Its areas of interest included the environment, health, narcotics, culture, and the border. Chaired by Matthew C. Nimetz, then counselor of the Department of State, and Alfonso de Rosenzweig Díaz, subsecretary of foreign relations, it brought together representatives of a half-dozen agencies of both the American and Mexican governments.[12] Among this group's most notable accomplishments was cooperation on a joint action against smugglers—the *coyotes* who assist approximately 40 percent of the illegal aliens entering the United States.

The Economic Group was headed by Richard Cooper, undersecretary of state for economic affairs, and Jorge de la Vega, Mexico's minister of commerce. So complex were the issues confronting this body that it spawned four subgroups to concentrate on trade; tourism; finance; and energy, minerals, industry, and investment.

Stephen Bosworth, then deputy assistant secretary of state, and Nathan Warman, the subsecretary of the Ministry of Patrimony and Industrial Development, chaired the Energy Subgroup. Despite the importance of energy-related matters, this group was relatively inactive. Its inactivity sprang from a decision by the two countries to handle the projected natural gas sale—the most controversial energy question in 1977 and 1978—outside the consultative mechanism in order to involve only those individuals and agencies possessing a specific interest in the transaction.[13] As a result the Finance Subgroup, headed by Undersecretary of the Treasury Anthony Solomon and Mexican Treasury Secretary David Ibarra Muñoz, played an indirect role in energy matters through its discussion of financial means to assist Mexico, including the economic resources to achieve the goals set for its petroleum sector.

The Trade Subgroup, chaired by Alan Wolff of the Office of the Special Trade Representative and Hector Hernández, his Mexican counterpart, managed in two years of periodic meetings to transform the approach in trade negotiations from emotionally charged confrontations to nonpolemical discussions. This change, which found U.S. representatives dealing with officials from five Mexican ministries (Patrimony, Agriculture, Treasury, Commerce, and Foreign Affairs), laid the groundwork for an agreement on tropical products, encouraged Mexico's affiliation with GATT, and produced an atmosphere for dispassionate treatment of such sensitive questions as charges that Mexico has "dumped" products in the U.S. market.

The working groups compiled a mixed record of achievement. With the exception of the Political Group, they were large, unwieldy ad hoc bodies with a changing cast of characters who, on each side, had competing interests and sometimes rather indefinite agendas. In the absence of a commitment by one or both nations to settle major disputes, the groups often busied themselves with peripheral matters. Generally speaking, the participants lacked authority to set policy, and, as noted, one of the most important bilateral issues—the terms for the shipment of Mexican gas to the United States—was largely taken up outside the consultative framework.

The groups did enable bureaucrats from a number of agencies to make trips abroad to acquire first-hand knowledge of the problem under discussion and become acquainted with the Mexican perspective. However, regularly assigned embassy personnel in constant communication with Washington might have handled such talks—at least in the preliminary stages—with appropriate civil servants of the host government. But many U.S. agencies, which in recent years have expanded their own international-affairs capabilities, view the State Department as a rival that does not serve their interests and should not act as their spokesman. Detractors, who argue that the Department can make but a limited contribution on energy, investment, and other matters discussed with Mexican counterparts, claim that Foggy Bottom is riven by competing stands on key issues. After all, over a dozen offices and more than fifty foreign-service officers, many in responsible positions, help fashion the Department's position on questions relating to Mexico. Wary of the State Department's pivotal role, other agencies sought to take up important measures outside the consultative mechanism. For its part the State Department, anxious to guard its bureaucratic "turf," has often appeared reluctant to turn matters over to the working groups where competing agencies had representatives. For example, the Department's Bureau of Economic and Business Affairs has tried to minimize the role of the Department of Energy in oil and gas issues affecting the two countries. Still, the mechanism provided an interagency forum for representatives of both countries to discuss selected issues and gain familiarity with each other and the difficult nature of the questions involved. It fostered the exchange of information and research and actually helped resolve several problems.

The activities of the working groups could not stem an erosion of U.S.–Mexican relations. This deterioration resulted from the collapse of negotiations over the sale of natural gas and efforts by the Carter administration to secure new legislation on illegal immigration, issues discussed in other chapters. President Carter visited Mexico in February 1979 amid this climate of recrimination and suspicion. If the American leader expected bouquets, he was in for a surprise. López Portillo treated his guest to frank, even bristling lectures. Typical of the tone was an allusion to the breakdown in natural gas negotiations, embedded in a luncheon toast: "Among permanent, not casual neighbors," said the Mexican president, "surprise moves and sudden deceit or abuse are poisonous fruits that sooner or later have a reverse effect."[14]

Carter soldiered on despite fatigue, scorching comments, and the strain of keeping abreast of developments in Iran and Afghanistan. He spoke Spanish to school children, lunched on tacos and beans with farmers in Ixtlilco el Grande, and attempted to reassure every audience that the United States had no imperialist designs on Mexico's hydrocarbons. Private conversations between the two men reportedly proceeded more amicably. Moreover, in a lengthy airport press conference after Carter's departure, López Portillo softened his language toward the president and the United States and attempted to give a more positive interpretation of the joint meeting. Mexico's ambassador to the United States, Hugo B. Margaín, later claimed that the negative image of the meeting projected by American newspapers was "totally false" and that relations between the two countries were "excellent, better than in a long time."[15] Nevertheless, the chief executive forcefully articulated the message that interdependence and mutual respect should replace paternalistic and patronizing attitudes in relations between the two countries. "For Mexico, a country deeply conscious of its past, this public airing of long-held grievances was almost obligatory before resentment toward the United States could be buried."[16] The leaders did agree to a resumption of negotiations over natural gas. They also urged a strengthening of the consultative mechanism to infuse it with "more dynamism, cohesion and flexibility for its more effective operation."[17]

What changes have been made in the consultative mechanism? According to a plan developed by the Office of Mexican Affairs of the State Department, the three original working groups have given way to nine, each co-chaired—on the U.S. side—by one State Department official and an appropriate representative of another agency. A new "Multilateral Working Group," consisting of the secretary of state and his Mexican counterpart, replaced the Political Group. An official at the assistant-secretary level heads each of the other eight. These, including the agency working in tandem with the State Department, are: Trade (Office of the Special Trade Representative), Energy (Department of Energy), Industry (Commerce), Finance (Treasury), Migration (Justice), Border (Commerce), Law Enforcement (Justice), and Tourism (Commerce). Representatives of U.S. and Mexican agencies that have an interest in the subject matter constitute the full membership of the nine bodies.

Because the working groups had not functioned as well as hoped, President Carter placed still another bureaucratic player on the binational stage. On June

22, 1979, he appointed a special coordinator for Mexican affairs to serve as the executive secretary for the U.S. side of the consultative mechanism. The name of Ambassador Patrick J. Lucey, envoy to Mexico, surfaced as the likely candidate for this assignment. After an article criticizing the governor-turned-diplomat appeared in the *Los Angeles Times*, [18] Carter warmly praised Lucey, emphasized that he was needed in Mexico City, and selected instead Robert C. Krueger, a forty-three-year-old former congressman from Texas who has good contacts on Capitol Hill and in the executive branch.

Krueger defies the image of a Texas politician, having been a Rhodes scholar and later a Shakespeare professor at Duke University before returning to New Braunfels to run his family's hosiery mill. He served in Congress from 1974 to 1978 and won accolades for his floor management of an industry-supported bill to deregulate natural gas. Low turnouts in black and Hispanic voting areas help explain his narrow loss to incumbent John G. Tower in a 1978 Senate contest. Presidential confidants Hamilton Jordan and Robert S. Strauss supported his appointment to the new post. The special coordinator has responsibility for the day-to-day conduct of affairs with Mexico. This mandate may permit him to cut across bureaucratic fiefdoms to fashion policy in a comprehensive rather than a fragmented manner—an approach strongly favored by Ambassador Lucey, who resigned on October 8, 1979, to join the Kennedy-for-President organization. He may even be in a position to explore linkages. In making decisions Krueger can solicit the counsel of senior interagency advisory groups whose composition varies according to the issue at hand.

Krueger, who seems interested in again seeking elective office, quickly gained a reputation for working effectively with the White House, Congress, and the business community. The anomalous character of his position and the absence of a clear mandate produced tension between his staff and the bureaus of Inter-American Affairs and Economic and Business Affairs. Unlike most diplomats, Krueger enjoys both an independent political base and the full backing of a key Carter adviser, Robert Strauss. Nonetheless, he remained at the margin of energy policy with regard to Mexico as high officials in the Department of Energy played an increasingly assertive role by cultivating personal contacts with Mexican counterparts. The first meeting of the Energy Group, held in Albuquerque in late August 1980, confined itself (at the request of the Mexicans) to such uncontroversial topics as technical cooperation, electricity exchanges between border cities, and the current status of the oil market.

Krueger had a difficult time strengthening the consultative mechanism because he and other officials insisted on making public comments on sensitive bilateral issues in lieu of assigning them to working groups for quiet discussions. His August 1979 statement on the need to consider compensating parties injured by the Ixtoc 1 oil spill, treated in chapter 9, was a prime example of ignoring the newly created communications channels. Officials on both sides of the border have indulged their taste for political rhetoric over the issue of Mexican tomato exports, as López Portillo indulged his when he gratuitously criticized Washington for freezing Iran's assets in the United States at the time of the embassy siege.

Krueger's undiplomatic statement combined with growing Mexican resistance to reparations for the maritime accident threatened the delay or cancellation of López Portillo's scheduled September 28–29, 1979, visit to Washington. An eleventh-hour natural gas agreement assured the holding of the summit meeting at which the two leaders focused on energy, environmental concerns, and the presence in Cuba of a Soviet combat brigade. Both men expressed satisfaction with the talks, which were conspicuously free of recriminations, accusations, and other intemperate behavior.

Group Interests

Governmental and private groups have evinced a keen interest in U.S.–Mexican energy relations. A great deal of activity has occurred on Capitol Hill. Senators and representatives from southern, Sunbelt, and eastern states served by the six transmission companies which have negotiated with PEMEX generally favor buying Mexican gas. Senators Lloyd Bentsen (D-Tex.), Richard Stone (D-Fla.), and Jacob Javits (R-N.Y.) most vocally attacked Secretary Schlesinger for vetoing the original agreement. In contrast, many legislators from the upper Midwest and Northwest, whose constituents depend on Canadian supplies, opposed the terms of the original agreement. An exception has been Sen. Frank Church, chairman of the Senate Foreign Relations Committee, who despite the use of Canadian gas by his Idaho electorate has suggested agreeing to the higher price as a kind of American aid program for Mexico.[19] While the Export-Import Bank was considering a $590-million credit to Petróleos Mexicanos to finance goods and services related to the gasline, Sen. Adlai E. Stevenson, III (D-Ill.) introduced a resolution disapproving the $2.60 price as "unreasonable" and a possible "dangerous precedent." Bentsen excoriated this legislation as an act of "economic coercion," stressing that oil and gas revenues were crucial to a "stable, secure and prosperous Mexico."[20] The measure, though never passed, sent a shock wave through the Export-Import Bank because Stevenson chaired the Senate subcommittee that oversees the agency's activities. The bank, which needed congressional approval to extend and renew its charter, announced postponement of final action on the loans pending an "indefinite additional review."

Sen. Edward M. Kennedy (D-Mass.) has included Mexican affairs among the many questions that his more than 150 committee and office staff members focus on. Kennedy, who chairs the Subcommittee on Energy of the Joint Economic Committee, became interested in Mexican energy questions on March 21, 1978, when Dr. Bernardo Grossling, a scientist at the U.S. Geological Survey, told his subcommittee that Mexico's holdings might be twenty times greater than its proven reserves, then totaling seventeen billion barrels.[21] The senator requested that the Congressional Research Service (CRS) conduct an in-depth study of Mexico's energy prospects. If Mexico did embrace large hydrocarbon reserves, he was anxious to pursue a "package deal," a comprehensive agreement in the mold of the SALT and Middle East settlements that covered trade, immigration, and energy. As he expressed it in the foreword to the CRS report:

> It is the flashpoints around the world which made foreign policy headlines, frequently alerting us, too late, to what might have been avoided with thoughtful concern in advance. In spite of brave words to the contrary, Mexico has been too close to home and too stable to evoke the genuine concern of our Government. We can be thankful that oil discoveries have alerted us to the true dimensions of the opportunities and pitfalls in the affairs of our two countries.[22]

The CRS analysis, which concluded that the United States would import increasing quantities of Mexican oil and gas deemed to total 30 to 50 billion barrels, reinforced Kennedy's belief that the Carter administration was badly handling bilateral relations. He berated Schlesinger for having "spurned the opportunity for negotiations with the Mexicans on price." He also took exception to the secretary's preference for favoring domestic gas production, including high-cost Alaskan supplies, over Canadian and Mexican imports. To the amazement of many observers, he expressed the unusual sentiment that Schlesinger's was a policy "for protecting domestic gas from competition," geared "to clear out the gas glut which embarrasses domestic producers."[23]

As noted, a great deal of Mexico's gas is associated with oil and must be reinjected, consumed at home, or exported to prevent loss through flaring. "Unless we encourage long-term gas supplies from Mexico," Kennedy warned, "Mexico will never maximize oil production."[24] In his view increased Mexican output will enable consuming nations, if not the United States, to shift "reliance away from the Middle East."[25] Even more important, sale of oil and gas will stimulate the economic growth required to solve the immigration problem, to which the Massachusetts Democrat now pays special attention as a ranking member of the Senate Judiciary Committee. With respect to Mexico as with a half-dozen other issues, Kennedy has painstakingly put distance between himself and the administration while cultivating an important constituency—in this case Mexican-Americans, who are well disposed toward the Catholic *político*.

Also concerned about relations with Mexico is the Hispanic Caucus in the House of Representatives. Chaired by Rep. Edward R. Roybal (D-Calif.), it also includes Henry "Hank" Gonzalez (D-Tex.), Robert Garcia (D-N.Y.), Eligio "Kika" de la Garza (D-Tex.), and Balthasar Corrado (P.R.), as well as 140 honorary members. Rep. Manuel Lujan, Jr., a Republican from New Mexico, is the only Hispanic-American congressman who has not affiliated with the group. The caucus, organized in late 1976, named its three-member staff in 1977 and held its first fund-raising dinner in September 1978. Its purpose is to establish an identity for Hispanic-American political leaders and sharpen the country's awareness of the needs of their people in housing, bilingual education, immigration, and employment. Energy forms a component of a comprehensive policy proposal on Mexico that was being elaborated in mid-1979.

Caucus members, along with other Hispanic-American leaders, have made a bid for greater influence with the Carter administration. They met with the president at his request on the eve of his February 1979 visit to Mexico and later advanced candidates to succeed Lucey when it appeared that he would leave his ambassadorial post. After Lucey did resign eight months later, Carter sought to ingratiate himself with the Hispanic-American community by naming Julian Nava as the new envoy. The son of an immigrant, holder of a Ph.D. from Harvard, and special assistant to the president of California State University at the time of his appointment, Nava had the backing of two powerful Hispanic leaders—Deputy Assistant Secretary of State Ralph Guzman and White House adviser Esteban E. Torres. The Mexican government initially opposed Nava's selection, but the Carter administration insisted on its choice.

Several factors—newness, small size, lack of resources, and disagreements with its constituency—now prevent the caucus from playing the role for Mexico that Jewish and Greek leaders play, respectively, for Israel and Greece. The Mexican government has kept a polite distance between itself and the Hispanic-American community. Still, López Portillo, well aware of the importance of ethnic politics in the United States, has met with a number of Chicano leaders, whose political influence can be expected to increase as Spanish-surnamed residents become the nation's largest minority, enjoy economic mobility, overcome internal divisions, and acquire skill in coalition-building.

Dozens of U.S. legislators have become acquainted with their Mexican counterparts through annual meetings held under the auspices of the Mexico–United States Interparliamentary Group.[26] This group, formed in 1961, was led for many years by former Senate Majority Leader Mike Mansfield, who developed an expertise in Mexican affairs. The American delegation, chaired by Senator Dennis DeConcini in 1980, may consist of as many as twelve senators and twelve representatives, although it has sometimes been difficult to fill the U.S. quota. The binational sessions, held alternately in Mexico and the United States, are essentially opportunities for travel, social contacts, and exchange of views. For example, the Americans were familiar with Enrique Olivares Santana, appointed minister of the interior in mid-May 1979, because he had served for several years as a leader of the group.

The insistence by American legislators that no member of the executive branch take a formal part in the debates has tended to diminish their policy-making effectiveness. Nonetheless, discussions on key problems confronting the two countries have had an impact. Interparliamentary group consideration of the Colorado River salinity issue paved the way for congressional approval of a settlement of the problem in the early 1970s. In 1978 the Mexicans made a convincing argument that railroad cars produced in their country should be admitted to the United States under the General System of Preferences. As a result American congressmen drafted, introduced, and backed a bill for such favored treatment. Energy dominated the 1979 Mexico City meeting and the 1980 Washington meeting. While neither country's delegation meets as a body except at the yearly conferences, the Mexicans have taken the interchange very seriously.

That U.S. state governments, which enjoy much greater autonomy than their Mexican counterparts, endeavor to shape policy further complicates bilateral relations and confuses opinion makers below the Río Grande. For instance, Texas governor William P. Clements argued publicly that a sensible energy strategy entails deregulating oil and gas, tumbling regulatory barriers, spurring the construction of nuclear plants, and acquiring Mexican hydrocarbons at the price fixed by PEMEX. To facilitate the last objective, he has proposed a conference of governors from adjoining U.S. and Mexican states. "The United States must not compromise its position as a world power," he said. "This is the moment that the United States must get off its knees and stand tall."[27] The news magazine *Proceso* interpreted this forceful statement to mean that "one of the most influential governors in the North American power structure" was prepared, if necessary, to dispatch armed forces to secure Mexico's oil and gas.[28] Other Mexican elites, whose familiarity with their own political system makes it difficult for them to comprehend the plurality of American politics and who are convinced that the U.S. Marine Corps eagerly awaits a return to the Halls of Montezuma, immediately assumed that the governor was working in concert with the White House and the Pentagon.

The Alaskan government, also acting independently of Washington, figured prominently in an initiative to send Prudhoe Bay crude to Japan in exchange for the shipment of Japanese-purchased Mexican oil to the U.S. Gulf Coast. Proponents claim the "swap" would save from sixty cents to two dollars per barrel in transportation costs, help diminish the four-hundred-thousand-bpd glut on the Pacific coast, and provide additional incentives for major oil firms to increase Alaskan production and enlarge the capacity of the eight-hundred-mile trans-Alaskan pipeline.[29] The Carter administration, wary of public reaction to the exportation of U.S. oil, has exhibited coolness toward this proposal that would require prior congressional approval before implementation.

Presidential candidates Edmund G. Brown, Jr., of California, who visited López Portillo in mid-1979, and John B. Connally of Texas have also urged closer energy ties with neighboring countries. They have been among the strongest advocates of a North American common market in energy, designed to increase the flow of hydrocarbons from Canada and Mexico to the United States and diminish the hemorrhage of dollars to OPEC nations. Though heady material for the political hustings, the concept has received a negative reception from both neighboring countries, where nationalists fear U.S. economic subjugation.[30]

Nongovernmental interest groups have also attempted to influence U.S.–Mexican policy. The American Federation of Labor and Congress of Industrial Organizations (AFL-CIO) has been particularly active in recent years. It has lobbied for (1) the termination of the *bracero* program, (2) an aggressive plan to halt the flow of illegal aliens, and (3) tariff protection against Mexican imports, including the products of the assembly plants along the border which the Mexican government has encouraged with special tax advantages. Organized labor has not yet taken an active part in energy-related issues.

Banks with a stake in the development of Mexico's hydrocarbons have been more conspicuous in this regard. The International Monetary Fund reported that Brazil and Mexico were responsible for 44 percent of the identifiable outstanding debt owed by non-OPEC nations to banks active in international markets in 1976. Loans to Mexico totaled $10.75 billion. Three years later the two countries accounted for 70 percent of the $100 billion in foreign debt of Latin American countries.[31] Among the principal lenders were the Bank of America and the Morgan Guaranty Trust Company which, in November 1977, helped put together a $1.2-billion, seven-year loan involving seventy-four international banks.[32] In 1979 the Bank of America syndicated a $2.5-billion credit to PEMEX in the form of bankers' acceptances, believed to be one of the largest commercial loans ever made. Such commitments enhance the interest of participating banks in U.S. policy toward Mexico, where they conduct much business. The banks have excellent contacts on Capitol Hill and also communicate their concerns through the Treasury Department and various chambers of commerce devoted to business activity in Mexico. The State Department and Office of Special Trade Representative also defend their interests.

The Mexican–United States Chamber of Commerce, composed of an equal number of members in both countries, is an important voice of private business with access to leading policy makers on both sides of the border. In October 1978 this organization, in cooperation with Petróleos Mexicanos, sponsored a tour of Mexico's oil and gas facilities for ten members, including Caterpillar Tractor Company, Rockwell International, and W. R. Grace and Company, which provide equipment, parts, and service to Mexico's oil industry. Ostensibly designed to give the firms an overview of Mexican operations, the activity may also have been intended to gain grassroots support for closer energy relations.

The chamber has also cultivated congressmen and senators. In January 1979 it made a PEMEX-prepared slide presentation to members and staff of the Senate Energy and Natural Resources Committee. Later in the year it developed plans for a second Mexican oil tour especially for key legislators and staff committee aides.

A number of corporations have individually attempted to influence the course of U.S.–Mexican affairs. For example, the gas-transmission companies fought hard to gain congressional and bureaucratic support for their deal with PEMEX. Articles appearing in influential industry-oriented publications reflect the business community's disenchantment with U.S. policy toward Mexico. A typical commentary, entitled "A Perplexing U.S. Coolness toward Mexican Oil," stated: "Washington has . . . done very little to boost development in southern Mexico, with one of the world's biggest non-OPEC oil reserves. In some ways, the U.S. has even hampered the Mexican efforts by feeding political ammunition to opponents of rapid development and early sale to American buyers. Washington policy has been neutral at best." The magazine also lamented the breakdown in the natural gas negotiations and observed that Mexico responded to the delaying action of the Export-Import Bank by swiftly turning to Europe and Japan for both steel and credits.[33] Other publications reiterated this view.

In an editorial entitled "Reviewing the Energy Debacle," the editors of the *Wall Street Journal* wrote: "The second huge cost has been the utter derangement of our relationship with Mexico. Because it embarrassed the lobbying effort with a $2.60 price, a deal to import huge quantities of Mexican gas was busted by the administration. The Mexicans understandably resent this, as they must resent the current claim that, oh well, they'll come back and sell it to us anyway." The *Energy User News* contended that obstacles placed in the path of Mexican oil development by the United States had forced the country to consider membership in OPEC.[34]

Ironically, friends of Israel who denounced Mexico's vote for a 1975 "Zionism is racism" resolution in the United Nations now see in that country's black gold an opportunity to reduce dependence on the Arab nations of OPEC. Most vociferous has been the *New Republic,* which in an August 19, 1978, editorial claimed that Mexican reserves stood at between 150 and 200 billion barrels and accused the U.S. government of concealing this fact from the American public and Congress, "apparently to avoid undermining energy policies premised on scarcity and foreign policies based on nuzzling the Arabs." It reserved special criticism for the "Mexico can't" school which contends that "PEMEX can't produce more than four or five million barrels a day by 1985." While denying acceptance of conspiracy theories, the piece stated that "both the Ford and Carter administrations seem to have made deliberate decisions to discount Mexican oil in charting U.S. energy and foreign policy." The *Near East Report,* a Washington-based newsletter with close ties to the American-Israel Public Affairs Committee, a lobbying organization for Israel, shares the *New Republic'*s perspective. It has argued that "the emergence of a Saudi-league oil power south of the Rio Grande may not be enough for us to tell the Arabs—who supply 18 percent of U.S. oil—to go jump in the lake, but it tends to diminish the importance of Saudi Arabia to the United States."[35]

Mexico's Foreign Policy

Mexican presidents employ foreign policy for domestic political purposes even more than most of their counterparts. As a perceptive scholar has remarked, the Mexican system must respond to four sets of interests: (1) the middle class, whose orientation is toward private enterprise and the United States; (2) the "government class," which favors state planning and coopera- tion with the Third World; (3) the well-organized and generally loyal working class, which exhibits slightly leftist leanings; and (4) the business and industrial community. Mexico's "progressive image . . . abroad has always been used to counterbalance a rather conservative domestic policy vis-à-vis its left-oriented critics."[36] Thus chief executives use foreign policy to bolster their domestic standing, reaffirm a commitment to the country's revolutionary tradition, and disarm critics.[37] In the face of desultory attacks, López Portillo adopted this strategy so often employed by his predecessors. In doing this he made clear that the United States would no longer be the only star in his country's international firmament.

López Portillo has shown his independence. For example, his ministers have emphasized that PEMEX would only ship "surplus" gas north of the border and then only at Mexico's price. If this amount were not paid, then Mexico would consume the fuel domestically, a possibility discussed in chapter 8. In addition, López Portillo turned down an invitation tendered Latin American heads of state to visit Washington on September 7, 1977, for the signing of the Panama Canal treaty agreement. Apparently he neither wanted to place his imprimatur on the document nor associate with the leaders of South American military regimes.[38] He did agree to fly to Panama to witness the June 16, 1978, exchange of ratification instruments; however, he made clear his disfavor over clauses that allow the United States to intervene militarily in Panama should the canal be closed for any reason. He manifested this disapproval by declining to attend a dinner for other Latin American leaders at the ratification ceremony, thereby forcing cancellation of the evening event. The Mexican president took advantage of his role as spokesman for all Latin American heads of state when he spoke at the canal transition ceremonies on October 1, 1979. He excoriated the legacy of colonialism that has beset Latin America, deplored "sickly and dependent industrialization," and saluted the victory of the Panamanian youth and their leader, Brig. Gen. Omar Torrijos, head of the National Guard. While conceding that the treaty served to bury the "opprobrium of colonialism and the abuse of force," he referred to the canal zone as a "military enclave," product of a "unilateral interpretation of a slanted and therefore partial neutrality."[39]

López Portillo further displayed his independence by making a fifteen-day visit to Eastern Europe in May 1978. He spent the bulk of his time in the Soviet Union, where he was praised for signing the "Tlaltelolco treaty" for prohibiting nuclear arms in Latin America. As a good-will gesture the Russians—who run a massive intelligence-gathering operation from their Mexico City embassy—pledged to share sophisticated fishing techniques and advanced petroleum technology.

The Mexican leader spent four days in Bulgaria, which promised to give agricultural lessons based on its experience in boosting food production 600 percent in four decades. Before and after the trip, reports suffused Mexico City that PEMEX, pursuant to an accord with the Soviet Union, would supply oil to Cuba in return for Russian shipments to Spain, a growing importer of Mexican crude. According to government spokesmen, the sales would begin at thirty thousand bpd, eventually climbing to seventy thousand. This move won loud approval from Mexico's small but recently recognized Communist party.

What will come of this overture to the left? Communist propaganda notwithstanding, technology in Azerbaijan pales in comparison to that available in Texas, and engineers, geologists, and equipment salesmen fill the nine daily flights between Houston and Mexico City. It is difficult to believe that the

Bulgarians have better know-how than that available in Mexico's own Agricultural Research Institute in Chapingo or in local projects supported by the World Bank and the International Maize and Wheat Improvement Center.

And despite the warm welcome accorded Castro when he visited Cozumel in May 1979, the quadrilateral deal with Spain, Cuba, and Russia may never become a major venture, despite obvious savings in shipping costs. This is in part because Venezuela has already joined such a four-party arrangement for a small amount of oil and in part because the Russians prefer the nuisance of dispatching tankers from the Black Sea to Havana rather than relinquishing the political leverage that comes from controlling their satellite's energy lifeline. In late August 1979 a spokesman for the Spanish government said that the "experiment" began two months before but involved only "very small quantities" of oil.[40] Well-informed sources in the U.S. government claimed that not a drop of Mexican oil had been sent to Cuba as of early 1980.

In any case Mexico's leaders will continue to publicize ties with Communist nations. López Portillo visited Peking in October 1978 "to strengthen relations and trade between the two nations," and a ranking official of the Foreign Ministry subsequently expressed his country's interest in increased contacts between Mexico and the Council of Mutual Economic Assistance (COMECON), the Soviet bloc's version of a common market. And during a September 1979 visit to Moscow, Gen. Félix Galván López, Mexico's defense minister, said it would be useful for Mexican officers to attend Soviet military academies. Meanwhile, the president of Hungary, the foreign minister of Poland, the chairman of Bulgaria's state council, and the vice president of Romania's parliament have visited Mexico City with a view to purchasing oil and promising closer economic cooperation. PEMEX stated its intention to begin shipping three thousand bpd to Yugoslavia in mid-1980.[41] López Portillo accepted Castro's invitation to visit Cuba between July 31 and August 2, 1980, in a move that showed solidarity with the Communist regime during a time of deepening economic problems illuminated by the flight of Cuban refugees.

But Communist nations have not been the monopoly's only interest. Political turmoil in Iran enhanced Mexico's attractiveness to energy-starved Japan. Two loan accords were signed in 1978: One committed a consortium of private banks to nearly $600 million, the largest credit without strings ever granted a foreign borrower by Japanese banks; the other amounted to $460 million. These funds will help finance a petrochemical complex, ports in Tabasco and Oaxaca, and the Alpha Omega road, rail, and pipeline link across the Isthmus of Tehuantepec.[42] At first the high Mexican price proved an obstacle to significant hydrocarbon sales to Japan; however, Japanese officials told an August 13, 1979, press conference that beginning in 1980, Mexico would sell their country a minimum of one hundred thousand bpd for at least ten years.[43]

Parallel to the oil agreement was a $500-million Japanese credit to PEMEX. This credit, the first negotiated by Mexico that is tied to hydrocarbon sales, provided an indirectly higher price for Mexican oil because of its preferential interest rate. Stung by Iran's demand for a $2.50-barrel premium, which Japan rejected, Prime Minister Masayoshi Ohira stated his country's desire to import 300,000 bpd from Mexico by 1982—about three fifths the volume obtained from Iran in early 1980 when he made the statement. Apparently anxious for greater economic assistance from the Tokyo government, Mexico reduced shipments to Japan to 27,500 bpd in the latter part of 1980.

Canada, whose supply of Iranian oil has been cut off, has rivaled Japan in strengthening energy relations with its North American neighbor. Canadian officials hoped that López Portillo's May 1980 trip to their country would lead to a mutually advantageous agreement. In particular they were anxious to begin buying crude from PEMEX to cover part of their 450,000-bpd import needs. Provided supplies are available, the Mexicans tentatively agreed to furnish 50,000 bpd by late 1980 and 100,000 bpd in 1981. The Canadians were also anxious to interest López Portillo in "Candu" reactors, which cost approximately $800 million and can be fueled with Mexican uranium without enrichment. Other items included in the talks were the possible sale of Canadian agricultural products to Mexico, the expansion of Canada's small program for accepting Mexican workers on a seasonal basis, mutual interests in relations with the United States, and steps to permit Mexico's use of a $1.58-billion Canadian credit offered the year before but delayed by bureaucratic problems and the failure to identify specific projects to which it could be applied.[44]

After the visit of King Juan Carlos I to Mexico, PEMEX increased contacts with Spain, which was to receive 160,000 bpd by the end of 1980. The Mexican state enterprise has also entered into an agreement with its Spanish counterpart (CAMPSA) to join a consortium that owns the 250,000-bpd Petronor refinery in Bilbao. Under the accord PEMEX obtained 15 percent of the capital in 1979 and an additional 18 percent the following year. Acquisition of this equity, formerly held by the Gulf Oil Corporation, should facilitate the sale of Mexican petroleum products in Europe.[45] Spain will have a 30-percent share in a shipyard to be constructed at Veracruz for tankers up to forty-four thousand tons.[46]

Even before Valéry Giscard d'Estaing's highly successful visit, France had signed an accord, effective January 1, 1980, to buy one hundred thousand bpd of Mexican oil. Pursuant to a protocol agreed to by Giscard and López Portillo, representatives of the two governments met to explore, among other things, the construction of both a factory to process uranium in Chihuahua, where rich deposits of the valuable mineral exist, and a nuclear power plant.[47]

During a May 1980 journey to Europe, López Portillo explored energy questions with officials in France, West Germany, and Sweden—countries that are eager for Mexican oil. He reportedly informed government leaders that because of growing inflation, his nation was not prepared to accept currency alone in exchange for crude, but expected a flow of technology as well. Like Canada and France, Sweden and West Germany are prepared to offer nuclear technology; however, López Portillo had not concluded a deal by mid-1980.[48] While the Mexican president did not fly to London, it is known, as noted earlier, that Great Britain has considered purchasing PEMEX crude to fill tankers that deliver North Sea oil to the Canadian and U.S. east coasts and, without another cargo, return empty.

Díaz Serrano has visited European and South American nations to investigate markets for oil and sources of technology. Over a dozen countries have entered into agreements with Mexico, and by January 1979 it had committed all of that year's production and most of the output for 1980. This activity formed part of an "export diversification plan" whereby the U.S. share of purchases will fall from 85 percent (1978) to 80 percent (1979) to between 60 and 66 percent (1980).[49]

During the first half of 1980, the U.S. portion of PEMEX sales abroad dipped to 74.1 percent (559,000 bpd) as exports averaged 754,400 bpd. Yet in July 1980 the United States received only 58.3 percent (536,000 bpd) of Mexico's shipments to foreign countries. It is impossible to know whether this figure represents a deepening of the diversification program or, as is likely, a monthly aberration explained by transportation bottlenecks, congested port facilities, inadequate storage capacity, adverse weather, delays in well completions, and—above all—surging domestic demand that have prevented PEMEX from attaining its projected export level, calculated to reach 1,468,000 bpd in early 1981. Canada, France, and Japan are among other importers discovering that the state monopoly cannot always deliver the volumes promised by high-ranking Mexican officials, particularly the minister of patrimony and industrial development.

Even as it picks and chooses among deals proposed by European nations and Japan, Mexico has lent a helping hand to Brazil, which with Mexican aid anticipates self-sufficiency in petroleum by the early 1980s. To achieve this goal, the Brazilian state oil company PETROBRAS concluded a working agreement with PEMEX and the Mexican Petroleum Institute in 1977.[50] Two years later Mexico was shipping 16,000 bpd of crude to Brazil.[51] PEMEX sold Brazil 40,000 tons of liquefied petroleum gas in 1979.[52] López Portillo also ordered shipment of 120,000 barrels to Costa Rica in late 1980 as an advance on the 7,000 bpd, one third of the republic's requirements, promised for 1980.[53] PEMEX officials ruled out the possibility of building a gas pipeline to Costa Rica to supply Central America because the region's low consumption would not assure sufficient sales to pay for the project.[54] In response to pleas from Central American and Caribbean regimes to follow Venezuela's lead and provide discounts or special arrangements,[55] Patrimony Minister Oteyza stated in 1979: "Although they are needy, priority in selling them our oil will be determined by the terms of international trade rather than by any other consideration."[56]

By early 1980 Mexico had moved away from this "strictly business" approach and was showing tangible evidence of its solidarity with revolutionary regimes. On January 24 López Portillo spent nine hours in Managua where he condemned the "satanic ambition of imperial interests" and suggested that the Nicaraguan revolution—like the Mexican and Cuban ones before it—offered a viable path for Latin American nations anxious to escape the problems besetting the hemisphere. He offered assistance to the country's fishing and communications industries and pledged that PEMEX would supply seven thousand bpd of oil, one half of the nation's consumption and an amount termed "indispensable" for the regime's survival.[57]

Two weeks later López Portillo welcomed Jamaican Prime Minister Michael Manley to Mexico. The leaders discussed Caribbean and Central American issues, emphasized the need for ideological pluralism and self-determination in the region, and announced increased economic cooperation. López Portillo agreed that Mexico would provide 10,000 of Jamaica's 27,000-bpd oil requirements in exchange for 420,000 tons of bauxite each year.[58] The oil deals to which PEMEX agrees in principle are subject to mutual acceptance of contract provisions and the availability of supplies. As table 10 shows, actual volumes shipped abroad do not necessarily correspond to the amount first announced after a meeting between Mexican and foreign dignitaries.

On August 3, 1980, López Portillo met with Herrera Campíns to announce a joint initiative designed to assist Central American and Caribbean countries. Under this plan Mexico and Venezuela together will ship up to 160,000 bpd to nations of the area on special terms. Specifically, they will grant credits to beneficiary countries amounting to 30 percent of their respective oil purchases for a period of five years at an annual interest rate of 4 percent. However, if the resources stemming from these credits are devoted to economic development, notably energy-related projects, the credits may be extended to twenty years at 2-percent yearly interest.

Supplies of petroleum will be shipped in accordance with commercial contracts signed by Venezuela or Mexico with the individual purchasing nation, and an effort will be made to export the oil in ships operated by the Caribbean Multinational Shipping Company. All countries of the area, including Cuba, are eligible to participate in the program, which may be renewed annually after an initial one-year period.

After a good deal of coaxing from Caracas, Mexico finally agreed to take part in the scheme lest Venezuela, which had indicated a readiness to increase its commitment of aid, gain political advantage in the region. In fact the amount of oil involved is small—Mexico's maximum contribution amounts to only 3.5 percent of its annual production—and assisting poorer countries with their petroleum needs demonstrates López Portillo's willingness to put into practice the tenets of his world energy plan.

Until the 1970s foreign affairs for Mexico meant, in essence, relations with its northern neighbor. Echeverría's move to change that relationship failed as economic difficulties made his country even more dependent on American financial assistance. Mexico's vulnerability was especially evident when U.S. and Canadian Jews launched a boycott of tourist facilities in the wake of the UN

TABLE 10
Mexico's Sales Abroad in 1980

Buyer	Quantity Announced for Shipment by End of 1980 (bpd)	%	Quantity Shipped in March 1980 (bpd)	%
United States[1]	730,000	58.3	550,000	71.7
Japan[2]	100,000	7.9		
France[3]	100,000	7.9	15,000	2.0
Spain[4]	160,000	12.8	60,000	7.8
Canada[5]	50,000	3.9		
Israel[6]	45,000	3.6	121,000	15.8
Brazil[7]	20,000	1.6	16,000	2.1
Jamaica[8]	10,000	0.8		
Nicaragua[9]	7,500	0.6		
Costa Rica[10]	7,000	0.6	5,000	0.7
Yugoslavia[11]	3,000	0.2		
India[12]	20,000	1.6		
Total	1,252,500	100.0	767,000	100.1

1. Information supplied by U.S. government.
2. *Daily Report* (Latin America), August 15, 1979, p. M-1.
3. Ibid., March 1, 1979, p. M-1.
4. *Excelsior*, March 11, 1979, p. 10-A.
5. Ibid., November 25, 1979, p. 28-A.
6. *Petroleum Economist*, 47 (January 1980):19.
7. *Daily Report* (Latin America), November 16, 1979, p. D-1.
8. *Proceso*, February 11, 1980, p. 26.
9. *Excelsior*, January 25, 1980, p. 1-A. The shipments were scheduled to begin in April.
10. *Petroleum Economist*, 47 (January 1980):19.
11. Sales will take place during the second half of the year.
12. Shipments will take place by the end of the year.

vote equating Zionism with racism. Yet despite his many problems, Echeverría whetted Mexican aspirations for a new, more assertive role on the world stage. The role that he couldn't secure with breathless globe-trotting and flamboyant speeches was assured by López Portillo, thanks to massive oil supplies. The attention lavished on Mexico by visiting monarchs, presidents, premiers, and foreign ministers has increased the confidence and boldness of the country's leaders while—notably in the case of the Communist bloc—defusing attacks from Marxist-oriented groups. It has also put Washington on notice that the United States is not the only important nation with which Mexico can cooperate on energy matters.

8

Mexico and the United States:
The Natural Gas Controversy

Background

The Natural Gas Policy Act of 1978, the Carter administration's compromise bill to deregulate gradually the price of natural gas, was one of the most controversial pieces of legislation in recent years.[1] The White House referred to it as the "centerpiece" of a plan vitally needed to curb energy imports, bolster the flagging dollar, brighten the image of the president, and inspire confidence in the nation both at home and abroad.[2] Supporters said that the measure, though flawed, would bring some degree of certainty to the natural gas market, but a spokesman for independent producers termed the proposal a bureaucratic nightmare whose passage might mean that "both intrastate and interstate supplies of gas would be dried up."[3] A leading consumer advocate dismissed the legislation as a "ripoff" that could cost Americans—notably the aged and the poor—as much as $35 billion.[4] And in a rare display of ideological bipartisanship, a coalition of ardent liberals and oil- and gas-state conservatives emerged in the Senate to block the bill's passage. Despite a fourteen-hour filibuster in the Senate and a last-minute effort to scuttle the bill in the House, a bleary-eyed Congress passed the legislation early on the morning of October 15, 1978, shortly before adjournment.

While passions rose steadily in Washington, López Portillo and Díaz Serrano carefully monitored the debate from Mexico City. They knew that the outcome of the legislative battle could determine the fate of projected natural gas exports, which could earn for their country $5 million per day in the early 1980s.[5]

Mexico produced 66 million cfd at the time of expropriation. This output, which rose more than tenfold to 720 million cfd during the Bermúdez years, averaged 2.92 billion cfd in 1979.[6] Meanwhile, gas reserves constitute approximately one third of the 60.10 billion barrels of proven holdings claimed by PEMEX on September 1, 1980. This amounts to almost 100 trillion cubic feet, using the monopoly's conversion figure of 5,000 Mcf to the barrel.[7] Over two thirds of the gas produced each day is "associated" with oil, with most of it coming from high-yield Reforma wells. Older wells register a gas-to-oil ratio

(GOR) of 1,000-to-one; that is, 1,000 cubic feet of gas are associated with each barrel of oil. At first GORs in the central Reforma fields (Sitio Grande, Cactus, Samaria, Río Nuevo, Iride, Cunduacán, and Nispero) maintained this level. Later 6,000 to 7,000 feet of associated gas per barrel issued from some wells. The average GOR for Reforma is somewhere in the neighborhood of 2,400-to-one, while 1,000-to-one prevails for the rest of the country. An exception is the offshore fields, where ratios of 250–500-to-one are found in the low gravity (20–25° API) fields of Akal, Bacab, Chac, Ha, Maloob, and Nohoch and ratios of 750–1,250-to-one in the medium gravity (30–35° API) fields of Abkatún and Ixtoc.[8] Table 11 provides an overview of natural gas production and reserves.

The ratios in Tabasco and Chiapas, extraordinary for Mexico, when added to the discovery of new fields in the North have led PEMEX to estimate that gas production will exceed 4 billion cfd in 1980 and climb to 5.4 billion cfd several years later.[9] Every projection thus far has proved conservative, and output may actually reach 8 to 10 million cfd according to a respected industry publication.[10] In view of the local market's inability to absorb this amount, what options exist to handle the surplus?

First, PEMEX could have expanded its petrochemical production even more. Once "stripped" of sulfur, the associated gas contains, by weight, 60 percent methane, the product that the state firm considered selling abroad.[11] Methane contains only two important raw materials for petrochemistry: am-

TABLE 11
Mexico's Natural Gas

Year	Reserves as of December 31 (millions of barrels)	Annual Production (Mcf)	Daily Production (Mcf)
1938	462.32	24,093,136	66,009
1946	371.78	26,052,618	71,377
1952	593.52	93,526,391	255,537
1958	1,558.13	262,626,152	719,524
1964	2,302.02	484,988,489	1,325,105
1970	2,279.13	665,005,369	1,821,932
1976	3,881,974.00	771,774,015	2,108,672
1977	5,573,651.00	746,863,000	2,046,200
1978	11,787,082.00	934,911,000	2,561,400
1979	12,243,344.00	1,064,559,000	2,916,600

Sources: The following volumes published by PEMEX: *Anuario estadístico 1975* (Mexico City: Petróleos Mexicanos, n.d.); *Anuario estadístico 1977* (Mexico City: Petróleos Mexicanos, 1977); *Anuario estadístico 1978* (Mexico City: Petróleos Mexicanos, 1978); *Memoria de labores 1977* (Mexico City: Petróleos Mexicanos, 1977); and *Memoria de labores 1978* (Mexico City: Petróleos Mexicanos, 1978). Also, *Petroleum Intelligence Weekly*, February 4, 1980, p. 11.

monia and methanol. The six-year, $20-billion development plan calls for a sharp increase in expenditures to triple petrochemical output by 1982. PEMEX will continue to enlarge the gas-sweetening operations at Cactus, designed to become the Western Hemisphere's largest with a 3.5-billion-cfd gas-handling capacity and a 1,300-ton-per-day sulfur output. Two additional 1,500-ton-per-day ammonia plants at Cosoleacaque will reportedly make this complex the world's foremost ammonia producer, with output rising from 3,000 tons per day (1976) to 13,000 tons per day (1982). Higher production would not only satisfy domestic demand but so saturate foreign markets as to depress prices severely. Mexico would produce 60,000 tons daily, more than half the world's consumption, if all the methane recoverable from Reforma were converted to ammonia. Facilities for such a transformation would have cost $4 billion in 1977. The folly of this expenditure became obvious to PEMEX when it reported that the international price of ammonia approximated that of natural gas—$2.60 per Mcf, by its estimates.[12] As Graciano Bello, then the company's southern zone manager, noted: "Building the plants would be no problem. The problem would be disposing of the products in an international market already facing more supply than demand in certain basic petrochemical building blocks."[13]

Second, the state oil firm might have implored industries and individuals to substitute natural gas for oil, thereby freeing additional crude for export. This strategy would have been especially attractive in Monterrey, where the use of oil contributes to pervasive air pollution. Several reasons militated against widespread conversion. Mexico lacked pipeline capacity to supply more gas to Saltillo, Torreón, Chihuahua, and other northern cities. Laying more mains would be extremely expensive and time-consuming, as would be the sweeping changeover of plants from oil to gas. In addition, a high subsidy—natural gas sold for 23 cents per Mcf to utility users and 34 cents to others, and PEMEX used significant quantities in internal operations, free of charge—meant that this fuel cost, on an equivalent basis, less than $2 per barrel in the domestic market, about one third the price of crude in 1977.[14] That government agencies charged with social functions bought much of the inexpensive fuel made a significant price increase politically difficult. Yet stimulating its use at home would have encouraged the wasteful consumption of a highly desirable, clean-burning energy source which could earn proportionally more foreign exchange per comparable unit than oil if sold in the international market.[15]

Third, the state monopoly could have pursued a conservationist line—either shutting in wells with major quantities of associated gas or extracting the gas and reinjecting it until this byproduct could be used efficiently. Petróleos Mexicanos has done both. Still, the pressing need for foreign exchange limited the number of wells that could be closed, and the reinjection of gas into the ground for storage cost approximately 80 cents per thousand cubic feet, exactly twice that of shipping it to the Texas border.[16]

Fourth, ever anxious to reduce its dependence on the United States, Mexico actively considered marketing Liquefied Natural Gas (LNG) abroad. Such a venture would entail processing the gas, lowering its temperature to 168 degrees (centigrade) below zero for transportation by special vessel, and reprocessing it after landing. The substantial costs of facilities for the liquefication cycle—about $7.5 billion—made this option unattractive. In 1977 the expenditures for LNG would have been $2.34 per Mcf compared to 40 cents for pipeline deliveries to the U.S. border. The former would have generated 27 cents in earnings compared to $2.21 for shipments to Texas, assuming a selling price of $2.60 per Mcf.[17] The 27 cents was roughly the amount derived by Algeria and Iran for deliveries to Europe. The absence of nearby markets required these countries to accept low profit margins; Mexico did not have to.

Two other liabilities attached to LNG: Purchase arrangements generally require long-term contracts, lasting at least twenty years after the first shipment is received (Mexicans resist binding themselves longer than a presidential *sexenio*); and construction requirements would mean that six years or more would elapse before sales began. PEMEX had included neither the value of the gas burned nor the cost of the delay in producing crude in the $7.5-billion estimate for building facilities. The time lag associated with LNG proved especially worrisome to Díaz Serrano, who observed that the world had only about twenty sure years "to continue to live the era of petroleum, and Mexico has this period to generate wealth by using the great demand and high prices that are currently being paid and that will surely rise over these 20 years that remain before the year 2000." The great distance to Yokohama—9,700 nautical miles compared to 5,250 to Rotterdam—meant that "the current possibilities of economically exporting methane gas to Japan are zero."[18]

Fifth, Mexico could flare the gas associated with the oil, which it so urgently needs to export. Díaz Serrano estimated that between 1957 and 1976, PEMEX burned off approximately $5.5 billion worth of gas.[19] The loss was $475 million in 1976 alone,[20] a year when the company managed to sell commercially only one third of the extracted gas. Officials deemed the waste of this important fuel intolerable and if continued, even at a diminished rate, an act that could engender scathing criticism of the government.

Sixth, the problems and limitations which beset each of the above-mentioned options lead to a sixth alternative. The surplus gas should be sold in the United States—a contiguous, affluent market that annually consumes twenty trillion cubic feet of the substance.

The "Deal" with the U.S. Firms

Mexico had no difficulty interesting U.S. corporations in its extensive gas reserves. American petroleum specialists had for several years been regular visitors to PEMEX's headquarters. The president and high officials of the Tennessee Gas Transmission Company, a subsidiary of Tenneco, had been among the most peripatetic travelers. They met at least ten times during the first half of 1977 with Díaz Serrano and other PEMEX executives[21] in sessions intended to devise a marketing agreement for the sale of Mexican natural gas to the United States.

On August 3, 1977, such an agreement crystallized into a "Memorandum of Intentions" signed between Petróleos Mexicanos and six pipeline firms serving thirty-one states. In addition to the Tennessee Gas Transmission Company, which agreed to take 37.5 percent of the gas, the consortium included Texas Eastern Transmission Corporation (27.5 percent), El Paso Natural Gas Company (15 percent), Transcontinental Gas Pipeline Corporation (10 percent), Southern Natural Gas Company (6.5 percent), and Florida Gas Transmission Company (3.5 percent).[22] A major consideration in the selection of these firms, among the many with which negotiations took place, was the broad distribution network they offered.[23] U.S. regulatory agencies had to approve a final contract before shipments could commence.

In a nutshell the memorandum provided for the initial delivery of approximately fifty million cfd, growing to two billion cfd by 1979—an amount exceeding 3 percent of American consumption. It contemplated a six-year accord, renewable for a second like term. Mexico reserved the option to lower or halt exports as required by domestic exigencies. The companies acceded to a Mexican request that the price of the gas be pegged to No. 2 fuel oil delivered to New York Harbor, an energy source that PEMEX considered comparable in terms of usage. This fixed the price of the first shipments at approximately $2.60 per Mcf, although the price could be renegotiated every six months.[24] The agreement also contained a "take-or-pay" provision whereby the American firms would be obligated to buy the Mexican gas irrespective of supplies available from domestic sources.

Several considerations cast doubt on the consortium's later claim to have engaged in "tough" bargaining over price.[25] To begin with, pipeline companies make money when volume or "throughput" is high and lose when their facilities are underused or idle; their primary concern is assured quantities of this premium fuel. Moreover, under regulations existing at that time, they could pass on their costs to customers by "rolling in" the price of relatively expensive imports with that of cheaper domestic gas, giving rise to a selling price somewhere between the two. Finally, Díaz Serrano praised the firms to the Mexican Congress. "The companies," he said, "accepted our prices and we enjoyed their full cooperation."[26] Nonetheless, Jack H. Ray, president of the Tennessee Gas Transmission Company, insists that the American firms got the best deal possible, especially in view of Mexico's original demand for "most favored nation" status—that is, a price equal to that paid for any gas, regardless of source, sold in the U.S. market.[27]

Most important of all, the tentative agreement anticipated construction of a gas pipeline (the *gasoducto*) from the Reforma fields to the U.S. border. All told, the forty-eight-inch-diameter conduit would stretch 735 miles along Mexico's alluvial plain, running from Cactus (Chiapas) through or near Cárdenas, Coatzacoalcos, Minatitlán, Poza Rica, Tampico, and Ciudad Madero to San Fernando (Tamaulipas). There it would branch, with one line reaching Monterrey and the other extending to Reynosa, the sister city of McAllen, Texas.

PEMEX also studied the possibility of a maritime line. It rejected this option, which might have been completed more rapidly, because of the problems involved in constructing and manning offshore platforms for the eighteen compressor stations, the additional costs involved, and the high component of foreign equipment and expertise required for a sea route. The state enterprise considered the creation of twenty-four thousand to thirty-five thousand construction jobs, many of which entailed modest skills, to be an important social and economic advantage of laying the pipes on land.[28]

The projected $1-billion cost of the *gasoducto* compared favorably with similar ventures. The trans-Alaskan oil pipeline from Prudhoe Bay to Valdez absorbed nine times this outlay, even though it had the same diameter and approximate length. The difficult nature of the terrain combined with low temperatures and environmental safeguards to increase the expenditures. Similar conditions along with greater distance have placed estimates for a gasline from Alaska's North Slope through Canada to the United States at $23 billion. Proponents of this project, yet to begin in 1980, hope that several billion cfd of gas will reach the upper Midwest and Pacific Northwest by the latter part of the decade.[29]

To assuage nationalist sensibilities, Díaz Serrano insisted that only "excess" gas would be sold to the United States. At the same time, PEMEX announced the route of the line only to San Fernando, a town two hundred miles north of Tampico, from which a spur would connect to the existing east-west pipeline serving Monterrey. Well-informed scholars have accused López Portillo of "tightrope-walking at the edge of truth" when he told Congress on September 1, 1977: "We have decided to construct a pipeline that will go from Cactus, Chiapas, to Monterrey with a branch that will go to Chihuahua and eventually loop back to the capital. Also, it will have another branch that will go to Reynosa for exporting to the United States."[30] Despite this treatment of the border link as almost an afterthought, the gas main was clearly designed as an export facility.

The enormity of the undertaking placed it beyond the capacity of Mexican industries, which could furnish but a limited amount of the required pipe, capital goods, and technology. Such North American firms as El Paso Natural Gas, Tenneco, and, to a lesser extent, the Bechtel Corporation, with experience in Alaska, worked closely with PEMEX in the engineering, construction, and testing.[31]

At first financing of the project appeared to be a problem. As we have seen, following the devaluation of the peso in 1976, the International Monetary Fund clamped a $3-billion annual debt restriction on Mexico in 1977 as part of an austerity program designed to rehabilitate the faltering economy. Inclusion of the $1-billion pipeline financing within this ceiling would have severely impeded the nation's ability to borrow for the development of other economic sectors. After weeks of deliberation, the IMF agreed that *gasoducto* credits would not count against the $3 billion. The organization realized that hydrocarbon exports were the major means of reducing Mexico's $30-billion foreign debt, three quarters of which had been incurred by the public sector.[32]

Private banks and official credit agencies of a half-dozen nations displayed a keen interest in assisting a scheme which Díaz Serrano optimistically concluded would pay for itself in the first two hundred days of operation.[33] At the head of the line stood the U.S. Export-Import Bank. It offered PEMEX a direct credit of $340 million to assist in financing the acquisition in the United States and exportation to Mexico of American goods and services. The bank anticipated that this credit, the repayment of which would commence in 1980 at an annual rate of 8.5 percent, would stimulate $400 million in purchases of American goods and services. The bank's board of directors also approved a second loan of $250 million on similar terms for equipment to accelerate production of oil and natural gas, increase refining capacity, and spur development of the petrochemical industry. This credit would induce $588 million in U.S. exports. The two loans would create forty thousand jobs in the United States.[34]

Reactions to the Proposed Gasoducto

President López Portillo's energy program has enjoyed smooth sailing in comparison to that of his North American counterpart, thanks in large measure to the centralized, authoritarian character of Mexico's political system. Still, the projected *gasoducto* excited stinging criticism, unexpected by the chief executive, within Mexico's usually pliant political parties and Congress. Not since the debate over the nation's Cuba policy in the early 1960s had a foreign policy decision created such public discussion.[35] For example, Francisco Ortíz Mendoza, leader of the legislative faction of the leftist Popular Socialist party (PPS), warned that the facility would heighten Mexican dependence on the United States. "Now [the Americans] want a gas pipeline," he said, "and soon they will ask for an oil pipeline. When they take all the gas and all the oil, they will leave us a large and useless pipeline 1,350 kilometers long and 48 inches in diameter."[36]

Víctor Manzanilla Schaffer, a member of the PRI and a federal deputy from Yucatán state, denounced the government's effort to secure the rapid passage, without sufficient committee consideration, of legislation designed to hasten construction of the gasline. At issue was an amendment to Article 27 of the Constitution, which codifies the diverse agrarian reform measures of the Mexican Revolution and regulates property rights. This specific proposal permitted PEMEX's temporary or permanent occupation or expropriation of land deemed necessary for investigation, exploration, or exploitation. The national oil firm planned to acquire a corridor 16.5 feet wide and 735 miles long for the pipeline itself as well as sites for roads, platforms, and compressor stations. In an unusual act of apostasy for the Revolutionary party, Manzanilla voted against the legislation, which he assailed as unconstitutional and contrary to the spirit of the country's hallowed agrarian reform.[37] "This move was also seen as another indication of the accumulation of vast power by PEMEX, deriving from the government's single-minded commitment to rapid develop-ment of the nation's petroleum."[38] Manzanilla's outburst followed by two weeks a move by 150 peasant families to block construction of the pipeline until their community received just compensation for the property condemned because of construction. Ironically, this *ejido,* located in Veracruz state, was named after Lázaro Cárdenas, father of Mexico's land and petroleum re-forms.[39]

Meanwhile, Senator Jorge Cruickshank García of the PPS censured an administration-endorsed mining bill, also under the purview of Article 27, as enhancing the likelihood that foreign firms could exploit the nation's newly discovered uranium deposits.[40] Despite these attacks Congress voted over-whelmingly to revise Mexico's fundamental law as requested by López Portillo.

Heberto Castillo, the most articulate critic of the government's energy policy, composed variations on the dependence theme first played by the PPS. He compared the *gasoducto* to the Panama Canal and warned that the structure could become an object of strategic significance, vulnerable to seizure by U.S. Marines. He urged instead a pipeline to a Caribbean port where the fuel could be liquefied for sale to the highest bidder. His National Front for the Protection of Natural Resources demanded an energy policy that would benefit the "have nots" who eke out a living at the base of Mexico's social pyramid.[41]

The press focused on opposition from deputies, senators, and left-wing parties. Often overlooked were "doubters" within the administration itself. These critics reportedly included senior cabinet members such as Jesús Reyes Heroles, then minister of the interior, ranking army personnel, and naval officers anxious to encourage greater maritime traffic, perhaps by constructing a port at Veracruz for LNG tankers.[42]

Predictably, the president and supporters of the *gasoducto* carried the day within official councils. They reiterated the enormous cost of liquefication facilities, tankers, and a deepwater port (the shallowness of the Gulf of Mexico now precludes the handling of supertankers at Veracruz). They observed that in the darkest days of the oil embargo, the United States neither dispatched troops nor endeavored to intimidate the Arab states militarily.[43] Díaz Serrano added that the Carter administration's commitment to human rights made the prospect of territorial aggression unthinkable. Moreover, he noted that supplying 33 percent of natural gas imports could hardly enhance dependence on a country that already provided two thirds of Mexico's trade, 70 percent of its foreign investment, and 90 percent of its tourism.[44] The president, whose response to critics "covered the gamut from didactic reasoning to something very close to angry damnation," argued that the gasline signified American dependence on Mexico.[45]

The persistent and unaccustomed carping proved irritating to the president and his energy advisers. Thus Petróleos Mexicanos sought to get a final contract signed with the consortium, complete construction of the pipeline, and to forge ahead with earning the generous supply of dollars that the duct promised to yield. A series of factors in the United States frustrated these plans.

As Mexico attempted to devise its own strategy, the United States Congress wrestled with an energy plan submitted by President Carter and Secretary of Energy Schlesinger. A key provision called for an immediate boost in the price of natural gas to $1.75 per Mcf, a figure based on the average price of the BTU equivalent of domestic crude. Deregulation of both newly discovered gas and that sold in interstate commerce would take place in stages, culminating with removal of government control on December 31, 1984. Paying Mexicans $2.60 per Mcf seemed unfair to many observers when domestic producers could look forward to just two thirds of this amount.

But that was only half the problem. Canadian gas, sold mainly in the "upper-tier" states of the Midwest, then entered the American market at $2.16. This charge derives from the price of OPEC crude oil delivered to Canada's east coast, plus transportation fees to a basing point (Toronto) minus the cost of shipping gas from Alberta to the U.S. border.[46] Should the Mexicans secure $2.60 for their supplies, would not Canadian producers, which supply one trillion cubic feet a year, demand parity and drive up the price to their American buyers? Yes, reasoned Sen. Adlai E. Stevenson, III (D-Illinois), chairman of the Banking Committee's Subcommittee on International Finance, which oversees the activities of the Export-Import Bank. One month after the bank notified Congress of the proposed $590-million loan to Mexico,[47] Senator Stevenson introduced a resolution that warned that Export-Import Bank financing of the "PEMEX natural gas project at such unreasonable prices for United States energy imports could set a dangerous precedent for prices of other U.S. energy imports." It further stipulated that such financing should await the secretary of energy's approval of the import price. Stevenson maintained that at $1.75 per Mcf, Mexico could recover the gas, operate the pipeline, finance its debt, and still enjoy an attractive profit. The additional 85 cents would constitute a "pure windfall" for PEMEX, adding $620 million annually to America's deficit in fuel imports. He also argued that by the time that the Mexican gas was integrated into the pipeline grid and transported to New York, it would cost about $3.60 per Mcf.[48]

Stevenson neither sought nor obtained co-sponsors for his resolution which, if passed, would not have bound the parties concerned. Having made his point with the bank and the administration, he did not press forward with legislation that had aroused scornful opposition from Sen. Lloyd M. Bentsen (D-Texas) and other spokesmen of gas-producing states.

As already discussed, Stevenson's action alarmed officials at the bank, which needed congressional approval of a three-month extension of its charter in 1977 and a four-year renewal of its charter in 1978.[49] On November 30, 1977, the bank announced postponement of final action on the loans pending an "indefinite additional review." Although both loans were eventually approved, this blatantly political move raised hackles in Mexico City.[50]

The gas deal also fell prey to governmental reorganization. A few months after assuming office, President Carter won congressional approval of a Department of Energy (DOE) to help devise and implement a national energy policy. With approximately twenty thousand employees, this cabinet-level body acquired the functions and staffs of the Federal Energy Administration, the Energy Research and Development Administration, and the Federal Power Commission, as well as portions of other departments and agencies. It came as no surprise when the transition precipitated bureaucratic infighting, blurred lines of responsibility, and outright confusion. Nowhere were the problems more acute than with regulatory functions. Within the energy field two bodies—DOE's Economic Regulatory Administration (ERA) and the Federal

Energy Regulatory Commission (FERC), successor to the Federal Power Commission—had grounds for claiming jurisdiction over the Mexican gas transaction. The former is concerned with the balance of payments, security of supply, proposed price at point of importation, national needs, and other issues of overall energy policy. The latter focuses on place of entry of the gas, siting, construction, operation of facilities, and sale price if the fuel enters the interstate market. Only after ERA approves a project does FERC become involved.

That aggrieved parties may challenge ERA and FERC actions at any point in the proceedings enhances the possibility of deadlock and delay. Studies reveal that FERC received 7,125 docketed filings in 1978. In addition, the commission carried over 6,833 cases that were unfinished at the end of the year.[51] Further complicating the picture was the possibility that an environmental-impact statement might be required of the *gasoducto*—even though the facility would rest solely on Mexico soil—before a contract could be signed for delivery of natural gas.

High-level officials of the departments of Energy and State circumvented the bureaucratic thicket by meeting directly with their Mexican counterparts, who—used to the dominant role of their chief executive—seemed baffled by the administrative process. "They thought that Carter could wave a magic wand and the deal would go through," confided a U.S. official who asked to remain anonymous. The first contact concerning a possible gas sale occurred in April 1977 when Díaz Serrano informed Schlesinger of Mexico's interest in exporting the fuel.[52] On June 1 Nathan Warman, undersecretary of Patrimony, wrote Secretary Schlesinger inviting an American delegation to Mexico City to discuss the utilization of gas from southern Mexico. Later in the month, Díaz Serrano and Patrimony Minister Oteyza met with Schlesinger and other U.S. representatives in Washington. At this June 27 session, the secretary of energy advised the visitors that (1) the U.S. government would have to approve the price aspect of any project, and (2) a problem would arise if the price exceeded the $2.16 per Mcf paid to Canada or if the escalation clause were tied to No. 2 fuel-oil prices in New York Harbor.

In a mid-July meeting of the Energy Subgroup of the consultative mechanism, a high-level official of the Federal Energy Agency (the DOE had not yet been created) carefully explained the intricacies of the regulatory process on natural gas imports. Another U.S. spokesman stated that the Natural Gas Act of 1939, which governed policy in this area, required that the price of imports be "just and reasonable." Mexican imports would probably not meet this criteria in view of the projected $1.75 price for domestic gas and the $2.16 charged by Canadian suppliers. Participants in this meeting agreed on a work program, which in part stated: "The two governments will seek appropriate means to accelerate discussions of the proposal for the exportation of natural gas to the United States."[53]

Despite the warnings from U.S. officials about possible difficulties, PEMEX and the six transmission companies signed the letter of intent on August 3, 1977. Eight days later the U.S. firms filed a preliminary application with the Federal Power Commission, seeking authorization to import the Mexican gas. The Americans believed that they had successfully conveyed to the Mexicans the political and bureaucratic problems that surrounded the pending sale. They even thought that they had hit upon a compromise; possibly the $2.60 per Mcf would apply only when large quantities of gas began to flow, two years hence.[54] By this time the U.S. price would be nearly as high. This plan would allow the Mexicans to save face by obtaining the highly publicized price of $2.60 while severing the tie to No. 2 fuel oil whose price, subject to marked fluctuations, is strongly influenced by OPEC. The Americans insisted that neither the cartel nor any other third party should determine the price in a bilateral arrangement.[55] In any case American officials, who claim that Mexican negotiators had agreed to abandon the tie to No. 2, believe that Díaz Serrano persuaded the president to veto such a move.[56]

Further complicating the picture is the fact that, unlike oil, a highly fungible substance, natural gas has no single world charge. Geographic propinquity, political relations, and alternative energy options are among the factors that determine the price in any given transaction. Before the shah fell, Iranians sold gas to Russia for 78 cents per Mcf in soft currency barter arrangements; Canadians were obtaining $2.16 from the United States in 1977 (a figure which had more than doubled by 1980); and Indonesians, who supply it to Japan for between $3.00 and $5.00, have concluded a deal for exports to the United States at $3.42, derived from a base of $1.25 and linked to the U.S. wholesale price index (50 percent) and the price of Indonesian crude (50 percent).[57]

In a pre-Christmas 1977 meeting between Díaz Serrano, Mexican Foreign Minister Santiago Roel García, and Schlesinger, the Mexicans again tried to link the price of natural gas to No. 2 fuel oil. The secretary of energy, in a manner perceived by the visitors as insulting, abrupt, and arrogant, adamantly rejected this demand. Instead, he called on PEMEX to accept the price of residual fuel oil (No. 6), then about $2.35 per Mcf, in order to secure an agreement.[58] The Mexicans held their ground, apparently believing that another cold winter would make the Americans more flexible (López Portillo had gained favorable publicity in the United States by making available a small additional quantity of gas during the frigid days of the previous winter).[59] The ensuing impasse prompted PEMEX to announce on December 22 that it would not renew the Memorandum of Intentions, scheduled to expire at the end of the year. The Mexicans justified the action on the grounds that their country, a champion of nonintervention, did not intend to meddle in the domestic-energy debate taking place in the United States. The gas companies, they felt, were using the pending deal as a lever to push up the fuel's price in the U.S. market.[60]

Roel García resisted charges that the American government was attempting to take advantage of the gas transaction to shape Mexican policy, especially with respect to illegal immigrants. "Mexico will neither accept pressures or blackmail, nor permit the White House to use the problem of laborers as a pretext to obtain eventual advantages in transactions dealing with raw materials," he stated.[61]

The transmission companies were livid that governmental interference had prevented their securing an assured supply of their economic lifeblood. That Schlesinger made them wait three weeks after a request for a meeting (December 17, 1977) before granting them an appointment (January 5, 1978) exacerbated their anger. In this session the secretary of energy explained his intransigence on the familiar grounds that the higher price would both inspire the Canadians to follow suit and undermine the pending energy legislation. He also argued that the lack of alternative markets would eventually cause Mexico to ship gas to the Texas border.[62]

Jack H. Ray believes that Schlesinger acted "illegally in trying the case in the newspapers" before a contract was even submitted to the regulatory authorities, at which time DOE's intervention would have been proper. As for the argument that Ottawa would raise its price to match that contained in an agreement with Mexico, Ray maintained then—and subsequent events support his thesis—that higher Canadian prices were inevitable, irrespective of a gas deal with Mexico. Ray also rejected the contention that linking the price of Mexican gas to fuel oil sold in New York Harbor would allow OPEC to set this charge, inasmuch as the price of No. 2 was only partly determined by cartel sales, with the price of U.S. supplies also playing a role.[63]

López Portillo Left "Hanging by his Paintbrush"

Mexican reaction to the American refusal to pay $2.60 was swift and loud. The president averred that he had been left "hanging by his paintbrush" when Schlesinger knocked over the ladder[64]—a phrase distinguished more by imagery than accuracy in view of the repeated warning signals received in Mexico City over potential pricing problems. Nonetheless, Mexican journalists, intellectuals, and officials vented the nation's collective anger at the secretary of energy. One of the most undiplomatic statements came from the country's chief diplomat: "Schlesinger is a liar," said Roel García, "and you can quote me on that."[65] In February 1978 the patrimony minister announced that Mexico would substitute the gas previously earmarked for its northern neighbor for oil in generating electricity and making petrochemicals. It would export the oil thereby made available.

Mexico's outcry was echoed in the United States on editorial pages, television programs, and Capitol Hill. Senator Kennedy accused Schlesinger of having "spurned the opportunity for negotiations with the Mexicans on price," suggesting an inflexibility belied by the record of contacts. Similarly, Sen. Jacob K. Javits (R-N.Y.) asked the secretary: "Isn't it a fact that we have very materially been poisoning the atmosphere between us and the Mexicans by the adamant position that you have taken on Mexican gas?" Meanwhile, Sen. Charles McC. Mathias, Jr. (R-Md.) wrote in January 1978 that the "whole fabric" of the agreement between the Mexicans and the six U.S. companies "was rent when an abrupt unilateral ruling was issued by the Department of Energy."[66]

Mexico suddenly discovered that it had no surplus gas to ship to Texas. As one scholar noted: "The *gasoducto* was miraculously rechristened the *Troncal Nacional* [National Trunk Line] and diverted from export service to the meeting of newfound domestic needs."[67] Díaz Serrano announced completion of the facility on March 18, 1979. It extends from Cactus to Los Ramones, Nuevo León; from there it continues as a forty-two-inch pipeline to China, Nuevo León, where it joins the Northern Pipeline System, linking Reynosa with consuming centers at Monterrey, Monclova, Torreón, and Chihuahua. Completion of the second phase of construction, scheduled for mid-1980, involves a seventy-four-mile, forty-two-inch pipeline from San Fernando to Reynosa to facilitate the export of gas to the United States. The Cactus–Los Ramones phase, completed ahead of the original timetable, required a total investment of $504 million, 60 percent of which was spent internally as the project provided employment to 18,500 construction workers. Installation of eighteen compressor stations would increase the line's capacity, 0.8 billion cfd when inaugurated, to 2.7 billion cfd. Laterals radiating from the main line will supply gas and spur economic development to outlying facilities. As PEMEX's director general boldly stated:

> Even if the United States were not a potential customer for the natural gas, we would still have had to construct the national trunkline to meet the energy needs of our population. The *gasoducto* is imperative to connect the northern and southern supply networks, link all known oil and gas fields, and supply natural gas to all the cities and future poles of development along the Gulf Coast.[68]

Can Mexico efficiently consume its natural gas internally? PEMEX officials stated that a conversion program would require two years, and it will be difficult to predict the outcome until that time elapses. The Department of State estimates that a concerted effort could increase domestic consumption from 1.5 billion cfd to 2.5 billion, a figure confirmed by foreign experts in Mexico City. As mentioned earlier, substitution entails formidable obstacles. PEMEX must build networks of distribution lines in each locality, and the costs of conversion are high. Mexican officials estimate that converting those fuel-oil users who can switch to natural gas will mean a $400-million investment. Contained in this figure are the transportation and distribution systems needed to deliver the gas to consumers (industrial users will bear a significant portion of this expense); additional equipment to improve refineries; and possible facilities in producing areas either to store the fuel or use it for field pressurization in lieu of water.[69] Moreover, powerful interests with a stake in existing energy systems will resist widespread changeover to gas lines and a metering system. Unions and their subcontractors drive the large trucks that supply factories with diesel and bunker oil and carry butane to homes and businesses; influential businessmen control storage centers; and well-entrenched politicians and their cohorts receive "commissions" on a series of related transactions. Too, a sharp rise in gas production would require the construction of new processing facilities. Government projections show that the gas sweetening and stripping capacity should be adequate to take care of planned output through 1981 or 1982, but greater production will necessitate designing and building additional plants.[70]

There may be an immediate need for such plants inasmuch as the 600,000 Mcf flowing from Cactus to northern Mexico in mid-1980 was reportedly "wet" and only partially sweetened, signifying that it contained light gasoline products and sulfur, which could rapidly corrode the new *gasoducto*. The Reforma gas winds up in Monterrey, with the 300,000 Mcf earmarked for the United States under the agreement discussed below supplied from the Monclova fields in the Sabinas Basin.

According to one observer, Mexico has achieved "considerable success" in absorbing its gas production.[71] PEMEX has reported a daily drop in flaring from 400,000 Mcf (1977) to 150,000 Mcf (1980), with further reduction when a thirty-six-inch line is completed to move gas ashore from the Campeche fields.[72] In addition, the Federal Electricity Commission claims that it alone could use 60 percent of the Reforma gas to replace oil as a boiler fuel in its generating plants, and the Rolls Royce Corporation has promoted the utilization of gas-driven turbines to produce electricity.[73] If unions can be coaxed into cooperating, opportunities abound to run gas lines to the Chrysler factory in

Toluca and the industrial area north of Mexico City where Ford, Dow Chemical, and Bacardí have operations. Monterrey's industrialists have been notably efficient in shifting from fuel oil to natural gas.

Insufficient data prevented an assessment of the nationwide conversion effort in late 1980. Although continued progress is likely, Mexico's climbing oil production suggests that increased quantities of gas will be available for shipment abroad.

Negotiations Resume

Once Congress passed the Natural Gas Policy Act of 1978, Ambassador Lucey tried to secure a reopening of bilateral contacts over the sale of this fuel to the United States. A presidential review memorandum (PRM-41), prepared for the National Security Council as a basis for impending discussions between Carter and López Portillo, urged completing the negotiations because the absence of an outlet for gas would limit oil production.[74] The final impetus came from the summit meeting: On February 16, 1979, it was reported that discussions would resume as soon as possible. This announcement proved the most significant development of a presidential visit marked by a host of blunt reminders that Mexico would not permit the "Colossus of the North" to exploit its hydrocarbons. As Carlos Fuentes, a world-acclaimed novelist, tersely expressed it: "Mexico is not an oil well."[75]

An American negotiating team headed by Julius L. Katz, assistant secretary of state, and Harry Bergold, assistant secretary of energy, arrived in Mexico City to resume talks on April 3, 1979. At that session the Mexicans indicated that conditions had changed since PEMEX and the transmission companies had signed the letter of intent. Now only eight hundred thousand Mcf might be supplied to the United States in contrast to the two billion cfd specified in the original memorandum. The U.S. negotiators, who kept in close touch with Ray and the American firms, took these and other new terms into consideration as they agreed to continue to work toward an accord. "Their calculation was that López Portillo, under fire from the left, was making a show of walking away from the deal. Their hope was that a subsequent meeting would yield better results."[76]

At the second session, held on May 4 in Washington, the Mexicans reported that domestic industries were absorbing most of the associated gas from the Reforma area. As a result no more than three hundred thousand Mcf could be sold, and this gas would come from old fields in the northern part of the country. The Mexicans were not even certain that this smaller amount would be available, casting doubt on the feasibility of a gas deal. The Americans were, in the words of one official, "genuinely disappointed. We thought Mexican industry was using so much natural gas that maybe no deal would go through. We saw Santiago Roel and Díaz Serrano and López Portillo making price a matter of national honor. We agreed to keep talking, but without much hope."[77]

Two events enhanced the possibility of an agreement when teams from Mexico and the United States next met on July 12 in Mexico City. In May the president had removed as foreign minister Roel, who had proved weak and ineffective in negotiations. Then, in its June session in Geneva, OPEC's oil ministers increased the basic price of crude oil by one fourth to $18.00 per barrel. This action led to an across-the-board increase in the prices of all grades of petroleum and its derivatives. During the negotiations, the Mexicans had insisted on pegging the gas price to that of No. 2 fuel oil delivered in New York Harbor. The equivalent price of this light fuel oil had risen sharply above the $2.60 figure that obtained when the talks collapsed in December 1977. For their part, the Americans insisted on a price linked to No. 6 residual oil, arguing that approximately 80 percent of the U.S. plants shifting from petroleum were using this cheaper heavier fuel, not Mexican gas for which industrial users would have to pay the full incremental cost. By that time the price of No. 6 had also risen sharply, making it possible for the Americans to agree to a figure well above the $2.60 originally sought by the Mexicans.

When the price proved an obstacle to agreement on July 12, the Americans decided to prepare a strong, well-documented justification for the use of No. 6 oil as a base price at the Washington meeting scheduled for July 27. At that session Assistant Secretary of State Katz presented the case for No. 6 and indicated that his country might agree to $3.30 per Mcf, and possibly a bit more. Yet the Mexicans were unwilling to come to terms.[78] One problem was that PEMEX had been excluded from the negotiations, which were now conducted principally by a representative of the Ministry of Patrimony, working closely with a colleague from the Ministry of Foreign Relations. Lacking the technical expertise to evaluate the market analysis offered by their American counterparts, the Mexican negotiators resorted to the political demand that the price had to exceed $4.00, a sum vaguely related to the anticipated higher price of No. 2. The Mexicans knew that any appearance of having "sold out to the gringos" would have strong repercussions at home. At least one American representative to the talks believed the Mexicans were stalling until Schlesinger's announced departure from the Department of Energy took place. Another U.S. negotiator spoke of an "impasse."[79]

So apparent was the reasonableness of exchanging American dollars for Mexican gas that a deal seemed inevitable. A study by the Congressional Research Service, figures from which appear in table 12, named "Mexican and Canadian gas as the least expensive supplemental source of gas available in the

United States." Moreover, influential voices such as those of the editorial writers of the *Washington Post* insisted that "the time has come to buy the gas."[80]

Still, the 1978 gas act spurred production so that a supply "bubble" existed in the United States during the 1979 negotiations. As a result the American representatives felt little pressure to hastily conclude a deal, even though many experts believed expanded gas sales to be a sine qua non of additional oil exports. The situation in the U.S. market made it difficult for the Mexicans to succeed with their intractable approach on the pricing question.

In view of the stalemate, Ambassador Lucey held a private meeting with López Portillo on August 3. He emerged from the session convinced that the Mexican chief executive would agree to a price of $3.40. Six days later the American team arrived in Mexico City in hopes of striking a bargain. Much to their dismay, the Mexicans refused the price. Moreover, Foreign Minister Castañeda, who had succeeded Roel García, took personal charge of the talks and explained that Lucey had misunderstood López Portillo. "The Americans concluded that Castañeda, resentful of the end run Lucey had made around him and persuaded that a higher price could be achieved, had caused López Portillo to reverse himself. At least some Americans left the fifth meeting certain that López Portillo had welshed on the deal."[81] The atmosphere surrounding the

TABLE 12
Estimated Supplemental Gas Prices, 1985

Country	Price per Mcf[1] (dollars)
Mexico[2]	$4.88
Canada[3]	3.56–4.88
Synthetic natural gas (coal)[4]	7.32–8.04
Synthetic natural gas (oil)[5]	5–6.75
Alaska gas[4]	5.81–6.11
LNG[6]	5.15

Source: U.S. Congress, Senate and House, *Mexico's Oil and Gas Policy: An Analysis,* Report prepared by the Congressional Research Service, Library of Congress, for the Committee.

1. Foreign sources of natural gas and LNG prices are for delivery to the U.S. border, and the price estimates for U.S.-produced SNG coal and oil are mainly production costs. The Alaska gas price of $5.81 is a U.S. border price estimate, but $6.11 reflects the large capital investment needed for delivery to the main U.S. distribution points in Dwight, Ill., and Antioch, Calif.

2. Reflects a 15-percent OPEC price increase in 1979 and thereafter assumes a 7-percent annual increase in No. 2 fuel-oil price.

3. CRS estimate based on Canadian and U.S. gas-pricing policy.

4. Prices for SNG coal and Alaskan gas from DOE intervention before FERC, ANG coal-gasification company proceeding, FERC docket Nos. CP75-278, et al., June 1, 1978, p. 5.

5. Estimate assumes naphtha as the feedstock and is based upon conversation with Bill Norman, J. Markowsky Associates, Boston, Mass.

6. Tenneco Atlantic Algerian project.

negotiations was further poisoned on August 22 when Ambassador Krueger publicly raised the question of compensation for the Ixtoc 1 oil spill. As discussed in chapter 7, this undiplomatic statement endangered the presidential summit meeting scheduled for September 28 in Washington.

The prospect that this meeting might be cancelled caused both sides to reassess their positions. The White House, sensitive to the domestic political rewards of good relations with Mexico, urged the departments of State and Energy to hammer out a compromise. Meanwhile, Castañeda informed Washington that he personally favored an agreement, provided an American official of suitable rank was dispatched to negotiate.[82]

On August 29 and 30 Deputy Secretary of State Warren Christopher met with Castañeda. Discussions focused on price in an effort described as "a bit like throwing darts at a board."[83] The Americans offered $3.50 per Mcf, which represented a weighted average of No. 2 and No. 6 prices in selected American cities; the Mexicans held out for $3.75. Finally, Castañeda, apparently tired of economic arguments, suggested they split the difference, giving rise to the mutually satisfactory price of $3.625. The Mexicans then proposed that they have the right to cancel the deal on ninety days' notice. Christopher went back to Washington to consider both this idea and a suitable escalator on which to base future price increases.

He returned to Mexico City on September 19 to consummate the agreement. The price was fixed at $3.625 as determined in late August. This base price could be reconsidered before January 1, 1980, if the price of gas "from comparable sources," i.e., Canada, surpassed $3.625. Future increases, to be considered every three months, would be geared to the international price of crude oil. (It was later decided that the revision would be based on a market basket of five crudes: Saudi Light, Mexican Isthmus, Algerian Saharan Blend, North Sea Forties, and Venezuelan Tía Juana Medium.)[84] The agreement was open-ended with respect to volume and duration: Either party could reduce the quantity involved or terminate the contract with 180 days' notice.[85]

The stipulated amount, far below the two billion cubic feet per day contemplated in the 1977 arrangement, represented only 8 percent of American imports and 0.5 percent of the country's total consumption. But observers close to both governments felt that the quarterly negotiating sessions would provide relatively unpublicized opportunities to expand the quantity. Slow progress in the Federal Electric Commission's conversion program could find Mexico with a two-billion-cfd daily exportable surplus in 1980.[86] In fact, Deputy Secretary Christopher indicated that an additional two hundred thousand Mcf could be agreed to in the near future.[87]

The president of Border Gas, the umbrella corporation for the six American transmission firms, doubts the possibility of a quiet accord with Mexico leading to expanded supplies. He is convinced that possible increases above three hundred thousand Mcf will find companies not party to the current agreement—the Panhandle Eastern Pipeline Company, United Gas Pipeline Company, and Northern Natural Gas Company, for example—going before regulatory bodies to seek a portion of any larger volume sold by PEMEX. "Every son of a bitch big enough to grow a moustache" will intervene in hearings that will be anything but low-keyed and diplomatic, he believes.[88]

With the opening of the San Fernando–Reynosa link, Mexico will be able to expand (perhaps double) shipments to McAllen. An ever greater increase may be anticipated because PEMEX incurred the additional expense of having its contractors lay double segments of forty-eight-inch pipe at all river, highway, and railroad crossings from Cactus to northern Mexico. In fact PEMEX's right-of-way is wide enough to accommodate three additional *gasoductos*.

In their first major bargaining venture over energy, the Mexicans failed to succeed with a "take-it-or-leave-it" approach—a fact that may lead to greater pragmatism in future talks. Still, their negotiators had nothing for which to apologize. Although giving ground on price in September, the Mexicans insisted on receiving the $4.47 price announced by Canada on January 17, 1980, two days after shipments began to McAllen, Texas. The Americans, eager to obtain a new source of supply, agreed to the higher figure. Further, the U.S. government wound up accepting an escalator tied to OPEC prices, something it had vigorously resisted in 1977. Such a link had always been implicit in Canadian prices, which the Mexicans now watch quite closely. The Carter administration was pleased to have a gas deal that betokened improved relations with a neighboring country whose importance transcends energy policy and international relations.

Former Secretary of Energy Schlesinger must have derived satisfaction from the terms of the contract. Had the 1977 agreement won approval with the price pegged to No. 2 fuel oil landed in New York Harbor, the Mexican gas would have cost over $6.00 in 1980. Thus by vetoing a price linked to this distillate, he had saved U.S. customers at least $1.53 per Mcf. Too, acceptance of the 1977 contract with its "take-or-pay" provision would have led to a shutting in of American production.[89] In any case Mexicans found a new bête noir against whom to level attacks in Robert Krueger, who suggested negotiations over compensation for the Ixtoc 1 oil spill, the subject of chapter 9.

The Ixtoc 1 Blowout

Introduction

As discussed in chapter 3, PEMEX's development emphasis had shifted by 1978 from the prolific Reforma trend to the offshore Campeche area. Not only did the maritime province offer the prospect of Tertiary, Cretaceous, and Jurassic pay zones compared with only the latter two onshore, but the gas-to-oil ratio was considerably lower—an important consideration in view of mounting criticism of flaring and the difficulty of fashioning an agreement to ship gas to the United States. In addition, PEMEX struck oil at 11,480 feet in Campeche Sound (Chac 1), while Reforma wells in 1978 averaged between 13,000 and 16,000 feet. Also, two giant Campeche wells that began production in October 1979 yielded sixty thousand (Akal 3) and fifty thousand (Akal 74-A) bpd, while the largest producer in the Reforma area was rated at eighteen thousand bpd and the average for Reforma's 170 wells was sixty-five hundred bpd.[1]

Maritime work could also proceed free of confrontations with peasants who increasingly resorted to violence to protect their farms and fishing grounds. Nor was such activity visible to the prying eyes of domestic and foreign journalists. Finally, such extremely costly operations, twice as expensive as those onshore and often requiring foreign expertise, gave rise to multimillion-dollar contracts and "commissions" for politicians, agents, union leaders, and PEMEX insiders.

PEMEX's goal in 1979 was the placement of a new offshore platform every two weeks, with twenty-six platforms under construction in the United States and Mexico. Maritime output reached eight hundred thousand bpd in August 1980, and some observers forecast a production potential as high as seven million bpd.[2]

Amid this rush of activity Ixtoc 1, an exploration well lying a few miles west of the Akal structure approximately fifty-five miles north-northwest of Ciudad del Carmen, suffered a blowout on June 3, 1979, sending an uncontrolled flow of dark brown oil into the blue waters of the Gulf of Campeche. The accident occurred when the crew of the semisubmersible rig *SEDCO-135*, operated by a private contractor (PERMARGO), encountered extremely high permeability at 11,800 feet and "lost circulation" of the drilling mud.[3] This mud, composed of water, clay, and certain chemicals, is pumped from the surface down the drillstring (a combination of thin-walled pipe called drillpipe and thick-walled pipe known as drill collars) to lubricate the drill bit, carry rock cuttings out of the hole, and control any formation pressure that might be encountered. Under normal conditions, it returns up the annulus, a space between the outside of the drillstring and the inside of the casing or hole, to the surface mud tanks from which recirculation takes place. Loss of circulation signals the escape of drilling fluid into the fractures of the rock at the bottom of the hole. The crew managed to regain circulation on the same day, June 1, and drilling resumed. They again lost circulation the next day when the drill bit hit a drilling break at 11,900 feet, some 656 feet from the programmed location of the reservoir. Such breaks or pockets, which cause the bit to drop or the drillstring to elongate a short distance, are not uncommon.

When circulation could not be regained by pumping mud through the drillstring, the senior SEDCO man on board—PERMARGO had contracted for a few SEDCO personnel to help maintain the vessel and provide advice as to its proper use—advised pumping salt water into the hole while more mud was mixed. PEMEX and PERMARGO supervisors vetoed the proposal. They attributed the circulation loss to fractures, anticipated because of experience with other wells in the area, and concluded it would be temporary, producing neither flow nor high pressure from the formation. "The view of the PEMEX and PERMARGO onshore senior personnel was that they had tried this filling procedure on prior occasions and found that filling the hole with salt water would render measures to correct the loss of circulation more difficult and was not necessary for safety," later stated Stephen Mahood, SEDCO's executive vice president, before committees of the U.S. Senate.[4]

After a high-level meeting in Ciudad del Carmen, the Mexicans decided to see if the situation remained stable for at least ten hours. If it did, they would pull the drillstring, remove the bit, and run the drillstring back into the hole open-ended. This move would allow them to pump in materials to seal off the fractures causing mud loss. "The SEDCO personnel advised against the procedure of removing the drill string without knowing the level of fluid in the hole. However, PEMEX elected to go ahead with such procedure," Mahood noted.[5]

Ten and one-half hours elapsed without any observable change in conditions. Thus at about 10:00 P.M. on June 2, the crew began "tripping out" the drillstring, stopping every nine hundred feet to make certain the well was filled with mud to compensate for the volume lost. Pursuant to standard operating practice, they tested the blowout preventers (BOPs), used to seal a hole, and the kelly cock, a safety valve on top of the drillstring above the rotary table.

The removal operation proceeded smoothly until the drillstring had been lifted to the point that the first drill collar, a heavy-walled cylinder of steel that concentrates weight immediately above the bit and maintains the drillpipe in tension, was visible at the rotary table on the rig floor vertically above the well. According to Mahood, the drill collars extended down about 630 feet and through the blowout preventer on the floor of the sea.

At 3:13 A.M. on June 3, all hell broke loose. Drilling mud flowed up the annulus and through the drillpipe (see figures 2 and 3) at a rapidly increasing rate. It proved impossible to install the drillpipe safety valve, designed to halt such a flow, because a key stand of drillpipe, unscrewed when conditions appeared normal, became cross-threaded. Moreover, workers were unable to plunge the drillstring back into the hole because the drill-collar safety clamp held the collars securely in place. Closing a device called a hydril stopped the flow up the annulus, but oil and gas continued to gush through the drill collars. An attempt to close the lower blowout preventer failed because shear rams could not cut the drill collars that jutted through the BOP.

The oil and gas, which soon ignited, began escaping at a prodigious rate, sending part of the rig's mast, rotary table, and thousands of feet of drillpipe to the bottom around the BOP stack. A helicopter rescue mission saved the sixty-four crew members, whose hopes of establishing a speed record for drilling were dashed by the accident.[6] The semisubmersible was moved on June 4 to prevent its sinking on the hole. Severely damaged, it was sunk five weeks later in a six-thousand-foot trench beyond Mexican territorial seas. This action, which later aroused criticism, was approved by the insurance underwriters who held salvage rights to the rig. Insured for $22 million, *SEDCO-135* was the first in a pioneering line of vessels and had been in service since 1965.

Other catastrophes have beset Mexico's oil industry. As we have seen, the first Dos Bocas well, drilled in 1908, spewed out of control for two months, sending thousands of barrels of oil each day into the Laguna de Tamiahua, from where it flowed to the sea, contaminating oyster grounds and fishing areas. Potrero del Llano 4, opened in 1910, also suffered a blowout for two months. Other major accidents included Well 6 in Poza Rica (1947), Santa Agueda 2 in the new Golden Lane (1953), Brasil 1, a gas well in the Reynosa area (1950s), two wells in the José Colomo fields in east Tabasco (early 1960s), and a well named Hormiguero or the "anthill" in Tabasco (late 1960s). In addition, a gasline explosion killed fifty-two people in a Tabasco oil camp in November 1978.

The Ixtoc 1 blowout was soon identified as Mexico's "worst ecological disaster" as 30,000 or more bpd escaped during the first few weeks. The renegade well released 3.1 million barrels of oil and approximately 3 billion cubic feet of natural gas before it was capped on March 22, 1980.[7] The spill

FIGURE 2—*The Ixtoc 1 Casing*

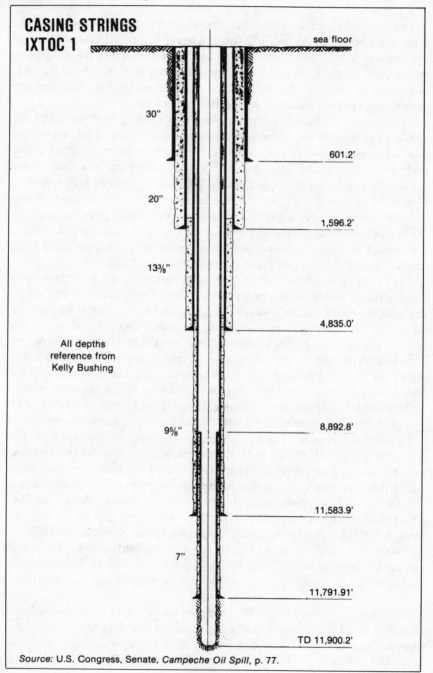

CASING STRINGS IXTOC 1

sea floor

30"

601.2'

20"

1,596.2'

13⅜"

4,835.0'

All depths
reference from
Kelly Bushing

9⅝"

8,892.8'

11,583.9'

7"

11,791.91'

TD 11,900.2'

Source: U.S. Congress, Senate, *Campeche Oil Spill*, p. 77.

FIGURE 3—*Stages of the PEMEX Ixtoc 1 Blowout*

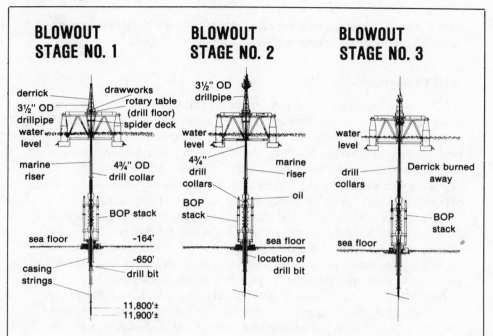

BLOWOUT STAGE NO. 1

derrick
3½" OD drillpipe
water level
marine riser

drawworks
rotary table (drill floor)
spider deck

4¾" OD drill collar

BOP stack

sea floor   -164'

casing strings
-650'
drill bit

11,800'±
11,900'±

BLOWOUT STAGE NO. 2

3½" OD drillpipe

water level

4¾" drill collars

BOP stack

marine riser

oil

sea floor

location of drill bit

BLOWOUT STAGE NO. 3

water level

drill collars

Derrick burned away

BOP stack

sea floor

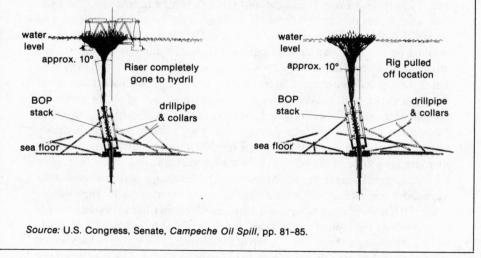

BLOWOUT STAGE NO. 4

water level

approx. 10°

Riser completely gone to hydril

BOP stack

drillpipe & collars

sea floor

BLOWOUT STAGE NO. 5

water level

approx. 10°

Rig pulled off location

BOP stack

drillpipe & collars

sea floor

proved to be the largest in the 120-year history of commercial oil development. Table 13 compares the Mexican tragedy with other oil spills.

PEMEX's Reaction

Differences reportedly emerged within PEMEX over whether to seek foreign assistance in taming the runaway well or draw exclusively on domestic resources.[8] Jesús Chavarría García, subdirector of exploration, is believed to have advanced the former position; Adolfo Lastra Andrade, subdirector of development, backed the latter, which enjoyed support from a group of young engineers who wanted the firm to demonstrate its technical prowess. Several previous blowouts had been "bridged" by sediment filling the hole, but it became immediately evident that no relatively simple solution existed for Ixtoc 1. Divers dispatched to survey the damage found themselves working in an underwater junkyard. For example, a six-ton beam from the substructure of the drilling equipment lay over the wellhead. One young diver returned from the scene to compare the morass of pipe to a platter of spaghetti.[9]

Any controversy over outside help quickly ended as PEMEX hired an American, Paul "Red" Adair, to assess the situation. Adair and his redoubtable "Red Devils" team had last attracted world attention when they capped the Ekofisk Bravo well in the North Sea. The seventy-one-year-old Texan, once the subject of a John Wayne action film, *Hell Fighters,* was no stranger to Mexico, where he had first fought a fire at the Chilipilla well near Ciudad PEMEX in the 1950s.[10] Adair had worked in the country several times since then. "I have seen many [hydrocarbon] releases similar to this one," he said, "but what makes this case particularly difficult is the huge quantity of gas and oil, and that it's found at the depth of 55 meters."[11] He also noted that the divers working on the well encountered poison gas, flames burning on the water's surface in a 295-foot diameter of the well, and gushing oil with a thickness of 66 feet that extended more than 656 feet from its source.[12] The murkiness of the water combined with healthy currents hindered the work of Adair and other experts who relied on underwater television cameras aboard an unmanned submersible for pictures of the damage. Even with strong lighting, visibility around the wellhead was only about one foot. Television viewers could see the approximate source of the oil current but were unable to determine the size of the various breaks in the piping. Moreover, the debris pile in the vicinity of the wellhead made subsurface containment of the blowout virtually impossible.

PEMEX and Adair soon agreed on the need to begin sinking two relief wells from carefully positioned jack-up rigs, the *Azteca* and *Interocean-II.*[13] These wells were angled to enter near the top of the deposit. There, one or both would serve as a conduit, first for salt water, then for drilling mud, which—because of

TABLE 13
The World's Major Oil Accidents

Well or Vessel	Type of Accident	Location	Date	Company Involved	Quantity (barrels)
Ixtoc 1[1]	Blowout	Gulf of Campeche	June 3, 1979–March 22, 1980.	SEDCO and PERMAGO (under contract to PEMEX)	3.1 million (and 3 billion cubic feet of gas)
Atlantic Express (Greek) and *Aegean Captain* (Liberian)[2]	Collision	Caribbean Sea	July 19, 1979	Mobil Oil Company	2.1 million
Amoco Cádiz[3] (U.S.)	Grounding	Brittany Coast	March 16, 1978	Amoco International Oil	1.6 million
Torrey Canyon[4] (U.S.)	Grounding	Brittany Coast	March 18, 1977	Union Oil Company	730,000 (est)
Bravo-14[5]	Blowout	North Sea (Ekofisk Field)	April 22–30, 1977	Phillips Petroleum Company	195,160
Ocean Eagle (Liberian)[6]	Grounding	San Juan Harbor (P.R.)	March 3, 1968	Trans Oceanic Tankers Corporation Northern Transatlantic Carrier Corporation	111,860
Union "A"[7]	Blowout	Santa Barbara, Calif.	January 24, 1969–February 14, 1969	Union Oil Company	77,400 (est)
Irenes Challenger (Greek)[8]	Tanker break-up	Pacific Ocean (220 miles southeast of Midway)	January 17, 1977	Tsakos Shipping and Trading Company	74,970

1. *Excelsior*, March 25, 1980, pp. 1, 18.
2. *Facts on File*, August 10, 1979, p. 592.
3. Ibid., March 9, 1979, p. 162.
4. Jeffrey Potter, *Disaster by Oil* (New York: Macmillan, 1973), p. 32.
5. *Facts on File*, May 7, 1977, p. 336.
6. Potter, *Disaster*, pp. 62, 107.
7. *Science*, November 19, 1976, p. 792.
8. *New York Times*, January 19, 1977, p. 14.

the suction of the Ixtoc 1 borehole—would flow into it, halting the rush of errant oil and gas. Although PEMEX claimed a world record in offshore drilling as it proceeded with the diagonal wells, problems plagued the venture. On June 11 an explosion on the *Azteca* platform from which the Ixtoc 1-A relief well was being drilled killed one welder and seriously injured four others.[14] And while the Ixtoc 1-B relief well made contact with Ixtoc 1 in late November, Ixtoc 1-A apparently missed its target, and the drill had to be backed up and a new attempt made, further delaying control of the well.

While the relief wells were being sunk, Adair's divers continued to try to close the preventers. After seawater and mud had been pumped into the well to reduce the pressure, they were able to shut down the flow on June 24. Within two hours it became evident that the damage was so extensive and the pressure so great that the seal would soon blow, creating the possibility of an unmanageable sea-floor crater. Thus the preventers were reopened and the fire reignited to burn off as much spilled oil as possible. This action unleased a flood of chocolate-colored oil capped by a fireball of burning gas. Although PEMEX made another unsuccessful attempt to control the well with the BOP on July 4, the incident convinced everyone of the crucial importance of the diagonal wells.[15]

Petróleos Mexicanos marshaled a small army of men and equipment to curb and contain the flow of oil. On hand at one time or another were one thousand Mexican technicians and workers, approximately five hundred experts from the United States, France, Norway, Canada, and other countries, five helicopters, a dozen planes, and twenty-two vessels, including tugboats, barges, lighters, and naval cranes.[16] One of the primary tasks was the installation of floating plastic barriers to contain the oil so that it could be skimmed from the water by collector ships. Rough seas and strong winds impeded this effort. According to official estimates, the relief operations combined with the lost oil cost PEMEX over $130 million.[17]

The monopoly also mounted a huge public-relations campaign that stressed the vigor and efficiency of its response to what it considered a freak accident caused by an unforeseen geological condition. This effort, which featured television programs and slick brochures, minimized foreign participation in decisions related to the relief work, the quantity of lost hydrocarbons, and the ecological damage resulting from the spill. PEMEX consistently set dates when it would regain control of Ixtoc and just as consistently failed to meet these targets. Its record led Senator Lowell P. Weicker, Jr., to conclude that "the Government of Mexico has withheld and covered information as to the continuing gravity and status of the Ixtoc 1 oilspill and has lied as to matters of fact relative to said disaster."[18]

PEMEX's campaign sprang from the firm's sensitivity to criticism in the media and before U.S. congressional committees concerning its handling of the crisis. For example, Donald Kash, chief of the U.S. Geological Survey's conservation division, claimed that the Ixtoc 1 operation failed to meet standards applied to drilling on the United States outer continental shelf.[19] Specifically, he cited (1) insufficient supplies of drilling mud at the well site, (2) removal of the drillstring without replenishing the drilling mud, (3) inexperienced and unqualified crew members, (4) inadequate supervisory personnel, and (5) inability to determine the bottom hole location.

Points (1), (2), and (3), based on newspaper and magazine articles, lack supporting evidence. Nonetheless, PEMEX and PERMARGO personnel can be criticized for removing the drillstring under uncertain conditions when they could have pumped Diesel-Bentonite cement through the bit, with the drill-string in place, to close off crevices in the formation. The decision to begin a sensitive undertaking at night raises further questions about the competence of the supervisors, as does the absence of a "check valve" near the bit to prevent the escape of materials up the drillstring. Point (4) will receive attention near the end of this chapter.

The most serious charge, point (5), relates to improper logging of Ixtoc 1. Anxious to drill as quickly as possible, PERMARGO failed to take readings to determine the direction and angle of the borehole. Such surveys must be run at prescribed intervals to avoid the hole's deviating more than three degrees from a vertical line from the sea bottom. Wells frequently follow a corkscrew-like path. However, failure to measure drift angles and take corrective action allows the drill bit to wander—possibly four, eight, or even twelve degrees from vertical—in pursuit of hydrocarbon deposits. The upshot in the Ixtoc 1 case was that neither the contractor nor PEMEX knew the bottom hole location—a conclusion given credence by the need to sink two relief wells to throttle the blowout.[20] After the accident occurred, the extreme pressures in the borehole prevented the insertion of a measuring device. Sensing tools, monitored by computers, assist the relief-well operators in locating a well such as Ixtoc 1. The smallness of the downhole target, seldom larger than one hundred feet in diameter, requires precise relief drilling. Specially designed instruments can detect a well's casing from a distance of two hundred feet, and it can be directionally located from fifty feet.[21]

Adding insult to injury, an official of Pollution Control Equipment, a Florida firm, reported that two months before the accident, Ing. Miguel Angel García Lara, head of PEMEX's Office of Environmental Control, decided against purchasing for $13 million security equipment to protect the environment from oil releases. The system, designed to encircle a well within thirty minutes after an emergency, included barriers, plastic fences, and recovery boats to capture and vacuum out of the water up to one thousand barrels each ninety minutes. PEMEX offered no reason for rejecting the devices, whose acquisition had been enthusiastically recommended by one of the firm's own technical advisers.[22]

Immediately after the spill, PEMEX relied on commercially available resources rather than soliciting or accepting the help of foreign governments to contain and clean up the oil. Already cited have been the activities of Red Adair and his team. The company bought or leased skimming vessels, suction devices, plastic fences, and oleophilic mops. Hercules 32R transports served as an air bridge between Houston and Ciudad del Carmen across which the Mott Oil Company shipped thousands of gallons of dispersants to combat contamination. The planes landed on a runway once used as a refueling stop for American B-29 bombers flying from the mainland to the South Pacific during World War II. Technicians from the Exxon Chemical Corporation arrived to supervise the spraying of Corexit, a foam designed to accelerate the degradation of the oil slick, thereby reducing damage to flora and fauna. Martech International of Houston provided remote-controlled surveillance equipment. Conair, British Petroleum, Shell Oil, Otis, Schlumberger, Eastman Whipstock, and Halliburton were among the many other foreign corporations that also supplied services or equipment. In addition, PEMEX purchased materials from national firms such as PERMARGO, PROTEXA, Guerrero Negro, and Proyectos y Construcciones, S.A.

Meanwhile, the U.S. National Response Team, an interagency body created by the National Oil and Hazardous Substances Pollution Contingency Plan, became concerned about PEMEX's inability to cap the runaway well.[23] At a June 19 meeting the task force sent, with permission of the Mexican government, a small group of U.S. scientists representing several agencies to the well site. This initial contact led to informal advice on the use of dispersants, priorities for clean-up operations, and other technical questions. On August 1 Mexican officials requested additional personnel and equipment from U.S. agencies, and the Response Team dispatched six people and two Coast Guard skimming barriers and associated auxiliary equipment. The equipment, which operated from August 12 to September 23 when PEMEX decided it was no longer needed, occasionally recovered five thousand bpd.[24]

Despite the importance of the international effort, approximately three quarters of which came from the United States, PEMEX continued to stress the predominance of its contribution. Ing. Luciano Flores Plauchú, superintendent of the Campeche Sound region, acknowledged the foreign presence but insisted that 98.8 percent of the specialists, workers, and experts attempting to control Ixtoc 1 were Mexican.[25]

The Mexicans did complement the outside help with their own ingenuity. PEMEX engineers chose to try a technique heretofore employed to diminish uncontrollable gas pressure in the catalytic cracking towers of oil refineries. Metal balls are poured into the rising column of gas or liquid so that they are partially suspended by the flow. The resulting friction reduces the fluid's speed and hence the pressure of the column. As the flow subsides, the balls fall to the base of the tube, further attenuating flow and pressure.[26]

PEMEX claimed that computer models and laboratory tests confirmed the possibility that metal balls might close down Ixtoc 1. Consequently, engineers began introducing two-inch balls through the kill line, a T-shaped structure that communicates with the wellbore. They used a French-made hose capable of withstanding pressures up to twenty thousand pounds per square inch. All told, the engineers injected 27,550 iron and 80,210 lead balls, some of which were bolted to large canvas squares in hopes that these patches would help seal the breaks.[27]

PEMEX officials insist that the injections helped to cut the flow to twenty thousand bpd by July 25 and ten thousand bpd by August 6.[28] Critics of the scheme argued that the success was more apparent than real—with the possibility that much of the flow simply began entering the Gulf waters as an emulsion deep below the surface. Jerome Milgram, professor of ocean engineering at the Massachusetts Institute of Technology and an inventor of equipment to clean up oil spills, concluded that the plan to use metal balls was "motivated more by emotion or desperation than by reason because straight-forward engineering calculations show that the balls will be blown right out of the well." He found that even the introduction of lead balls proved inadequate "because control of the well has not been achieved after many months of ball pumping."[29] In any case injections were suspended when the tube used to channel the flow of metal balls broke. "It is suspected that many of them simply swept through the large gaps in the pipe and were lost."[30]

Another herculean effort involved the deployment of an inverted steel funnel atop Ixtoc 1 to recover much of the ten thousand bpd gushing from the wellhead. Once in the 310-ton device, called a *sombrero* by oilmen, the petroleum could be piped to a nearby platform for separation into oil, gas, and water and subsequent transmission to onshore refineries. Mechanical flaws in handling equipment and bad weather delayed "operation sombrero" until October. Shortly before noon on October 18, Mexican and U.S. workers on a platform above the intractable well began to turn valves. They watched intently as the ten-foot high flames from burning natural gas receded and the churning, oil-filled waters calmed. PEMEX claimed that the octagonal steel funnel, estimated to cost between $40 and $50 million,[31] collected 60 to 80 percent of the increasingly irregular flow. Milgram was less sanguine about its success. He said that the hurriedly built sombrero was designed for thirty thousand Mcf of gas, but that the actual daily flow at the time of installation was about one hundred thousand Mcf. The device collected only about 10 percent of the escaping oil, about half of which was rejected by the separation system. He claimed that the oil flow, which had drastically subsided according to Mexican officials, was about fifty thousand bpd as of October 20.[32] Milgram's estimate may be more accurate than PEMEX's as evidenced by the fifty-four-thousand bpd production of Cantarell 94-A, another offshore well with properties similar to those of Ixtoc 1. Although it was built by Brown & Root of Houston, Díaz Serrano hailed the sombrero as a "very important product of the Mexican Petroleum Institute, under the direction of exploration and production engineers."[33]

The relief wells eventually entered the reservoir. Through them were pumped nine to ten million barrels of seawater which, along with the water existing in the formation, substantially reduced the flow of oil and, especially, gas. It was then possible to place a drillship, the *Río Pánuco,* above the well and insert a one-hundred-foot length of pipe through the BOP stack as a conduit for cement and heavy mud used to fill and control the well. Once "dead" the well was closed with an "up-hole plug" formed by five hundred sacks of quick-setting cement.[34]

Environmental Damage

Throughout the long summer and fall of 1979, officials of the Mexican government consistently dismissed assertions that Ixtoc 1 would render major damage to Mexico's environment. On June 10, just a week after the blowout, Eugenio Echeverría Castellot, a petroleum engineer and the PRI's candidate for governor of Campeche, called the accident "lamentable but not alarming because the dangers of contaminating the marine ecology are minimal." Of every one hundred barrels released, he added, fifty are burned off, twenty-five evaporate, and twenty-five remain on the surface of the sea and are carried away.[35] On June 19 Díaz Serrano repeated the same argument, although his figures were even more optimistic. He claimed that fire consumed fifteen thousand bpd, evaporation took care of twenty-five hundred barrels, skimmer boats recovered another seventy-five hundred, and only five thousand escaped into Campeche Sound. The application of a biodegradable solvent further diminished the danger of environmental damage, he said. As early as June 14, PEMEX contended that it had "total control" of the oil slick.[36]

Nonetheless, two days later large quantities of oil reached Tabasco, where the fine sands of Miramar beach were converted into a gummy mass covered by blackened vegetation washed ashore. (The beaches of Tuxpan, Laura Villar, Tamaulipas, Campeche, and Yucatán were also affected.) Hundreds of dead fish covered both the beach and the mouth of the Grijalva River as Díaz Serrano met with local governors who greeted with skepticism his assurance that schools of shrimp would not be injured. Also unconvinced were the fishermen's unions, which prepared a manifesto to López Portillo and Díaz Serrano demanding both reparations for losses and a halt to offshore drilling until the most comprehensive safety precautions were taken. Heretofore, they contended, the equipment used by the company had been old and inadequate.[37]

Shrimp had the greatest economic significance of all the thirty commercial fish and shellfish caught in the Gulf of Mexico. Ciudad del Carmen, the principal port in the stricken area, accounts for 30 percent of the total national production of an item whose exports to the United States generate over $130 million annually.[38]

Biologists cite three immediate consequences of an oil spill for aquatic life. *First,* a slick impedes the photosynthesis of flora by preventing the sun's rays from penetrating the water. *Second,* it impairs the natural interchange of gases (oxygen and carbon dioxide) between the water and the atmosphere, preventing plant and animal organisms from breathing. *Third,* it causes metabolic and possibly genetic changes in organisms that filter their food and whose gills or other filtering organs encounter petroleum. All told, oil has five hundred or more components which affect sea creatures. The director and two department heads at the Escuela Superior de Ciencias Marinas in Ensenada, Baja California, accused PEMEX of concealing information related to the oil spill, thereby making it more difficult to ascertain the effects of the blowout.[39] Despite this perceived lack of cooperation, by mid-summer 1979 researchers from a half-dozen institutions had begun investigating the extent of damage, especially to female shrimp.[40]

The government did install barriers to prevent the oil from entering such fragile spawning grounds as Campeche's Laguna de Términos, the Laguna de Tamiahua, and the Laguna de Tampamachoco. Scientists from the National Polytechnic Institute, who said it would take years to ascertain the impact of the oil spill on fauna and flora, reported in early August that the accident had partially paralyzed the fishing industry in Campeche Sound.[41] For months the Mexican navy closed an area around the spill to all boats.

On July 22, forty-eight days after Ixtoc 1 caught fire, PEMEX announced it would assume full responsibility for damage caused by the burning well in the waters off Campeche.[42] Yet as late as October 8, José Luis García Luna, the company's manager of petrochemicals and coordinator of the anticontamination work, rejected the idea that Ixtoc had harmed the environment because petroleum is "a great ally of nature" which "increases the yields of marine species." Any damage, he added, was "aesthetic because the crude which reaches the beaches in small quantities dirties the feet of tourists."[43] Four days later the state company finally admitted that the errant well was a source of contamination;[44] however, no mention was made of steps to indemnify persons injured by the accident—a subject reportedly considered "taboo" at PEMEX.[45]

As described above, the United States government was actively involved in combating the spill within weeks of its occurrence. In addition to on-site efforts, the Coast Guard and other federal agencies sought to determine the danger posed to the Gulf Coast by the oil slick, which by mid-July had broken into seven major patches as it moved toward northern Mexico and southern Texas. The Coast Guard cutter *Valiant,* carrying scientists from the National Oceanic and Atmospheric Administration (NOAA) and the Environmental Protection Agency (EPA), zigzagged through the Gulf water, using dye markers and special instruments to gauge the direction and velocity of the currents. The data collected were fed into a NOAA computer in Seattle to resolve conflicting theories on currents. At the same time, specially equipped Coast Guard planes flew over the slick to assay its size and growth.

Meanwhile, the U.S. Fish and Wildlife Service helped its Mexican counterparts airlift thousands of young Kemp's Ridley turtles to safe water from their oil-menaced hatching grounds at Rancho Nuevo, north of Tampico. The creatures, which are dime-sized at birth but reach eighty pounds as adults, are among the rarest of turtles.[46]

Efforts along the Texas coast complemented those taken in Mexican waters. Dr. John Robinson of NOAA headed a federal interagency scientific team of toxicologists, chemists, biologists, geologists, meteorologists, and oceanographers attempting to avert a disaster. Of particular concern was a thin finger of land, the country's longest barrier island, that stretches 137 miles from Port Aransas in the North to South Padre Island in the South. The Texas barrier strand, itself teeming with beaches, hotels, restaurants, condominia, and summer homes, has become an increasingly popular area, especially for tourists who spend hundreds of millions of dollars each year. However, behind the island lies an even greater treasure. This is Laguna Madre and Corpus Christi Bay, whose water, marshlands, and mudflats serve as breeding and nurturing grounds for many species of aquatic life, including sports fish, blue crabs, oysters, sea turtles, and much of the shrimp eaten by Americans. This area also constitutes a rookery for many bird varieties, including the endangered brown pelican, as well as a stopover point for such threatened migratory species as the peregrine falcon and redhead duck. A local wildlife refuge provides a winter home for most of the world's one hundred whooping cranes.

In Texas the U.S. Fish and Wildlife Service held bird clean-up workshops for 125 government employees and volunteers, set up seven oiled-bird rehabilitation centers along the Gulf Coast, convened numerous briefings and planning sessions, developed contingency plans, and recovered a total of twenty-six contaminated birds—eight of which were blue-faced boobies, an open sea specie. The service also collected and analyzed several oiled sea turtles.[47]

The Coast Guard towed long "floating fences," extending two feet below the surface, to block the four "passes" or cuts through the barrier island. Oil penetrating these barriers would be virtually impossible to clean up and could destroy parts of the wetlands that are the "area's chief glory."[48] The Coast Guard also installed two 500-foot anti-oil booms and skimmers to guard the 1,550-foot-wide Aransas Pass, the regularly used channel into the port of Corpus Christi.[49] Yet August, the month when the slick first reached the Texas waters, is the height of the hurricane season, and tidal surges generated by a major tropical storm could simply push the oil across the thin barrier islands into the grassy waters and lush marshes.

Fortunately no such tragedy took place. Still, "patch oil slicks and tarry globs of crude as big as baseballs" began invading the white beaches of south Texas on August 7, 1979. The beaches were repeatedly covered with oil for several days along a 140-mile stretch.[50] Unlike tropical storms—brief, discrete events familiar to local residents—the Campeche slick, compared by journalists to "a huge expanse of pancakes and mousse," occasioned a prolonged vigil because it had broken into hundreds of ribbons, streaks, and patches of varying thickness, texture, and toxicity.

Beginning in September the prevailing winds off the Gulf of Mexico—the winds which pushed the oil over five hundred miles—were countered by the "blue northers," strong, cold air currents from Canada that carried the slick away from shore. This naturally induced departure did not take place until over 132 tons of debris—mixed oil, water, and sand—had been strewn along the barrier islands, soiling at least 25 percent of the beachfront.[51] As early as mid-August, the resort beaches at South Padre Island reported a 30-percent decline in tourism.[52] Crews finally succeeded in cleaning the beaches. Nevertheless, months and even years will pass before scientists, optimistic in 1980, can assess the long-term damage to the ecological system.

Compensation

The Ixtoc 1 disaster has precipitated litigation that will eclipse Charles Dickens's *Jarndyce* and *Jarndyce* in matters of complexity, legal fees, and judicial delay. Anticipating that suits would be filed against it, SEDCO, the Dallas-based firm which leased *SEDCO-135* to PERMARGO, initiated a "liability-limitation" action in the Federal District Court in Houston on September 11, 1979.[53] Under this 1853 admiralty statute, SEDCO attempted to exonerate itself from liability for damages from the disaster or, alternatively, place a ceiling of $300,000 on the overall award to plaintiffs should the court find in their favor. Pursuant to this proceeding, all parties with claims against SEDCO were instructed to come forth.

The Justice Department, acting on behalf of the U.S. government, charged that SEDCO had violated the Federal Water Pollution Control Act and the Rivers and Harbors Act (commonly called the "Refuse Act"). It alleged that (1) the drilling rig was "unfit and unsuitable" for its mission, (2) those responsible for the vessel were "incompetent," and (3) supervisory personnel were "negligent in carrying out their responsibilities." The United States sought to recover $6 million in clean-up costs and "an amount yet undetermined" for damage to the nation's property and resources. The "ultra-hazardous" nature of oil-drilling led Justice Department attorneys to assert that SEDCO should be held "strictly liable" for all injuries, not simply those up to $300,000, related to the blowout and the ensuing "public nuisance."

Mark White, the Democratic attorney general of Texas, also filed suit against SEDCO, which was founded by William P. Clements, Jr., the Republican governor of Texas and a political rival of White's. Clements, who placed his SEDCO shares in a blind trust upon assuming public office, at first dismissed the possible damage of the oil spill as "much ado about nothing." He also incurred White's wrath by opposing litigation against Mexico until the State Department had tried to negotiate a settlement. The attorney general insisted that those responsible for injuries arising from the accident should definitely be taken to court and that the governor's statement would diminish chances for fruitful negotiations with Mexico.[54] White's determination to sue was reportedly strengthened by the sinking of *SEDCO-135* following the accident. This, he implied, might have been an attempt to conceal important evidence.[55] The state of Texas sought to recover more than $60 million for damage to the environment, harm to fish and animal life, loss of tax revenues, and clean-up costs.

In addition to the federal and state government, claimants include the Texas AFL-CIO, Exploration Logging, S.A., Padre Island National Seashore Company, local political subdivisions (class action), Gulf Coast businessmen and property holders (class action), seafood interests (class action), commercial fishermen (class action), and a group of fifty operators of motels, restaurants, marinas, and related activities. Total stated claims exceeded $427 million in May 1980.

At that time the Justice Department had not decided whether to add PERMARGO or PEMEX as defendants, although it appeared likely that the government would sue PERMARGO, as had some other claimants. SEDCO had filed a motion to implead or bring suit against PERMARGO and PEMEX on the grounds that it had a written contract with the former and an implied relationship with the latter (as a third-party beneficiary) to hold SEDCO harmless should an accident occur. Both PERMARGO and PEMEX deny any such agreement. Acceptance of SEDCO's motion to implead would allow the U.S. government to press a claim against the two Mexican corporations without formally suing either one. Should the court render a judgment against PEMEX, which claims protection under the Sovereign Immunities Act, it remains to be seen how the Justice Department would enforce its award, if it sought to, against the firm's property in the United States without igniting a political explosion.

Years will elapse before the case is settled. An example of just one of the many preliminary, time-consuming issues is the jurisdictional question, raised by the claimants in a motion for summary judgment, of whether *SEDCO-135* even qualifies as a "vessel" under the Liability Limitation Act. In arguing the negative, claimants' attorneys cited the rig's lack of self-propulsion, its tendency to remain in a fixed position for long periods, its attachment to the ocean floor by drilling gear, and a trend perceived in recent court decisions to replace a mechanistic "locality" test with a flexible "traditional maritime activity" standard. On the other hand, SEDCO's lawyers stressed the broad definition of "vessel" found in the Jones Act and the Longshoremen's Act, Congress's failure to modify this definition, *SEDCO-135*'s registration as a vessel by the U.S. government, its mobility, and the fact that the rig had made eleven long voyages, twice crossing the Atlantic, as it logged a total of 15,947.5 nautical miles.

The United States might also seek relief under the Convention on the Continental Shelf, which obliges a coastal state in safety zones established around continental shelf installations "to undertake, in such zones, all appropriate measures for the protection of the living resources of the sea from harmful agents."[56] Also relevant is Article 24 of the Convention on the High Seas, which states:

Every State shall draw up regulations to prevent pollution of the seas by the discharges of oil from ships or pipelines or resulting from the exploitation and exploration of the seabed and its subsoil, taking account of existing treaty provisions on the subject.[57]

The most promising precedent for resolving an international environmental dispute occurred in the *Trail Smelter Arbitration*. Canada was held liable to the United States for damages resulting from sulfur dioxide fumes from a lead-zinc smelter at Trail, British Columbia, which injured farms across the border in

Stevens County, Washington. The two countries involved submitted the dispute to an international arbitration court which set a monetary compensation in 1941. The tribunal found that

> under the principles of international law, as well as of the law of the United States, no State has the right to use or permit the use of its territory in such a manner as to cause injury by fumes in or to the territory of another or the properties of persons therein, when the case is of serious consequence and the injury is established by clear and convincing evidence.[58]

Both the United States and Mexico signed the Stockholm Declaration on the Human Environment, which sprang from a 1972 United Nations conference. Article 21 of this document embraces the tenets of the *Trail Smelter Arbitration:*

> States have, in accordance with the Charter of the United Nations and the principles of international law, the sovereign right to exploit their own resources pursuant to their own environmental policies, and the responsibility to ensure that activities within their jurisdiction or control do not cause damage to the environment of other States or of areas beyond the limits of national jurisdiction.[59]

All of the international declarations, documents, and principles come to naught if there is no satisfactory forum for handling the conflict. The International Court of Justice in the Hague provides such a forum; however, reservations that the United States and other nations attached to accepting this tribunal at the time of its 1946 creation have virtually eliminated compulsory jurisdiction. And Mexico has shown no interest in negotiating over compensation, much less submitting claims to this or any other international body. This position, adamantly set forth by López Portillo in his third state of the nation address (September 18, 1979), sparked a standing ovation from members of all parties in Congress. He argued that international law established no responsibility for accidents, a point reiterated after Ixtoc 1 was capped. Moreover, the Mexican chief executive pointed out that the United States refused to compensate his countrymen whose cotton farms suffered damage because of the increasing salinity of the Colorado River, which flows from Nevada to Baja California, and from which Mexico is guaranteed 1,500,000 acre-feet of water annually by a 1944 treaty. The salt entered the water because the Wellton-Mohawk Reclamation District in Arizona began rinsing soil with methods requiring the pumping of highly saline waters into the Gila River near its confluence with the Colorado. This practice, which continued for a decade, contaminated over 123,000 acres valued at $100 million in the Mexicali Valley, according to a leader of the Independent Peasant Confederation.[60] Under a 1973 agreement, the United States agreed to cleanse the water to a greater degree before sending it below the border.[61]

Mexican scholars observed that while the Ixtoc 1 blowout was an accident, the United States had intentionally used Mexico and its coastal waters as its "trashcan for more than 40 years."[62] In support of this thesis, Lic. Leopoldo González Aguayo, coordinator of the Center for International Relations of UNAM's Faculty of Social and Political Sciences, cited industrial discharges in the Río Grande, the use of fertilizers and pesticides which reach the Gulf of Mexico and Sea of Cortés in the form of "red tide," and the dumping of radioactive waste and the flushing of ships' bilges in the Gulf of Mexico.[63] Spokesmen for the Communist, Popular Socialist, and Worker's Socialist parties agreed that because of their pollution of Mexico and the Gulf waters, Americans had no right to seek reparations. These left-wing parties further alleged that the United States would use the oil spill as a means to exert pressure on Mexico in the Carter–López Portillo talks scheduled for late September in Washington. "This [Ixtoc 1] has been converted into a favorable card for the United States," according to a Communist party representative.[64]

Controversy over the issue almost caused cancellation of the presidential meeting. On August 23, 1979, Ambassador Robert C. Krueger, recently named to coordinate U.S.–Mexican affairs, bluntly announced his country's interest in negotiating questions of responsibility and payment for the accident. This statement so infuriated the Mexican government, which had received an inquiry about negotiations only a few hours before Krueger's pronouncement, that López Portillo considered calling off his visit to the White House. "We were disconcerted because a special ambassador-designate, in one of his first actions and without waiting for a reply, makes public a measure of this kind," said the Mexican president. "It is without precedent and constitutes an attitude that is disconcerting." López Portillo further characterized the publication of the U.S. diplomatic note as "some concession to an internal pressure group"[65]—a conclusion that was probably correct in light of Krueger's strong ties to Texas politics.

As a result of his statement, Krueger replaced former Energy Secretary James R. Schlesinger as the principal American target for nationalistic outrage. Nonetheless, agreement on a natural gas deal assured the convening of the two-day presidential meeting on September 28. In a joint communiqué issued the following day, the two leaders "agreed on the need for both countries to prevent events or actions on one side of the land or maritime boundary from degrading the environment on the other side."[66] Officials in Washington interpreted these words differently from their Mexican counterparts. For example, National Security Council adviser Guy Erb said that discussions about the oil spill would include the issue of compensation, although negotiations would have to precede any agreement.[67] This interpretation, also carried by the *New York Times,* elicited a scathing statement from Mexico's Ministry of Foreign Relations, which repeated the text of the joint communiqué and emphasized that "the presidents decided to negotiate agreements to prevent further damage to the environment of both countries and deliberately excluded any reference to possible damages caused in the past by one country or the other and particularly the question of compensation and indemnities that could derive from them."[68]

As a result of this communiqué, bilateral meetings began in mid-January 1980 that U.S. officials hoped would lead to (1) a joint contingency plan for spills of oil and other noxious substances (such as the plan that took effect with Canada in 1974), (2) a long-range environmental agreement embracing such questions as offshore drilling and tanker traffic, and (3) discussion of damages arising from the Ixtoc 1 spill. These deliberations, co-chaired by Matthew Nimetz and Alfonso de Rosenzweig Díaz, took place under the auspices of the Border Working Group of the consultative mechanism. While negotiators made rapid progress toward fashioning a contingency plan, a subject previously under discussion for several years, the Mexicans refused to talk about broader environmental issues lest such discussions draw more attention to the Ixtoc 1 embarrassment. State Department observers believed that capping the well would make it easier for Mexicans to consider mutual environmental concerns. The inclusion of Canada in future sessions, not necessarily conducted under the Border Working Group, would seem like a reasonable step. U.S. diplomats never raised the extremely sensitive question of compensation, preferring to leave it to the judicial process.

Responsibility for the Blowout

Oscar Flores Sánchez, Mexico's attorney general, exonerated "the workers, employees, technicians and professionals of PEMEX and of the drilling firms" from any criminal behavior related to the Ixtoc 1 catastrophe. The blowout, he said, resulted from the unforeseeable presence of caverns that the drillpipe entered more than 11,500 feet below the sea bottom. The "phenomenon of nature" gave rise to a "serious accident, but in no sense [was it] a national disgrace." After all, the state firm had previously sunk 114 wells in the continental shelf without experiencing a similar problem. This official pronouncement supposedly followed a study of the disaster conducted by members of the attorney general's office; however, the lack of expertise and sources of independent data forced reliance on information provided by PEMEX. A presentation of the report's technical details was called incomprehensible by a reporter for *Excelsior*.[69]

That no finding of criminal behavior emerged must not obscure the conditions under which the blowout took place. For reasons stated at the beginning of this chapter, Petróleos Mexicanos has placed extraordinary emphasis on lifting as much oil as rapidly as possible from the Campeche zone. This breathless venture has found contractors, whose earnings closely correlate with distance drilled and wells completed, cutting corners and thus enhancing the possibility of more incidents like Ixtoc 1. But ultimate responsibility rests with PEMEX, whose personnel either acquiesce in the shortcuts, perhaps for personal financial gain, or lack the experience to provide competent and thorough supervision.

Under Dovalí Jaime PEMEX had three rigs for coastal drilling (one owned by the state firm, two by Perforadora México) and six inland water-drilling barges (two owned by the state firm, four by private companies). Supervising the day-to-day operations of the offshore units was a group of eight to ten extremely competent marine experts including Antonio Garrillo Miranda, Juan Medina Ruíz, Frankemberg Velasco, and Miguel Angel Benitez Hernández. By 1980 the number of rigs drilling in Campeche Sound alone had increased to twenty, and PEMEX was reportedly looking for two additional "jack-up" rigs. In addition, there were five construction barges working at the Cantarell complex, with another on order.[70] Meanwhile, PEMEX had transferred most of its seasoned engineers with maritime experience to office jobs in either Mexico City or the superintendencies without having adequately trained new ones to replace them. The result was fewer skilled engineers to supervise a vastly accelerated offshore program.

Díaz Serrano managed to find a silver lining in the dark cloud of disaster. In a September 20, 1979, report to Congress, he said that the oil spill should help "restore the confidence [of Mexicans] in our future." Certainly the several million barrels spilled must not be considered a loss inasmuch as they revealed the presence of a deposit containing eight hundred million barrels. He recalled that the director of PETROBRAS, the Brazilian national oil company, recently announced the termination of petroleum exploration in the Amazon Basin after having invested more than $2 billion in the enterprise. "How much would many countries give to have just one Ixtoc!" he proclaimed.[71]

Díaz Serrano's full presentation to Mexico's 370 deputies, delayed two hours because of last-minute consultations with the president, consisted of his reading a sixty-four-page document after which he answered ninety-seven questions posed by representatives of the country's seven registered parties. This was the most intense interrogation to which a spokesman for Mexico's executive had ever been subjected. The director general stood during the entire session, which lasted eight hours and twenty-seven minutes. Most queries of PRI deputies appeared designed to make their compatriot look good. But thunderous applause greeted his agreement, in response to a question posed by a member of the opposition National Action party, to permit an audit of PEMEX's books. The left focused on potentially embarrassing matters concerning his financial links to PERMARGO, which he claimed to have severed on October 16, 1975, and his relationship to former CIA director and Republican presidential contender George Bush, whom Díaz Serrano claimed was a PERMARGO board member for only two months before beginning his political career. Although the engineer occasionally fumbled questions (such as the unaccounted-for loss of 28.2 percent of oil and 27.8 percent of gas produced by PEMEX), his marathon presentation was viewed as dignified, controlled, forceful, frank—even *macho*. *Excelsior* called the dialogue with Congress "spontaneous, and its results, convincing."[72] López Portillo's overt, enthusiastic, and renewed support for his friend and appointee following this virtuoso performance signaled to elites in the country's authoritarian political system that the time had come to end both attacks on the national oil company and carping about Ixtoc 1.

10

The Prospects for Mexican Oil

International: A New Saudi Arabia?

Is Mexico a new Saudi Arabia? All of the information emanating from PEMEX and American publicists notwithstanding, any such comparison is at best premature. By its official estimates, challenged by some analysts, Mexico has about 60.10 billion barrels of proven reserves; Saudi Arabia, where only limited exploration has taken place in recent years, has nearly three times that amount or 163.35 billion barrels, and neither King Khalid nor Petroleum Minister Zaki Yamani have released calculations of what are believed to be vast potential holdings. Moreover, its production costs are in the neighborhood of thirty-five to fifty cents per barrel, a fraction of the Mexican figures. Mexico's output, over 2.3 million bpd in late 1980, may climb to 3.5 or 4 million by mid-decade. The very steps that López Portillo could take to accelerate the flow even more—grant franchises to multinational oil firms, lay an oil pipeline parallel to the *gasoducto* to the Texas border, impose stringent consumption controls on his citizens, purge the venal Oil Workers' Union—are politically impossible.[1] Meanwhile, the Saudis, who expanded production from 8.5 million bpd in 1978 to 9.5 million in 1979, could have a capacity in the neighborhood of 12 million bpd by the mid-1980s, assuming investments are made in additional production facilities. A more realistic figure might be a 10.5-million-bpd capacity by 1983.[2]

Mexico's large, fast-growing population now consumes six out of ten barrels produced in the country, a figure that will fall only to five out of ten by the end of López Portillo's term. In contrast, Saudi Arabia, with fewer than eight million people in a country twice the size of Western Europe, can export almost all of its output. Saudi Arabia also differs from Mexico, an independent producer, in its pivotal position in OPEC, where it accounts for 30 percent of total production and one third of reserves. Riyadh's influence is not confined to petroleum. The Saudi government had amassed over $19 billion in reserves as of early 1980, while placing about half of its $60 billion invested abroad in the United States. Reluctance to invest or hold vast quantities of dollars would have a devastating effect on the U.S. currency. Just as it can lift more oil to meet demand, the kingdom can reduce yields as the situation dictates. This "balancing" role has contributed to OPEC's strength and longevity by obviating the need to assign production quotas.

It is true that a severe cut might limit Saudi Arabia's $142-billion development program, announced in 1975. Yet modernization poses severe hazards to this conservative Moslem state. At June 1979 prices, Saudi Arabia's earnings from production at current levels and income from investment abroad were sufficient to fund development plans and foreseeable foreign-policy needs through 1982.[3] An important segment of the ruling family, alarmed at the shah's ouster from the Peacock Throne, believes that the country has already moved too far too fast and worries about the influx of foreign workers necessitated by rapid economic growth. Also of concern is inflation, under control in 1980, which afflicted the country for four years following the rapid increase in expenditures from oil revenue. "Indeed, it is reasonable to assume that the Saudi royal family is so wealthy that its primary objective is not maximizing the wealth of the nation it governs, but rather assuring its own survival as the ruling family."[4] Thus the Saudis could retrench if necessary to preserve the monarchy and safeguard their prized cartel. Table 14 compares Mexico and Saudi Arabia in key areas relating to petroleum.

If not a Saudi Arabia, Mexico is nonetheless an increasingly important player on the world petroleum stage. The extent of its importance depends upon

TABLE 14
Mexico and Saudi Arabia: A Comparison

	Population (millions)[1]	Gross National Product[2] (millions of dollars)	Per Capita Income (dollars)	Proven Oil Reserves[3] (billions of barrels)	Oil Production[4] (million bpd)	Maximum Sustainable Production Capacity[5]	Natural Gas Reserves[4] (trillion cubic feet)	International Reserves[6] (millions of dollars)
Mexico	67.7	$83,800	$1,238	40.00	2.00	2.7	100.00	$ 1,927
Saudi Arabia	8.1	55,400	6,840	163.35	9.52	10.3	93.2	19,175

Sources: *Encyclopedia Year Book 1980* (Danbury, Conn.: Grolier, 1980,) pp. 339, 441; *Oil & Gas Journal* (December 31, 1979), pp. 70–71; Department of Energy, *Monthly Energy Review* (June 1979):98; and International Monetary Fund, *International Financial Statistics* (April 1980):269, 335.
  1. Estimated for 1979.
  2. Estimated for 1977.
  3. January 1, 1980.
  4. Estimated average for 1979.
  5. March 1979 for Saudi Arabia; for Mexico this is a level to be reached by 1982.
  6. January 1, 1979 for Saudi Arabia; February 1980 for Mexico.

a host of factors: (1) oil production in Mexico, (2) economic growth in the United States and elsewhere, (3) progress in energy efficiency, (4) oil output in non-OPEC nations, (5) oil production in Saudi Arabia and other cartel members, (6) availability of such other "conventional" energy sources as natural gas, coal, and nuclear power, and (7) availability of such unconventional sources as shale oil, solar, tar sands, geothermal, tidal, and wind.[5] Estimates of the free-world demand for oil in 1985 vary greatly. The Central Intelligence Agency has projected that needs will run between 68.3 and 72.6 million bpd, with U.S. imports in the neighborhood of 12 to 15 million bpd. To satisfy this demand, OPEC would have to supply between 46.7 and 51.2 million bpd, with Saudi Arabia's contribution being at most 18 million. The Organization for Economic Cooperation and Development estimated that in the absence of a major conservation program, the United States would import 9.7 million bpd in 1985, while the mean estimate for the same year of fourteen studies prepared in 1977–1978 was 11 billion bpd.[6]

Two events in 1979 may cause a reassessment of these projections. The first is President Carter's July 15, 1979, speech, in which he pledged that U.S. imports would never again exceed their 1977 level of 8.8 million bpd. Implementation of this promise would further curb the growth in world consumption, not only because of limited American imports, but because U.S. abstinence has a powerful "echo effect" on the economies and, therefore, demand in Western Europe and Japan.[7] Second is the decision taken at year-end 1979 by a number of OPEC members—Libya, Nigeria, Iran, and Algeria, for example—to raise prices to almost $35 per barrel. This increase contributed both to conservation and a recession, thereby further restraining demand within consuming countries. Optimists hold that Mexico may export 5 million bpd of oil and natural gas by mid-decade. While to do so is technically possible, reaching this figure would wreak such havoc on the inflation-ridden economy as to be politically untenable. The specter of runaway inflation aided conservative bankers and the economic cabinet in early 1980 as they successfully battled Díaz Serrano's proposal to raise production from 2.25 to 4 million bpd by the end of the *sexenio*.

Elements of the private sector, alarmed over spending policies that could overheat the economy, have increasingly shared the position of the left in opposing greater output. PEMEX was permitted to proceed with only a modest elevation of the production plateau for the 1980–1982 period (2.7 million bpd), and the change of presidents, next scheduled for December 1, 1982, militates against a new, even more ambitious development program before the next chief executive is securely ensconced in office. Too, a national energy program, released in late 1980, proposes to limit exports while stressing conservation and economic balance. If the plan takes effect, Mexico will contribute only 1.5

million bpd to the world market. While such quantities should help restrain OPEC prices, they would hardly diminish the power of the cartel or hasten its dissolution. Since excess productive capacity of as much as 8 million bpd in recent years did not shatter OPEC, PEMEX's additional output would hardly provide a coup de grâce. As discussed in chapter 6, Mexican leaders have forsworn actions that would prove injurious to the thirteen-member organization and the attractive prices it has secured.

United States-Mexican Relations in the 1980s

Media reports have stressed the negative aspects of United States–Mexican relations. Newspaper, magazine, and television commentaries have paid inordinate attention to prickly toasts, artless references to "Montezuma's revenge," and recriminations over the natural gas negotiations. Largely ignored has been the remarkable degree of cooperation in energy matters between Mexico and private and public groups in the United States. PEMEX bought $658 million worth of equipment, parts, and supplies, chiefly from U.S. manufacturers, in 1979, and it plans to spend $5.7 billion between 1979 and 1986; it relies on American firms for drilling rigs and crews for offshore exploration and development; and it hires American scientists and technicians for important geological and seismological work. Many Americans work for Schlumberger, a French corporation that takes and evaluates borings that indicate the productivity of wells. Symbolic of this need for U.S. expertise was the decision to have DeGolyer and MacNaughton certify the larger reserve estimate announced early in the López Portillo administration.

Exports to the United States will continue to grow in volume as they decrease as a percentage of Mexico's sales abroad. Assuming the United States purchases 60 percent of its total shipments (1.8 to 2.5 million bpd), Mexico may supply 10 to 15 percent of its neighbor's imports by 1985; the percentage would double if imports are kept at the 1977 level. Yet, adherence to the new energy plan would limit sales to America to half of all exports or 750,000 bpd, assuring that PEMEX would supply less than 10 percent of U.S. imports. Little coverage has been given Mexico's readiness to sell crude to the Department of Defense for the National Strategic Petroleum Reserve. This reserve, begun in 1975 and projected to reach 750 million barrels, is stored in Texas and Louisiana caverns to be used as a cushion against future oil embargoes or shortages. PEMEX's two sales for this purpose were made in June (5.4 million barrels) and July (13 million barrels) 1978 at a total price of $242.7 million.[8] Other Mexican oil, first purchased by U.S. companies, provided 40 percent or 25,000 to 30,000 bpd of the crude entering the reserve.[9] According to the *Washington Post,* the Carter administration stopped buying oil for the reserve, which contained 97.1 million barrels in January 1980, when the Iranian revolution disrupted world production patterns. After purchases had halted, the Saudis warned that they would cut back output if Washington continued stockpiling supplies.[10]

The Pentagon purchases emphasize the advantages of Mexican oil supplies in terms of proximity and security. Richard B. Mancke has developed this argument, stressing Mexico's distance from "the numerous sources of conflict" that could disrupt the critical flow of oil from Africa and the Persian Gulf. All told, he identifies "six sets of factors"—minority dissidence, boundary disputes, disruption of feudal societies, possibility of external intervention, a local arms race, and the Arab-Israeli conflict—that could curtail shipments to the United States. Such events as the Iraq-Iran conflict, which followed the completion of his book, lend support to his argument as does the fact that a half-dozen wars, a dozen revolutions, and innumerable assassinations and territorial disputes have afflicted the Mideast since the end of World War II.[11] In contrast, Mexico is a stable, peaceable nation connected to America by sea lanes that are short—three days steaming time from Mexican oil ports compared to forty-five days from the Persian Gulf[12]—and would be relatively easy to defend in the event of conflict. As one American hyperbolically expressed it: "I'd rather have it [petroleum] across the river than try to get my tankers through Soviet sub patrols."[13]

Mancke also said that increased imports of PEMEX oil would improve the U.S. trade balance because Mexico would spend most of its new dollars on American goods and services. Such imports would also obviate or at least minimize a large-scale effort in the production and consumption of such "still uneconomic domestic fuels" as solar power, oil shale, synthetic oil, and gas made from coal. Related to this would be reduced capital outlays because the lead time to produce this petroleum and gas would be short and the investment small compared to excavating new coal mines, producing oil in the Baltimore Canyon, or constructing major coal- or nuclear-powered electric generating plants. Moreover, "if because of the greater security of imports of Mexican petroleum, the United States decides that it can increase its total imports of crude oil and natural gas, then there will be a substantial reduction in the total energy-related damage to the environment and to public health and safety."[14]

Thus six arguments have been advanced for encouraging expanded production and greater U.S. consumption of Mexican oil: (1) proximity, (2) security, (3) trade balance, (4) postponed use of exotic fuels, (5) reduced investment, and (6) environmental protection. Some of these points are more compelling than others. For example, the Ixtoc 1 blowout raises doubts with respect to environmental advantages flowing from an acceleration of PEMEX production. A number of specialists also applaud expenditures for solar power, sometimes considered an exotic fuel, on the grounds that the next century will see the last accessible oil field drained, and the sun's rays offer a clean, abundant energy source that cannot be manipulated by a private or public cartel.

Quibbling over justifications aside, virtually every U.S. policy maker recognizes the importance of access to Mexico's oil and gas. Differences emerge, however, as to whether the United States should encourage massive production—say, six or seven million bpd by 1985 and ten million bpd by 1990, with the bulk being sold in the American market. The United States has little influence on such decisions. But, for the sake of balance, we should point out liabilities that would attend a sharp acceleration in output. To begin with, Mexico's economy could not absorb the newly generated monies without mounting inflation—possibly to "Chilean" proportions. By mid-1980 the rise in the cost of living, calculated on an annual basis, exceeded 25 percent; and many economists expected an even greater increase in 1981. In the unlikely event that Mexico pursued such an expansionist policy, the quid pro quo could be formidable: the acceptance of a large flow of illegal immigrants, with marked political, economic, and environmental consequences for the United States. Projections made by the Environmental Fund indicate that undocumented immigrants and their dependents will add 40 million people to the U.S. population, increasing it to 306 million in the year 2000. These "illegal" residents will consume approximately 2.3 billion barrels of oil annually.[15] Even a Spanish-speaking Quebec, activated by the drumbeat of secession, could emerge in the Sunbelt by the end of the century. Then the T-shirts now on sale in San Antonio, which proclaim "You are now in occupied Mexico," may not appear so facetious.

As discussed in chapter 5, the industrial development plan offers little hope for creating a sufficient number of jobs. Moreover, the decision to stay out of GATT together with a reluctance to devalue the peso (a move that might unsettle lenders) points to a persistent policy of tolerating, if not encouraging, inefficient, capital-biased production. Continued inflation will sharpen the uncompetitiveness of Mexican goods, and we can expect increased reliance on foreign credits and oil sales to generate the revenues required to cover balance of trade deficits in nonpetroleum items. The emphasis on hydrocarbon production, which gives rise to relatively few jobs, instead of labor-intensive manufactures darkens the employment picture.

Indeed an underlying, if unstated, assumption of the industrial plan is that energy dependence and domestic political considerations will force the United States to absorb massive numbers of Mexicans who cannot find suitable work at home. For that reason Washington should act at once to stem the current of undocumented immigrants into the country. Reducing this flow is more a political than a technical problem, and a combination of methods—ubiquitous electronic sensing devices, spotlight-equipped helicopters, noncounterfeitable work cards, sturdy fencing, criminal penalties for employers knowingly hiring illegals, and expansion of the U.S. border patrol—merits consideration. This is not a punitive measure against the migrants, whose courage, determination, and capacity for hard work evoke praise. Rather, such a policy is designed to make powerful constituencies in Mexico—key industrialists, businessmen, professionals, bureaucrats, intellectuals, labor leaders, and political chieftains—aware that only by stimulating employment can they avert massive social unrest and continue to enjoy a reasonably comfortable life. Sensitivity and flexibility must accompany the closing of the border escape valve. During a five-to-ten-year transition period, temporary migration visas could be issued to vitally needed workers, and the quota for legal immigrants should be raised.

Support for such action will grow, notably during periods of recession in the United States as television, magazines, and newspapers move past sensationalist accounts heralding Mexico as the new Saudi Arabia to in-depth coverage of the enormous wealth accumulated by a small percentage of Mexicans while the country's poor live in subhuman conditions. In view of Mexico's population explosion, postponement of an assertive border policy will turn a flood into a tidal wave within a decade or two.

The creation of employment in Mexico will allow that proud country to avoid the path taken in the United States, where the promotion of capital- and energy-intensive agriculture eliminated the jobs of hundreds of thousands of poor blacks and whites in the South. Descendants of many of these uprooted people now live without work in major American cities, dependent on welfare and—in some cases—drugs.

Apart from the illegal immigration question, growing dependence on Mexico would expose America to pressures exerted by future presidents who may not be as moderate as López Portillo. In addition, easy access to a new oil cornucopia would allow politicians to delay the development of an intense energy conservation program. Such a program could cut U.S. energy usage by up to 40 percent and thereby diminish the country's international vulnerability.[16]

Oil is fungible. And the important thing about Mexican crude is not who buys it, but that it is sold. Its availability on the world market will not shatter OPEC, but it should relieve the pressure on North Sea and Persian Gulf sources. A backyard reserve would be useful to the United States in case of a Mideast explosion, world war, or interdiction of the shipping lanes. In the event of an international crisis, Mexico might also find it convenient to have a contiguous market for its exports. Even then the United States would be treaty bound to share available supplies with fellow members of the International Energy Agency. Although agriculture is not the subject of this book, it is interesting to note that the imperative to buy an increasing volume of basic farm products abroad may pose as serious a problem for Mexico as energy does for America. Mexico's need was especially evident in 1979 when, after a poor harvest, it purchased a major portion of the American grain denied Russia in the aftermath of the Afghanistan invasion. This food-energy complementarity, replete with opportunities for linkage, further illustrates a growing coincidence of interests that means neither country should again have to approach the other with hat in hand.

The Future in Mexico

López Portillo and Díaz Serrano have painted an attractive picture of the prosperity that awaits their oil-rich country. But they have yet to reveal a convincing plan to convert the hydrocarbons into jobs, without which Mexico may follow the same route as Venezuela, Ecuador, and even Iran. Appropriate technology appears to make sense in a country where unemployment or underemployment besets half the work force, many members of which live in the shadows of the machines that deprive them of a livelihood.[17] Yet Mexican entrepreneurs and transnational corporations continue to erect modern plants, replicas of those found in the United States and Western Europe. To many Mexicans, appropriate or labor-intensive technology smacks of imperialism wrapped in "small is beautiful" prose—an effort to preserve Third World nations as "hewers of wood and drawers of water."

The World Bank, which has demanded that developing nations commit themselves to population control as a prerequisite for assistance, should apply a similar requirement for appropriate technology. This would stimulate the activities of small- and medium-sized producers who generally have labor-intensive operations. Among the needs of these entrepreneurs are: (1) trade associations to lobby, fund common research, and gather and share information; (2) access to inventory and working capital on a competitive basis; and (3) the acquisition of management skills. In making loans international financial agencies should stipulate that Mexico earmark a significant amount of the monies for these producers. Even then few labor-intensive projects will be undertaken if unreasonable social benefits and inflexible labor laws make human workers expensive and virtually impossible to discharge.

To answer the common complaint that appropriate technology is unavailable, the Department of Commerce, in conjunction with local groups, should sponsor "technology fairs" in Mexico as well as visits of small businessmen and industrialists to the United States. America should also support creation of a capability within the United Nations Development Program to catalogue existing appropriate technology, help develop new applications of these techniques, and disseminate this information to interested nations. Finally, Congress might change U.S. tax laws to encourage investment abroad by smaller American firms whose technology is more likely to be labor-intensive than that of giant multinationals.

One third of Mexico's population still resides in the *campo,* where work is often available fewer than one hundred days per year. Additional jobs will spring from road-building, water and soil conservation, and irrigation work. Such endeavors, which the World Bank encourages through the PIDER program in 106 impoverished areas of Mexico, should be emphasized over the capital-devouring agroindustries endorsed by many politicians and industrialists.

Increased trade offers another means to facilitate development. Unfortunately, domestically produced goods are quite expensive, and only a limited number can compete internationally. It is a severe indictment of Mexican producers that Asian nations, which have emphasized exports in the absence of oil reserves, undersell them in scores of manufactured items in the U.S. market. López Portillo wants Americans to buy more of his country's tomatoes, shoes, textiles, and other goods whose domestic producers are already up in arms over external competition. The United States might consider greater access for Mexican light manufactures at the expense of Asian states only if its neighbor vigorously shifts from import licenses to tariffs to stimulate competition at home while steadily dismantling this tariff protection on a wide array of goods. Affiliation with GATT would have been a step in the right direction, but this move would have necessitated a gradual lowering and eventual elimination of the country's high tariff wall. Although López Portillo was unwilling to take this bold step, his successor may be more interested in encouraging efficient production and foreign trade.

Officials should also invite Mexico to take greater advantage of existing U.S. import preferences, available through the General System of Preferences, on paper goods, dried fruits, chemicals, and metal products. Japan, whose 116 million inhabitants boast an average income of approximately $6,000, may be a promising market for Mexican goods. Mexico might persuade the Tokyo government to relax its fierce protectionism as part of a political arrangement linked to oil sales. Tourism, an extremely labor-intensive industry, could expand greatly with careful planning and skillful promotion. The nation's magnificent countryside, scenic coastlines, cultural riches, and archaeological treasures offer virtually untapped opportunities. At present only a few large cities and resorts have competent programs to attract and accommodate foreign as well as Mexican tourists. Many jobs in tourism, which currently earns about $2 billion for the country, require relatively little skill.

Although Mexico has a great deal of petroleum, attention must be given to husbanding these reserves. The government now sets low energy prices, thereby encouraging energy- or capital-intensive production. Conservation practices such as increased prices, automobile mileage standards, efficiency criteria for appliances, and prudent electricity-pricing techniques should be encouraged. Technical assistance might come both from the increasingly conservation-minded Canadian and American experts and through the U.S.–Mexican working groups responsible for energy and economic questions.

Because oil will last only a few decades, diversification of Mexican energy sources should be emphasized. There are major deposits of commercial-grade coal in Coahuila which provide 11.1 percent of the nation's energy needs. Mexico is one of the only countries with identified geothermal resources, located near Mexicali on the U.S. border, which can easily be developed. But next to petroleum, Mexico's greatest potential lies in solar energy, in which López Portillo has shown a strong interest. As one of the first countries to take part in the Solar Energy Project of the Organization of American States under the Mar del Plata Resolution of 1974, Mexico has purchased solar water pumps for arid areas in Sonora, Durango, San Luis Potosí, and northern Baja California. Also under investigation is the widespread use of solar photovoltaic cells for hot-water heating and heat generation, and a model all-solar village for 250 people, Sonntlan, is being constructed at Las Barrancas, Baja California, as part of a German-Mexican cooperative program.

Solar power offers many advantages for a sun-drenched country like Mexico. One of its often overlooked benefits is decentralized power generation. Mexico could become a world leader both in petroleum production and in the advancement of nonconventional energy sources. Solar energy could also be used to produce liquid and gaseous fuels such as alcohol, methane, and hydrogen. Success with solar power might induce Mexico City's policy makers to give this energy source as much financial support as has been accorded a fledgling nuclear program, under which the first reactor will come on line in the early 1980s.

The United States and international bodies can have but a limited impact on Mexico. The oil presents a unique opportunity: It immunizes the country from the world energy crisis while providing an assured source of income that can be devoted to development. Without oil, the nation's economy would be in disastrous shape. But the black gold may tempt the elites to buy short-term support through some conspicuous social expenditures and the awarding of congressional seats to opposition parties while avoiding the sacrifices required by genuine reforms. Unless major changes are made in the next few years, the country could exhaust its most accessible reserves without achieving the populist goals of the Mexican Revolution. Moreover, a country whose leaders occasionally blame their problems on "U.S. imperialism" or the machinations of transnational firms will be denied a convenient scapegoat.

Older Mexicans have been long-suffering, fatalistic, and quiescent. Their children and grandchildren may prove less docile. In recent years relatively few people have endorsed violence. In the 1960s only 9 percent of migrants from the countryside to cities believed protest demonstrations an effective way to influence government decisions.[18] However, in a 1979 public-opinion survey conducted in Mexico by the *Los Angeles Times,* 30 percent of the respondents in the eighteen- to twenty-nine-year-old category said progress would only come "with total change and perhaps with violence." Although optimistic about the future, Victor Urquidi, president of the Colegio de México, says that the government must begin to bridge the chasm between "haves" and "have nots": "I believe the next 10 to 15 years are our last chance. We must finally produce a significant change in income distribution." Failure to provide young people with a stake in the system may have serious consequences for a regime whose commitment to "revolutionary" goals often seems more rhetorical than real. If Mexico does not succeed, Urquidi predicted, "social tensions will grow difficult. There will be invasions of property, insecurity in the cities, new political leaders."[19]

The Aztecs believed that they had to provide the mighty sun god Huitzilopochtli with the blood of human hearts to strengthen him for battle against the night, the stars, and the moon so that he could rise and bring the day. It remains to be seen whether President José López Portillo and other Mexican elites—the latter-day priests, who manage prodigious oil reserves—will demand the sacrifices necessary to uplift the country's masses.

Appendix: PEMEX Statistics

PEMEX: Reserves and Production
(Millions of Barrels)

Year	Total Hydrocarbon Reserves	Index of Reserves (1938=100)	Total Production of Hydrocarbons	Index of Production (1938=100)	Accumulated Production	Ratio of Reserves to Production[1]
1901–1937					1,874	
1938	1,276	100	44	100	1,918	29
1939	1,190	93	50	114	1,968	24
1940	1,225	96	51	116	2,019	24
1941	1,225	96	50	114	2,069	24
1942	1,236	97	41	93	2,110	30
1943	1,257	99	40	91	2,150	31
1944	1,548	121	43	98	2,193	36
1945	1,515	119	49	111	2,242	31
1946	1,437	113	55	125	2,297	26
1947	1,388	109	64	145	2,361	22
1948	1,367	107	67	152	2,428	20
1949	1,650	129	71	161	2,499	23
1950	1,608	126	86	195	2,585	19
1951	1,919	150	96	218	2,681	20
1952	2,241	176	98	223	2,779	23
1953	2,233	175	93	211	2,872	24
1954	2,549	200	104	236	2,976	25
1955	2,751	216	115	261	3,091	24
1956	2,959	232	119	270	3,210	25
1957	3,374	264	125	284	3,335	27
1958	4,070	319	153	348	3,488	27
1959	4,348	341	172	391	3,660	25
1960	4,787	375	177	402	3,837	27
1961	4,990	391	189	430	4,026	26
1962	5,008	392	196	445	4,222	26
1963	5,150	404	206	468	4,428	25
1964	5,227	410	227	516	4,655	23
1965	5,078	398	231	525	4,886	22
1966	5,357	420	241	548	5,127	22
1967	5,486	430	264	600	5,391	21
1968	5,530	433	276	627	5,667	20
1969	5,570	437	290	659	5,957	19
1970	5,568	436	311	707	6,268	18
1971	5,428	425	306	695	6,574	18
1972	5,388	422	317	720	6,891	17
1973	5,432	426	327	743	7,218	17
1974	5,773	452	387	880	7,605	15
1975	6,338	497	452	1,027	8,057	14
1976	11,160	875	482	1,095	8,539	23
1977	16,002	1,254	546	1,241	9,085	29
1978	40,194	3,150	672	1,527	9,757	60
1979	45,803	3,590	803	1,825	10,560	57
1980[2]	60,100	4,710	N.A.	N.A.	N.A.	N.A.

Source: Petróleos Mexicanos, *Anuario estadístico 1978* (Mexico City: Petróleos Mexicanos, 1978), p. 5.

1. This figure represents total proven oil and gas reserves divided by annual production.
2. Estimated by PEMEX.

PEMEX: Natural Gas Production

Year	Annual Production Cubic Meters (millions)	Annual Production Cubic Feet (millions)	Daily Average (production) Cubic Meters (millions)	Daily Average (production) Cubic Feet (millions)
1938	682	24,093	1.9	66
1939	906	32,004	2.5	88
1940	926	32,713	2.5	90
1941	883	31,190	2.4	85
1942	836	29,509	2.3	81
1943	675	23,851	1.8	65
1944	689	24,334	1.9	67
1945	747	26,363	2.0	72
1946	738	26,053	2.0	71
1947	930	32,871	2.5	90
1948	1,008	35,596	2.8	98
1949	1,270	44,853	3.5	123
1950	1,762	62,213	4.8	170
1951	2,422	85,533	6.6	234
1952	2,649	93,526	7.3	256
1953	2,645	93,387	7.2	256
1954	2,659	93,904	7.3	257
1955	3,392	119,771	9.3	328
1956	3,534	124,777	9.7	342
1957	4,568	161,325	12.5	442
1958	7,438	262,626	20.4	720
1959	9,328	329,364	25.6	902
1960	9,665	341,265	26.5	935
1961	10,210	360,506	28.0	988
1962	10,516	371,308	28.8	1,017
1963	11,371	401,515	31.2	1,100
1964	13,735	484,988	37.6	1,329
1965	13,965	493,157	38.3	1,351
1966	14,983	529,128	41.1	1,450
1967	16,221	572,832	44.4	1,569
1968	16,355	576,871	44.8	1,580
1969	17,247	609,056	47.2	1,669
1970	18,832	665,026	51.6	1,822
1971	18,220	643,426	49.9	1,763
1972	18,696	660,232	51.2	1,809
1973	19,164	676,750	52.5	1,854
1974	21,089	744,673	57.8	2,040
1975	22,271	786,458	61.0	2,155
1976	21,855	771,774	59.9	2,114
1977	21,149	746,863	57.9	2,046
1978	26,474	934,911	72.5	2,561
1979	30,181	1,065,800	82.7	2,920
1980[1]	39,977	1,411,735	109.5	3,857

Sources: Petróleos Mexicanos, *Anuario estadístico 1978* (Mexico City: Petróleos Mexicanos, 1978), p. 10; and *Petroleum Intelligence Weekly*, February 4, 1980, p. 11.
1. Estimated by PEMEX.

PEMEX: Production of Crude Oil, Condensates, and Liquids
(Millions of Barrels)

Year	Crudes and Condensates	Gas Liquids	Total	Daily Average
1938	38,482	336	38,818	106
1939	42,891	415	43,306	119
1940	44,045	403	44,448	122
1941	43,031	354	43,385	119
1942	34,826	322	35,148	96
1943	35,153	306	35,459	97
1944	38,197	306	38,503	105
1945	43,543	335	43,878	120
1946	49,240	293	49,533	136
1947	56,289	828	57,117	156
1948	58,520	1,254	59,774	164
1949	60,902	1,325	62,227	170
1950	72,422	1,459	73,881	202
1951	77,308	1,471	78,779	216
1952	77,278	1,629	78,907	216
1953	72,433	1,665	74,098	203
1954	83,651	1,579	85,230	234
1955	89,395	1,975	91,370	250
1956	90,660	3,437	94,097	258
1957	88,266	3,931	92,197	253
1958	93,533	7,108	100,641	276
1959	96,393	9,365	105,758	290
1960	99,049	9,722	108,771	298
1961	106,784	10,036	116,820	320
1962	111,830	9,733	121,563	333
1963	114,867	10,962	125,829	345
1964	115,576	13,928	129,504	355
1965	117,959	14,182	132,141	362
1966	121,149	13,872	135,021	370
1967	133,043	16,881	149,924	411
1968	142,360	18,126	160,486	440
1969	149,860	18,519	168,379	461
1970	156,586	21,013	177,599	487
1971	155,911	21,361	177,274	486
1972	161,367	23,644	185,011	507
1973	164,909	26,573	191,482	525
1974	209,855	28,416	238,271	653
1975	261,589	32,665	294,254	806
1976	293,136	34,149	327,285	897
1977	358,090	38,136	396,226	1,086
1978	442,607	42,689	485,297	1,330
1979	536,926	53,644	590,570	1,618
1980	757,839	N.A.	N.A.	2,071[1]

Source: Petróleos Mexicanos, *Anuario estadístico 1978* (Mexico City: Petróleos Mexicanos, 1978), p. 9.
1. Estimated by PEMEX.

Taxes Paid by PEMEX
(Millions of Pesos)

Year	Taxes and Duties Paid to Federal Government[1]	Taxes Paid to State Governments
1938	41	1
1939	63	2
1940	87	2
1941	76	3
1942	76	3
1943	93	4
1944	92	4
1945	109	5
1946	137	6
1947	181	8
1948	231	9
1949	230	9
1950	293	11
1951	312	12
1952	327	13
1953	339	14
1954	399	16
1955	459	30
1956	493	40
1957	540	50
1958	508	51
1959	576	63
1960	694	65
1961	778	70
1962	840	72
1963	864	77
1964	955	85
1965	1,015	93
1966	1,097	102
1967	1,134	110
1968	1,301	122
1969	1,455	133
1970	1,585	145
1971	1,776	141
1972	1,921	186
1973	2,283	393
1974	3,800	532
1975	7,674	583
1976	9,682	374
1977	19,764	506
1978	30,283	554
1979	47,000	N.A.

Sources: Petróleos Mexicanos, *Anuario estadístico 1978* (Mexico City: Petróleos Mexicanos, 1978), p. 3; and *Excelsior*, March 19, 1979, p. 13-A.

1. As of 1978 PEMEX's federal tax consisted of a 17-percent levy on total income from the sale of petroleum products and one of 12 percent on sales of petrochemicals; in addition, the firm pays duties on imports and exports of crude and its derivatives equal to current tariff levels. The state tax is an additional 2-percent levy on gasoline sales.

PEMEX: Personnel and Salaries

Year	Average Number of Workers[1]			Expenditures on Personnel (millions of pesos)
	Permanent	Transitory	Total	
1938	14,786	2,814	17,600	69
1939	16,278	3,823	20,101	98
1940	17,464	4,476	21,940	100
1941	16,238	3,524	19,762	96
1942	15,880	4,691	20,571	104
1943	16,498	4,737	21,235	109
1944	17,088	5,779	22,867	148
1945	17,660	7,986	25,646	187
1946	18,576	10,612	29,188	233
1947	20,025	8,797	28,822	250
1948	20,135	8,949	29,084	283
1949	20,138	8,948	29,086	341
1950	22,117	11,987	34,104	338
1951	23,166	13,387	36,553	421
1952	24,255	11,533	35,788	465
1953	24,579	12,350	36,929	504
1954	25,354	14,577	39,931	645
1955	26,537	16,815	43,352	785
1956	27,148	15,319	42,467	870
1957	27,934	16,495	44,429	991
1958	28,668	16,864	45,532	1,098
1959	29,324	16,371	45,695	1,241
1960	30,018	16,739	46,757	1,371
1961	31,134	15,024	46,158	1,380
1962	31,830	15,535	47,365	1,529
1963	32,858	16,747	49,605	1,683
1964	33,472	16,900	50,372	1,852
1965	34,315	19,658	53,973	2,080
1966	35,377	22,362	57,739	2,414
1967	38,448	24,224	62,672	2,724
1968	39,904	27,813	67,717	3,180
1969	41,789	26,610	68,399	3,416
1970	43,053	28,009	71,062	3,841
1971	44,153	31,345	75,498	4,389
1972	44,697	31,051	75,748	4,787
1973	45,633	31,023	76,656	5,048
1974	47,735	29,938	77,673	6,157
1975	49,166	32,037	81,203	8,037
1976	51,049	37,003	88,052	8,841
1977	52,669	39,011	91,680	11,546
1978	54,632	41,023	95,655	14,194

Source: Petróleos Mexicanos, *Anuario estadístico 1978* (Mexico City: Petróleos Mexicanos, 1978), p. 48.
1. Beginning in 1970, employees in PEMEX's projects and construction branch were not included in these figures.

PEMEX: Five Year Budget
(Millions of Dollars)

	1978	1979	1980	1981	1982
Marine programs	$258.64	$ 271.05	$127.73	$ 64.14	$ 64.14
Geophysical	117.86	185.05	161.13	208.33	254.32
Exploration & Production	735.26	1,356.98	997.48	1,110.40	1,211.34
Refining	470.37	448.08	270.47	214.92	270.26
Petrochemicals	500.48	809.44	263.26	153.04	143.31
Administration & Social programs	50.12	57.35	42.28	44.50	47.22
Total	$3,132.50	$3,802.65	$2,166.13	$2,054.99	$2,217.94

Source: Offshore, 39 (May 1979):146.

Notes

Introduction

1. John Hickey, "Pemex: A Study in Public Policy," *Inter-American Economic Affairs,* 14 (Autumn 1960): 73.
2. Quoted in the *Christian Science Monitor,* August 8, 1979, p. 22.

Chapter 1: Development of Mexico's Oil Industry

1. Pan American Petroleum and Transport Company, *Mexican Petroleum* (New York: PAP&TC, 1922), p. 23; cited in Richard B. Mancke, *Mexican Oil and Natural Gas: Political, Strategic, and Economic Implications* (New York: Praeger, 1979), p. 20. (Chapter 2, "History and Future of the Oil Industry in Mexico," is the text of an address delivered by Edward L. Doheny at the second annual meeting of the American Petroleum Institute, Chicago, December 6, 1921.)
2. Edgar Wesley Owen, *Trek of the Oil Finders: A History of Exploration for Petroleum* (Tulsa: The American Association of Petroleum Geologists, 1975), pp. 246–47.
3. Pan American Petroleum, *Mexican Petroleum,* p. 25; cited in Mancke, *Mexican Oil,* p. 20.
4. José López Portillo y Weber, *El petróleo de México: Su importancia/sus problemas* (Mexico City: Fondo de Cultura Económica, 1975), pp. 17–18.
5. Henry Bamford Parkes, *A History of Mexico* (Boston: Houghton Mifflin, 1960), p. 286.
6. Mancke, *Mexican Oil,* p. 27.
7. Howard F. Cline, *The United States and Mexico* (New York: Atheneum, 1965), p. 52.
8. Parkes, *History of Mexico,* p. 294.
9. Cline, *United States and Mexico,* p. 53.
10. Carleton Beals, *Porfirio Díaz: Dictator of Mexico* (Westport, Conn.: Greenwood Press, 1971), p. 29.
11. Mira Wilkins, *The Emergence of Multinational Enterprise: American Business Abroad from the Colonial Era to 1914* (Cambridge, Mass.: Harvard University Press, 1970), pp. 116–20.
12. Pan American Petroleum, *Mexican Petroleum,* p. 17; cited in Mancke, *Mexican Oil,* p. 23.
13. Ernest Gruening, *Mexico and Its Heritage* (New York: Appleton-Century, 1934), p. 604.
14. Daniel Cosío Villegas, *Historia moderna de México,* VII, part 2 (Mexico City: Editorial Hermes, 1965), p. 1127.
15. Mancke, *Mexican Oil,* p. 24.
16. Owen, *Trek of the Oil Finders,* pp. 250–51.
17. Ibid., p. 399.
18. Lorenzo Meyer, *Mexico and the United States in the Oil Controversy, 1917–1942* (Austin: University of Texas Press, 1977), p. 23.
19. López Portillo y Weber, *El petróleo de México,* pp. 19–20.
20. Alfred Tischendorf, *Great Britain and Mexico in the Era of Porfirio Díaz* (Durham: Duke University Press, 1961), p. 67; cited in Mancke, *Mexican Oil,* p. 25.

21. Ibid.

22. Owen, *Trek of the Oil Finders*, p. 251.

23. J. Fred Rippy, *British Investment in Latin America, 1822–1949* (Minneapolis: University of Minnesota Press, 1959), p. 101.

24. Antonio J. Bermúdez, *The Mexican National Petroleum Industry* (Palo Alto, Calif.: Institute of Hispanic American and Luso-Brazilian Studies, Stanford University, 1963), p. 3.

25. Meyer, *Oil Controversy*, p. 67; and Mancke, *Mexican Oil*, pp. 45–46.

26. Owen, *Trek of the Oil Finders*, p. 389.

27. Ibid., pp. 255, 398–400.

28. Ibid., pp. 398–400.

29. Charles C. Cumberland, *Mexican Revolution: The Constitutionalist Years* (Austin: University of Texas Press, 1972), p. 246.

30. Meyer, *Oil Controversy*, p. 27.

31. Thomas A. Bailey, *A Diplomatic History of the American People*, 7th ed. (New York: Appleton-Century-Crofts, 1964), p. 559. Professor Bailey identifies this intervention as primarily an effort to overthrow Victoriano Huerta, who had ousted and executed Francisco I. Madero, the leader of the revolution and president until February 1913.

32. Quoted in Cline, *United States and Mexico*, p. 155.

33. Meyer, *Oil Controversy*, p. 50.

34. Owen, *Trek of the Oil Finders*, p. 961.

35. Franklin Tugwell, *The Politics of Oil in Venezuela* (Stanford: Stanford University Press, 1975), pp. 183, 11.

36. Ibid., p. 11.

37. J. Richard Powell, *The Mexican Petroleum Industry: 1938–1950* (New York: Russell & Russell, 1972), p. 15.

38. Owen, *Trek of the Oil Finders*, pp. 966–67.

39. Meyer, *Oil Controversy*, p. 12.

40. William Cameron Townsend, *Lázaro Cárdenas, Mexican Democrat* (Ann Arbor, Mich.: George Wahr, 1952), pp. 44–45.

41. Mancke, *Mexican Oil*, p. 51.

42. Arthur W. MacMahon and W. R. Dittmar, "The Mexican Oil Industry Since Expropriation," part I, *Political Science Quarterly*, 57 (March 1942):33.

43. Townsend, *Lázaro Cárdenas*, p. 257.

44. Cline, *United States and Mexico*, p. 242.

45. Josephus Daniels, *Shirt-Sleeve Diplomat* (Chapel Hill: University of North Carolina Press, 1947), p. 258.

46. *New York Times*, April 21, 1943, p. 10.

47. Daniels, *Shirt-Sleeve Diplomat*, pp. 249–50.

48. Powell, *Mexican Petroleum Industry*, p. 129.

49. *New York Times*, January 6, 1940, p. 5.

50. Powell, *Mexican Petroleum Industry*, p. 143.

51. Owen, *Trek of the Oil Finders*, pp. 972–73.

52. Ibid., p. 970.

53. *New York Times*, March 2, 1944, p. 5.

Chapter 2: Forging a National Oil Company

1. *New York Times*, October 28, 1946, p. 6.

2. Powell, *Mexican Petroleum Industry*, pp. 20, 147.

3. Cline, *United States and Mexico,* p. 253.

4. Ibid.

5. Powell, *Mexican Petroleum Industry*, p. 149.

6. Interview with Joaquín Hernández Galicia, Ciudad Madero, May 31, 1978.

7. Upon his request these monies were returned to Bermúdez when he left office; interview with Antonio Vargas MacDonald, Mexico City, June 2, 1979.

8. Interview with Gustavo Roldán Vargas, member, Executive Committee, STPRM, Mexico City, May 29, 1978.

9. Petróleos Mexicanos, *Anuario estadístico 1978* (Mexico City: Petróleos Mexicanos, 1978); unless otherwise indicated the *Anuario* provides statistical information found in the remainder of this book on such matters as employment, reserves, drilling, petrochemicals, and production.

10. *Time,* June 16, 1947, p. 42.

11. Bermúdez, *Mexican Petroleum Industry*, p. 80.

12. Antonio J. Bermúdez, *La política petrolera mexicana* (Mexico City: Cuadernos de Joaquín Mortiz, 1976), p. 38.

13. Ibid., p. 54.

14. Bermúdez, *Mexican Petroleum Industry*, p. 168.

15. Hickey, "Pemex," p. 73.

16. Bermúdez, *Mexican Petroleum Industry*, pp. 66–75.

17. Hickey, "Pemex," p. 73.

18. *New York Times*, March 7, 1949; and Bermúdez, *Mexican Petroleum Industry*, pp. 32–34.

19. *New York Times*, March 5, 1952, p. 49.

20. Bermúdez, *Política petrolera mexicana*, p. 25.

21. Interview with Pascual Gutiérrez Roldán, Mexico City, May 24, 1978.

22. These writings have been collected in Antonio Vargas MacDonald, *Hacia una nueva política petrolera* (Mexico City: Editorial Promoción, 1959).

23. *New York Times*, January 25, 1959, p. 10.

24. Ibid.

25. Interview with Antonio Vargas MacDonald, June 2, 1979.

26. *Business Week,* April 11, 1959, p. 80.

27. Interview with Antonio Vargas MacDonald, June 2, 1979.

28. Interview with Gustavo Roldán Vargas, May 29, 1978.

29. Interview with Pascual Gutiérrez Roldán, May 24, 1978.

30. Ibid.

31. Ibid.

32. Petróleos Mexicanos, *Informe del director general, 18 de marzo de 1960* (Mexico City: Petróleos Mexicanos, 1960), p. 12.

33. Petróleos Mexicanos, *Informe del director general, 18 de marzo de 1964* (Mexico City: Petróleos Mexicanos, 1964), p. 14.

34. PEMEX, *Informe del director general . . . 1960*, p. 8.

35. PEMEX, *Informe del director general . . . 1964*, p. 14.

36. Interview with Pascual Gutiérrez Roldán, May 24, 1978.

37. PEMEX, *Informe del director general . . . 1964*, p. 7.

38. Interview with Pascual Gutiérrez Roldán, May 24, 1978.

39. Bermúdez, *Política petrolera mexicana*, pp. 69–70.

40. *New York Times*, June 5, 1960, p. 5.

41. Interview with Antonio Vargas MacDonald, June 2, 1979.

42. *New York Times*, January 26, 1964, p. 24.

43. Bermúdez, *Política petrolera mexicana*, pp. 69–70.

44. Ibid., p. 71.

45. Jesús Reyes Heroles, the next director general, announced that PEMEX alone would construct the plant.

46. Jorge Echaniz, "Petróleos Mexicanos, veintecinco años de vida de la industria nacionalizada" (tesis profesional, 1963), p. 159; cited in Francisco Alonso González, *Historia y petróleo, México: El problema del petróleo* (Madrid: Editorial Ayuso, 1972), pp. 242-43.

47. This party was called the National Revolutionary party (Partido Nacional Revolucionario—PNR) from 1929 to 1937. From that year until 1945, it was the Party of the Mexican Revolution (Partido de la Revolución Mexicana—PRM). Since then it has been known as the Institutional Revolutionary party (Partido Revolucionario Institucional—PRI).

48. Petróleos Mexicanos, *Petroleum Policy: Reports of Mr. Jesús Reyes Heroles, Director General, Petróleos Mexicanos, 1965-1966-1967* (Mexico City: Petróleos Mexicanos, n.d.), p. 71.

49. Ibid., pp. 10, 11.

50. Bermúdez, *Política petrolera mexicana*, p. 75.

51. *Latin America*, July 4, 1969, p. 211.

52. Petróleos Mexicanos, *Petroleum Policy: Report of the Director General of Petróleos Mexicanos, Lic. Jesús Reyes Heroles, March 18, 1970* (Mexico City: Petróleos Mexicanos, 1970), pp. 11, 21.

53. Petróleos Mexicanos, *Política petrolera: Informes del director general de Petróleos Mexicanos, Lic. Jesús Reyes Heroles, 1968-1969-1970* (Mexico City: Petróleos Mexicanos, 1970), p. 72.

54. PEMEX, *Petroleum Policy . . . 1965-1966-1967*, pp. 10, 13.

55. PEMEX, *Petroleum Policy . . . 1970*, pp. 25, 9.

56. PEMEX, *Petroleum Policy . . . 1965-1966-1967*, pp. 39-40.

57. Petróleos Mexicanos, *1938-1978 Petróleos Mexicanos* (Mexico City: Petróleos Mexicanos, 1978), p. 79.

58. PEMEX, *Petroleum Policy . . . 1965-1966-1967*, p. 7.

59. Alfonso Martínez Domínguez, quoted in *Latin America*, April 26, 1968, p. 135.

60. Petróleos Mexicanos, *Petroleum Policy: Report of the Director General of Petróleos Mexicanos, Lic. Jesús Reyes Heroles, March 18, 1968* (Mexico City: Petróleos Mexicanos, 1968), pp. 19-20.

61. *Latin America*, July 19, 1968, p. 226.

62. PEMEX, *Petroleum Policy . . . 1970*, p. 5.

63. Ibid., p. 6.

64. Ibid., p. 7.

65. Interview with Pascual Gutiérrez Roldán, Mexico City, June 1, 1977.

66. PEMEX, *Petroleum Policy . . . 1965-1966-1967*, p. 16.

67. Interview with Joaquín Hernández Galicia, May 31, 1978.

68. Jaime Aguilar Briseño, *"La Quina"* (Privately printed, n.d.), pp. 206-07.

69. Interview with former Secretary General Terrazas Zozaya, Mexico City, June 6, 1978.

70. Ibid.

71. PEMEX, *Petroleum Policy . . . 1970*, p. 15.

72. Ibid., p. 28.

73. Petróleos Mexicanos, *Report of the Director General Ing. Antonio Dovalí Jaime, March 18, 1971* (Mexico City: Petróleos Mexicanos, 1971), p. 8; and *Latin America*, February 26, 1971, p. 67.

74. *World Oil*, August 15, 1979, p. 82.

75. Petróleos Mexicanos, *Report of the Director General Ing. Antonio Dovalí Jaime, March*

18, 1976 (Mexico City: Petróleos Mexicanos, 1976), p. 9; and *Petroleum Economist, Latin America and Caribbean Oil Report* (London: Woodcote Press, 1979), pp. 118-19.

76. PEMEX, *Report of the Director General . . . 1976*, pp. 5, 17.

77. Petróleos Mexicanos, *Report of the Director General Ing. Antonio Dovalí Jaime, March 18, 1974* (Mexico City: Petróleos Mexicanos, 1974), p. 11.

78. PEMEX, *Report of the Director General . . . 1976*, p. 16.

79. Ibid., p. 20.

80. PEMEX, *Report of the Director General . . . 1974*, pp. 16-17.

81. Petróleos Mexicanos, *Report of the Director General Ing. Antonio Dovalí Jaime, March 18, 1972*, (Mexico City: Petróleos Mexicanos, 1972), p. 21.

82. Ibid., p. 17.

83. Ing. Ernesto Verdugo, quoted in *Excelsior*, October 14, 1977, p. 4-A; and interview with Verdugo, Mexico City, May 23, 1978.

84. PEMEX, *Report of the Director General . . . 1976*, p. 23.

85. PEMEX, *Report of the Director General . . . 1971*, pp. 9-10.

86. PEMEX, *Report of the Director General . . . 1976*, pp. 23-24.

87. PEMEX, *Report of the Director General . . . 1974*, p. 8.

88. PEMEX, *Report of the Director General . . . 1976*, p. 33.

Chapter 3: The Oil Policy of José López Portillo

1. *Proceso*, August 13, 1979, p. 29; and *Latin America Political Report*, July 13, 1979, p. 212.

2. "Mexico: Crisis of Poverty/Crisis of Wealth," Supplement to the *Los Angeles Times*, July 15, 1979, p. 10.

3. The *Wall Street Journal*, August 8, 1977, p. 1. For a description of conditions in Mexico at the time López Portillo took office, see George W. Grayson, "Mexico's Opportunity: The Oil Boom," *Foreign Policy*, No. 29 (Winter 1977-78):65-66; and Frank E. Niering, Jr., "Mexico: A New Force in World Oil," *Petroleum Economist*, 46 (March 1979):109. Niering's is a thorough, balanced, informative analysis of Mexico's oil wealth.

4. *Excelsior*, October 27, 1977, p. 10-A.

5. Díaz Serrano, quoted in the *New York Times*, July 16, 1978, p. F-5.

6. *Ocean Industry*, 13 (May 1978):41.

7. *Proceso*, September 24, 1979, pp. 11-12.

8. Richard Fagen, quoted in "Mexico: Crisis of Poverty," *Los Angeles Times* supp., July 15, 1979, p. 16.

9. Interview with Walter Friedeberg Merzbach, Williamsburg, Va., March 27, 1980.

10. Petróleos Mexicanos, *Linea troncal nacional de distribución de gas natural; El director general de Petróleos Mexicanos ante la H. Cámara de Diputados* (Mexico City: Petróleos Mexicanos, 1977), p. 11.

11. William D. Metz, "Mexico: The Premier Oil Discovery in the Western Hemisphere," *Science*, December 22, 1978, p. 1262.

12. Bruce Netschert, "Special Survey of Mexico's Oil and Gas Potential," in *Petroleum Economist, Latin America and Caribbean Oil Report*, p. 136.

13. *Offshore*, 38 (January 1978): 44; and Mancke, *Mexican Oil*, p. 62.

14. *Los Angeles Times*, May 18, 1979, p. 6.

15. "Mexico: Crisis of Poverty," *Los Angeles Times* supp., July 15, 1979.

16. *Ocean Industry*, 13 (May 1978):41.

17. Interview with Walter Friedeberg Merzbach, March 27, 1980.

18. *New York Times*, January 10, 1979, p. D-1.

19. For a perceptive examination of the six-year plan, see *Petroleum Economist*, 44 (March 1977):94, 96.

20. *New York Times*, January 10, 1979, p. D-1.

21. Mancke, *Mexican Oil*, p. 2.

22. *New York Times*, January 10, 1979, p. D-3.

23. Ibid.

24. See Mancke, *Mexican Oil*, p. 88; and *Oil & Gas Journal*, February 20, 1978, p. 20.

25. *Oil & Gas Journal*, November 19, 1979, p. 209.

26. Ibid., July 31, 1978, p. 118.

27. Fernando Hiriart, cited in U.S. Congress, Senate, *Mexico: The Promise and Problems of Petroleum*, Report prepared for the Committee on Energy and Natural Resources, 96th Cong., 1st sess., 1979, pub. no. 96-2, p. 26.

28. *Excelsior*, March 19, 1980, p. 12-A.

29. These definitions are found in Howard R. Williams and Charles J. Meyers, *Oil and Gas Terms*, 4th ed. (New York: Bender, 1976), and are cited in U.S. Senate, *Mexico: Promise and Problems*, p. 16. Díaz Serrano discusses these definitions in "¿En que consiste una reserva petrolera?" *Ciencia y desarrollo*, No. 13 (March-April 1977):5-9.

30. *Excelsior*, March 19, 1980, p. 12-A.

31. *World Oil*, February 1, 1979, p. 40; and Niering, "Mexico: A New Force," p. 111.

32. The following section relies heavily on E. N. Tiratsoo, *Oilfields of the World* (Houston: Gulf, 1976), pp. 290-95; and Philip R. Woodside, "Comments on Oil and Gas Reserves in Mexico," Paper presented at the Workshop on Oil and Gas Supply from Foreign Sources, Reston, Virginia, June 12-13, 1979.

33. Niering, "Mexico: A New Force," p. 106.

34. *Excelsior*, March 19, 1980, p. 12-A; and *Petroleum Intelligence Weekly*, January 21, 1980, p. 11.

35. Mancke, *Mexican Oil*, pp. 69-70.

36. *Oil & Gas Journal*, June 5, 1978, p. 70.

37. *Excelsior*, September 30, 1979, pp. 1-A, 10-A, and November 1, 1979, pp. 1-A, 8-A.

38. *Offshore*, 38 (January 1978):44; cited in Mancke, *Mexican Oil*, p. 65.

39. PEMEX, *Linea troncal de gas natural*, pp. 13-14.

40. *World Oil*, February 1, 1979, p. 41, and August 15, 1979, p. 83.

41. Ibid., August 15, 1979, p. 83.

42. *Excelsior*, August 24, 1980, p. 31-A.

43. *Petroleum Economist*, 45 (December 1978):527-28.

44. *World Oil*, August 15, 1979, p. 84.

45. Ibid.

46. U.S. Congress, Senate and House, *Mexico's Oil and Gas Policy: An Analysis*, Report prepared by the Congressional Research Service, Library of Congress, for the Committee on Foreign Relations and the Joint Economic Committee, 95th Cong., 2nd sess. (Washington, D.C.: Government Printing Office, 1979), p. vii.

47. Metz, "Mexico: Premier Oil Discovery," p. 1263, and Mancke, *Mexican Oil*, p. 69.

48. Mancke, *Mexican Oil*, p. 74.

49. *Los Angeles Times*, May 18, 1979, p. 6.

50. Interview with Walter Friedeberg Merzbach, March 27, 1980.

51. *Los Angeles Times*, May 18, 1979, p. 1.

52. *World Oil*, August 15, 1979, p. 84.

53. "Mexico: Crisis of Poverty," *Los Angeles Times*, supp., July 15, 1979.

54. *Proceso,* December 4, 1976, p. 6.
55. *World Oil,* August 15, 1979, p. 40.
56. *Virginian-Pilot* (Norfolk), November 24, 1979, p. A-21.
57. *Proceso,* May 21, 1979, pp. 6–8; see also Ernest Duff, "Mexico's reforma política," Paper presented at the annual meeting of the Middle Atlantic Council of Latin American Studies, University of Delaware, Newark, Delaware, April 17–19, 1980.
58. *Virginian-Pilot,* November 24, 1979, p. A-21.
59. U.S. Foreign Broadcast Information Service, *Daily Report* (Latin America), February 9, 1979, p. M-1, and February 27, 1979, p. M-1.
60. *Offshore,* 38 (January 1978):43; cited in Mancke, *Mexican Oil,* p. 86.
61. *World Oil,* August 15, 1979, p. 81; and Petróleos Mexicanos, *Memoria de labores 1978* (Mexico City: Petróleos Mexicanos, 1979), p. 63.
62. *World Oil,* February 15, 1979, p. 135.
63. Ibid., February 1, 1979, p. 40; *Petroleum Economist,* 47 (January 1980):19; and *Petroleum Intelligence Weekly,* March 17, 1980, p. 12.
64. Christopher Buckley, "Mexico's Oil Boom and What's in It for Us," *Esquire,* December 19, 1978, p. 49.
65. *Proceso,* July 17, 1978, p. 8.
66. *New York Times,* January 6, 1979, p. 25.
67. *Proceso,* June 19, 1978, pp. 16, 17.
68. *Excelsior,* November 19, 1979, pp. 1-A, 8-A.
69. *Proceso,* June 19, 1978, pp. 17, 19, and July 9, 1979, p. 11.
70. Ibid., July 16, 1979, p. 22.
71. Ibid., p. 18.
72. *Excelsior,* December 19, 1977, p. 1-A.
73. Ibid., December 20, 1977, p. 4-A.
74. *Newsweek,* February 19, 1979, p. 30.
75. *Proceso,* July 19, 1979, p. 11.

Chapter 4: The Oil Workers' Union

1. The following discussion of the labor movement in Tampico has profited greatly from "Coyuntura y conciencia: Factores convergentes en la fundación de los Sindicatos Petroleros de Tampico durante la década de 1920," an excellent paper written by S. Lief Adleson, a doctoral candidate in history at the Colegio de México, delivered at the Fifth Mexican–North American Historians' Conference, Pátzcuaro, Michoacán, October 1977.
2. Ibid., p. 6.
3. Joe C. Ashby, *Organized Labor and the Mexican Revolution under Lázaro Cárdenas* (Chapel Hill: University of North Carolina Press, 1967), p. 195.
4. Adleson, "Coyuntura y conciencia," p. 8.
5. Ibid., pp. 5–6.
6. Ibid., p. 8.
7. Ibid., p. 12.
8. Ibid., p. 20.
9. Ibid., p. 21.
10. Ibid., pp. 21–22.
11. Ibid., p. 24.
12. Ibid., pp. 26–27.
13. Ashby, *Organized Labor,* p. 195.

14. For the names of those attending this session and the declaration of principles they adopted, see Sindicato de Trabajadores Petroleros de la República Mexicana, *Acta constitutiva y estatutos generales* (1977), pp. 9–18.

15. Ibid., p. 76 ff.

16. Ibid., p. 86.

17. Interview with Samuel Terrazas Zozaya, June 6, 1978.

18. STPRM, *Acta constitutiva*, p. 111.

19. Powell, *Mexican Petroleum Industry*, p. 149.

20. Interview with Pascual Gutiérrez Roldán, June 1, 1977.

21. *Proceso*, December 11, 1978, p. 14.

22. Interview with Antonio Vargas MacDonald, June 2, 1979.

23. *Excelsior*, December 6, 1975, p. 22-A.

24. Briseño, "*La Quina*," p. 21 ff.

25. Interview with Joaquín Hernández Galicia, May 31, 1978.

26. Interview with Brigado Piñeyro de los Santos, secretary of organization and statistics, Mexico City, May 29, 1978.

27. Interview with Samuel Terrazas Zozaya, June 6, 1978; Hebraicaz Vázquez Gutiérrez, Carlos Ibarra, and Manuel Limón, quoted in *Excelsior*, October 23, 1979, p. 18-A.

28. *Excelsior*, March 3, 1977, p. 28-A.

29. Interview with Samuel Terrazas Zozaya, June 6, 1978; and *Excelsior*, March 6, 1977, p. 1-A, 13-A.

30. Briseño, "*La Quina*," pp. 185–96.

31. Petróleos Mexicanos, *Anuario estadístico 1976* (Mexico City: Petróleos Mexicanos, 1976); and Secretaria de Programación y Presupuesto, *Trabajo y salarios industriales, 1975* (Mexico City: Dirección General de Estadística, 1977), p. 1.

32. Interview with Brigado Piñeyro de los Santos, May 29, 1978.

33. *News* (Mexico City), January 12, 1959, p. 12-A.

34. *Proceso*, October 24, 1977, p. 13.

35. Interview with Antonio Vargas MacDonald, June 2, 1979; *News*, January 12, 1959, p. 12-A.

36. Interview with Pascual Gutiérrez Roldán, June 1, 1977.

37. Interview with Hebraicaz Vázquez Gutiérrez, Mexico City, May 31, 1978; for charges of job-selling brought by two workers in 1977, see *Proceso*, October 24, 1977, p. 13.

38. Vázquez Gutiérrez, quoted in *Excelsior*, October 14, 1977, p. 11-A.

39. *Proceso*, October 29, 1979, p. 33.

40. Interview with Joaquín Hernández Galicia, May 31, 1978; and *Excelsior*, October 13, 1977, p. 12-A.

41. Interview with Francisco Alonso González, August 17, 1980.

42. *Excelsior*, January 16, 1978, pp. 10-A, 17-A.

43. For further information on the López Díaz case, see *Excelsior*, January 14, 1978, pp. 1-A and 8-A, and January 16, 1978, pp. 1-A and 10-A. The most extensive—though sensationalistic—reporting of the incident was done by Roberto Blanco Moheno in four issues of the weekly magazine *Impacto*, March 8 (p. 8), March 15 (pp. 8, 69), March 22 (pp. 8, 9), and March 29, 1978 (p. 89).

44. Interview with Antonio Vargas MacDonald, June 2, 1979.

45. *News*, January 13, 1959, p. 12-A, January 14, 1959, p. 15-A; and *Proceso*, May 28, 1979, pp. 12–13.

46. *Proceso*, May 28, 1979, p. 13.

47. Interview with Walter Friedeberg Merzbach, March 27, 1980.

48. *El Universal,* October 7, 1958, p. 1, October 8, 1958, p. 17, October 9, 1958, p. 21, and October 12, 1958, p. 1.

49. Interview with Gustavo Roldán Vargas, June 5, 1979.

50. Interview with Francisco Alonso González, Mexico City, June 1, 1979.

51. *Proceso,* December 11, 1978, p. 14.

52. Ibid., p. 15.

53. *Excelsior,* November 17, 1977, p. 5-A.

54. *Proceso,* October 29, 1979, pp. 33–34.

55. *Excelsior,* November 22, 1978, p. 11-A.

56. Ibid., June 26, 1978, p. 25, and January 26, 1978, p. 4-A.

57. Ibid., June 22, 1978, p. 4-A.

58. Interview with Hebraicaz Vázquez Gutiérrez, May 31, 1978.

59. Interview with Ernesto Verdugo, May 23, 1978.

60. *Excelsior,* October 15, 1977, p. 14-A.

61. Ibid., October 27, 1977, p. 10-A, and January 26, 1978, p. 4-A; and *Impacto,* March 15, 1978, p. 8.

62. *Excelsior,* October 15, 1977, p. 14-A.

63. Samuel Terrazas Zozaya, quoted in ibid.

64. *Excelsior,* January 16, 1978, p. 10-A.

65. Interview with Hebraicaz Vázquez Gutiérrez, May 31, 1978.

66. Roger D. Hansen, *The Politics of Mexican Development* (Baltimore and London: Johns Hopkins Press, 1971), p. 125.

67. "Survey of Mexico," the *Economist,* April 22, 1978, p. 24.

68. Victor Manuel Negrete Cueto, quoted in *Excelsior,* January 26, 1978, p. 4-A.

69. Ing. Tavare Azcona Pavon, general manager for sales, PEMEX, quoted in *Excelsior,* November 12, 1977, p. 1-A.

70. Susan Kaufman Purcell and John F. H. Purcell, "State and Society in Mexico: Must a Stable Polity Be Institutionalized?" *World Politics,* 32 (January 1980):200.

71. Ibid.

72. *Excelsior,* November 17, 1977, p. 1-A.

73. Ibid., November 24, 1979, p. 14-A.

74. Interview with Joaquín Hernández Galicia, May 31, 1978.

75. La Quina (ibid.) stated that the union's goal was to reduce the percentage of transitory jobs that are sold from twenty to five. In a meeting with President López Portillo, Secretary General Oscar Torres Pancardo pledged to fight corruption; see *Ultimas noticias,* May 27, 1977, p. 8.

76. *Excelsior,* January 11, 1978, p. 20-A.

77. Ibid., November 24, 1979, p. 14-A.

78. Ibid., October 27, 1977, p. 10-A; and November 14, 1977, p. 1-A.

79. Ibid., October 28, 1979, pp. 4-A, 7-A.

Chapter 5: Oil, Jobs, and Economic Development

1. Quoted in *Daily Report,* February 8, 1979, p. M-1; and *Excelsior,* October 27, 1977, p. 1-A.

2. Raymond Vernon, *The Dilemma of Mexican Development* (Cambridge, Mass.: Harvard University Press, 1963), p. 9.

3. Leopoldo Solís, "The External Sector of the Mexican Economy," Paper presented at the Center for International Affairs, Harvard University, March 1977 (mimeo.), p. 3.

4. Gustav Ranis, "Se está tornado amargo el milagro mexicano?" *Demografía y economía,* 8

(1974):20–23, cited in John S. Evans and Dilmus D. James, "Increasing Productive Employment in Mexico," Paper presented at the Brookings Institution–El Colegio de México Symposium on Structural Factors in Migration in Mexico and the Caribbean Basin, Washington, D.C., June 28–30, 1978, p. 5.

5. David Felix, "Income Inequality in Mexico," *Current History*, 72 (March 1977):112.

6. It fell from 17.7 percent in 1950 to only 9.5 percent in 1969; see Solís, "External Sector," p. 6.

7. David Gordon, "Mexico: A Survey," the *Economist*, April 22, 1978, p. 18.

8. Susumu Watanabe, "Constraints on Labour-Intensive Export Industries in Mexico," *International Labour Review*, 109 (January 1974):27, 31.

9. Evans and James, "Increasing Employment in Mexico," p. 5.

10. Watanabe, "Constraints on Labour-Intensive Export Industries in Mexico," p. 32.

11. Lic. Jesús Puente Leyva, federal deputy for District 6 of Nuevo Leon, quoted at the second annual Symposium on Mexico–U.S. Economic Relations, Xochimilco University, Mexico City, May 23–25, 1979.

12. Wayne A. Cornelius, "Mexican Migration to the United States: Causes, Consequences, and U.S. Responses," Report prepared for the Migration and Development Study Group, Massachusetts Institute of Technology (Cambridge, Mass.: Center for International Studies, 1978), pp. 41–42.

13. Saul Trejo Reyes, "El desempleo en México: Características generales," *El Trimestre económico*, 42, pp. 677–78; cited in Evans and James, "Increasing Employment in Mexico," p. 2; and Manuel Gollás, "El desempleo en México: Soluciones posibles," *Ciencia y desarrollo*, No. 14 (May–June 1978):74; cited in Evans and James, ibid., p. 2.

14. *Latin America Economic Report*, May 13, 1977, p. 72.

15. Ibid.

16. *Washington Post*, August 17, 1978, p. A-19.

17. *Excelsior*, November 18, 1979, p. 11-A.

18. This assumes a population of seventy million and a 2.8 percent growth rate; see *Proceso*, October 15, 1979, p. 23.

19. Gordon, "Mexico: A Survey," p. 4.

20. Based on data contained in Consejo Nacional de Población, *México demográfico: Breviario 1975* (Mexico City: Consejo Nacional de Población, 1975).

21. *Proceso*, October 15, 1979, p. 24.

22. Alfredo Gallegos, Jorge García Peña, José Antonio Solís, and Alan Keller, "Recent Trends in Contraceptive Use in Mexico," *Studies in Family Planning*, 8 (August 1977):203.

23. *Latin America Regional Reports: Mexico & Central America*, March 21, 1980, p. 7; *Proceso*, March 3, 1980, pp. 6–10.

24. *Latin America Regional Reports: Mexico & Central America*, March 21, 1980, p. 7.

25. Ibid.; and *New York Times*, March 9, 1979, p. A-6.

26. Interview, Mexico City, June 1, 1977.

27. *Proceso*, October 15, 1979, p. 22.

28. David S. North, "The Migration Issue in U.S.-Mexico Relations," Paper presented at the second annual Symposium on Mexico–U.S. Economic Relations, Xochimilco University, Mexico City, May 23–25, 1979, p. 1.

29. Cornelius, "Mexican Migration," pp. 19–21.

30. Ibid.

31. Ibid.

32. Provided in telephone conversation with INS, April 9, 1980.

33. North, "Migration Issue," p. 9.

34. American Federation of Labor and Congress of Industrial Organizations, "Statement of Rudy Oswald, Director, Department of Research, American Federation of Labor and Congress of Industrial Organizations, Before the Senate Committee on Judiciary on S. 2252, 'Alien Adjustment and Employment Act of 1977,' " Washington, D.C., May 17, 1978 (mimeo.).

35. National Association for the Advancement of Colored People, "NAACP 68th Annual Convention Resolutions," adopted at convention held in St. Louis, Missouri, June 27–July 1, 1977, p. 16.

36. Sen. Lloyd Bentsen, "Survey of President Carter's Illegal Alien Problem," 1978 (mimeo.). Interpretations of these findings must be made with caution inasmuch as only 3.2 percent of those persons selected from a list of 114,000 Spanish-surnamed Texans responded.

37. For an environmentalist's perspective, see Michael Christiano, " 'A Tidal Wave' of Illegal Aliens," Letter to the editor, *Washington Post*, December 28, 1978, p. A-22.

38. *U.S. News & World Report*, April 25, 1977, p. 35.

39. *Facts on File*, February 23, 1979, p. 127.

40. *New York Times*, November 7, 1978, p. A-22.

41. Roberto Jaramillo Flores, quoted in *Excelsior*, October 29, 1977, p. 14-A.

42. North, "Migration Issue," p. 2.

43. *Excelsior*, June 2, 1979, p. 1-A.

44. *Daily Report*, June 26, 1979, p. M-1.

45. "Survey of Venezuela," *Economist*, December 27, 1975, pp. 1–28.

46. Stephen C. Dodge, "Venezuela's Bright Future," *Current History*, 70 (February 1976):65.

47. *Encyclopedia Yearbook 1980* (Danbury, Conn.: Grolier, 1980), pp. 547, 575.

48. *Americana Annual* (Danbury, Conn.: Grolier, 1979), p. 580.

49. *Bank of London & South America Review*, 13 (April 1979):250.

50. On March 25, 1980, Jesús González Escobar, director of the Domestic Hydrocarbon Market, asserted that the $1-million daily subsidy paid by the state to gasoline consumers in the form of artificially low prices deprived the country of funds that could be used to construct nearly three thousand schools each year. He also observed that distilled water is three times more expensive in Caracas than gasoline; see *El Nacional* (Caracas), March 26, 1980.

51. *Latin America Economic Report*, December 22, 1978, p. 397.

52. *New York Times*, October 25, 1974, p. 16. The World Bank reported that the wealthiest 10 percent of households increased their portion of total income from 41.2 percent in 1962 to 51.2 percent in 1971; see Jeffrey A. Hart, "Industrialization and the Fulfillment of Basic Human Needs in Venezuela," Paper prepared for an Institute of World Order Colloquium, revised October 1979 (mimeo.), p. 29.

53. *Bank of London & South America Review*, 13 (April 1979):250.

54. Ibid., p. 251.

55. *Latin America Weekly Report*, April 18, 1980, pp. 4–5.

56. Hart, "Industrialization and Human Needs in Venezuela," p. 26.

57. Ibid., p. 14.

58. *Bank of London & South America Review*, 13 (April 1979):252.

59. Ibid., p. 250.

60. *Latin America Economic Report*, September 14, 1979, p. 287, and October 19, 1979, p. 321.

61. *New York Times*, July 23, 1978, sec. IV, p. 4; and U.S. Arms Control and Disarmament Agency, *World Military Expenditures and Arms Transfers 1967-76*, Pub. 98 (1978), p. 41. In constant dollars, the GNP increased from $2.8 billion (1971) to $4.4 billion (1976).

62. *Washington Post*, September 9, 1975, p. 17.

63. U.S. Arms Control and Disarmament Agency, *World Military Expenditures*, p. 41.

64. Phillip Berryman, "Ecuador's Oil: No Bonanza for the People," *Christian Century,* June 11, 1975, pp. 604–05.

65. *Latin America Economic Report,* December 1, 1978, p. 370.

66. *Latin America Regional Report: Andean Group,* February 29, 1980, p. 8.

67. *Latin America Economic Report,* August 26, 1977, p. 132.

68. *Bank of London & South America Review,* 13 (May 1979):305.

69. *Latin America Regional Report: Andean Group,* February 29, 1980, p. 8.

70. *Latin America Economic Report,* December 22, 1978, p. 398.

71. *Bank of London & South America Review,* 13 (May 1979):306.

72. James H. Street, "The Mexican Paradox, Other Models and the Technological Challenge," Paper Presented to the North American Economic Studies Association, Mexico City, December 27, 1978, p. 14.

73. Ibid., p. 12.

74. *Americana Annual,* p. 580.

75. *Washington Post,* January 31, 1977, pp. A-1, A-11.

76. John J. Bailey, "Presidency, Bureaucracy, and Administrative Reform in Mexico: The Secretariat of Programming and Budget," Paper presented at the annual meeting of the American Society for Public Administration, San Francisco, April 14–17, 1980, p. 18.

77. Ibid., p. 36.

78. Ibid., pp. 8, 17; and *Mexico: Latin America Special Report,* March 25, 1977, p. 4.

79. Gordon, "Mexico: A Survey," p. 18.

80. *Comercio exterior de México,* 23 (January 1977):1–2.

81. Bailey, "Presidency, Bureaucracy, and Administrative Reform," pp. 19–20.

82. Ibid., p. 21.

83. An analysis of the reform appears in *Comercio exterior de México,* 23 (January 1977):1–10.

84. Bailey, "Presidency, Bureaucracy, and Administrative Reform," p. 29.

85. *Comercio exterior de México,* 23 (January 1977):13; a tentative assessment of the reform can be found in ibid., pp. 38–40.

86. Ibid., p. 13.

87. *Latin America Economic Report,* September 30, 1977, p. 154.

88. *Comercio exterior de México,* 23 (February 1977):53.

89. Ibid., (June 1977):227.

90. *Latin America Economic Report,* August 12, 1977, p. 121.

91. Ibid., September 2, 1977, p. 135.

92. Ibid., May 13, 1977, p. 70.

93. Ibid., February 10, 1978, p. 48.

94. Ibid., March 30, 1979, p. 104.

95. *Latin America Regional Reports: Mexico & Central America,* January 11, 1980, p. 5.

96. The English version of the four decrees, issued between February 2 and March 19, 1979, which implement the plan is published in the *Quarterly Economic Report* (Mexico City), March 1979. See also: *Financial Times* (London), April 10, 1979, p. 4; *Latin America Economic Report,* May 4, 1979, p. 132; *Daily Report,* March 16, 1979, p. M-1; and *Comercio exterior de México,* 25 (June 1979):105–13.

97. *Excelsior,* March 13, 1979, p. 11-A.

98. "Federal Executive Decree of March 6, 1979," in *Quarterly Economic Report,* March 1979, n.p.

99. Reportedly, Mexican trains average only 7.5 miles per hour because of the condition of roadbeds and equipment. The Monterrey chamber of commerce complained that two hundred

factories were operating at less than half capacity for lack of raw materials at the same time that one hundred thousand tons of freight collected dust in the city's railyards. Meanwhile, reports from Sinaloa alleged that fifty thousand to seventy thousand tons of agricultural produce were immobilized for want of rail transport; see *Latin America Economic Report,* February 2, 1979, p. 38.

100. *Daily Report,* March 16, 1979, p. M-1.

101. The decision to stay out of GATT reportedly sprang from an unwillingness to embark upon a controversial international policy simultaneously with the introduction of a 10-percent value-added tax at home. In addition, three other considerations may have affected the stance: (1) an unwillingness to devalue the peso in order to spur exports; (2) the collapse of agricultural output in 1979, giving rise to additional government protection and subsidies that clash with GATT policy; and (3) the fear that Mexico would lose flexibility in setting oil production levels in view of article 20G of the membership protocol, requiring that a proportional reduction in domestic consumption accompany any decrease in oil sales abroad. Both leftists concerned about the growing influence of transnational firms in Mexico's economy and right-wing businessmen alarmed over potential competition dominated the opposition to membership, believed originally to be supported by López Portillo. Observers identified a majority of the cabinet as in support of the final decision. These members were Oteyza (Patrimony), David Ibarra Muñoz (Treasury), Francisco Merino Rábago (Agriculture), Pedro Ojeda Paullada (Labor), and Jorge Castañeda (Foreign Relations). In favor of entry were Jorge de la Vega Domínguez (Commerce), Miguel de la Madrid Hurtado (Planning), and Enrique Olivares Santana (Interior); see *Latin America Weekly Report,* March 28, 1980, p. 4.

102. *Latin America Economic Report,* February 17, 1978, p. 52.

103. *Latin America Weekly Report,* December 14, 1979, p. 77.

104. *Latin America Economic Report,* June 22, 1977, p. 190.

105. *Excelsior,* March 16, 1979, pp. 1-A, 8-A.

106. *Daily Report,* April 17, 1980, p. M-1.

107. *Latin America Regional Reports: Mexico & Central America,* January 11, 1980, p. 6.

108. *Proceso* published a draft of the program on October 29, 1979, pp. 6–12; see also *Latin America Weekly Report,* December 14, 1979, pp. 76–77.

109. *Comercio exterior de México,* 25 (June 1979):208.

110. *Proceso,* October 29, 1979, p. 10.

111. *Latin America Weekly Report,* March 28, 1980, p. 4.

112. *Comercio exterior de México,* 25 (June 1979):209.

113. *El Universal,* March 28, 1979; quoted in ibid., p. 213.

Chapter 6: Mexico and the Organization of Petroleum-Exporting Countries

1. For a discussion of the Mossadeq period, see Peter Avery, *Modern Iran* (New York: Praeger, 1965), pp. 416–39.

2. David Wise and Thomas B. Ross, *The Invisible Government* (New York: Random House, 1964), p. 110; and the *New York Times,* March 6, 1967, p. 33.

3. Anthony Sampson, *The Seven Sisters* (New York: Viking Press, 1975), p. 156.

4. Tugwell, *Politics of Oil in Venezuela,* p. 61.

5. Zuhayr Mikdashi, *The Community of Oil Exporting Countries* (Ithaca: Cornell University Press, 1972), p. 33.

6. Tugwell, *Politics of Oil in Venezuela,* p. 62.

7. Although the term *cartel* is widely applied, OPEC does not, according to the textbook definition of such an association, regulate the purchasing, production, or marketing of goods by its

members. In recent years, it has acted as a consultative group for setting reasonable minimum prices.

8. Dankwart A. Rustow and John F. Mugno, *OPEC: Success and Prospects* (New York: New York University Press, 1976), p. 21.

9. Edward J. Williams, "Mexico, Oil and OPEC," *Latin American Digest,* 11 and 12 (Fall–Winter 1977–1978):6.

10. Ibid., p. 4.

11. Ibid., p. 5.

12. *Excelsior*, September 10, 1977, p. 4-A.

13. Jorge Díaz Serrano, quoted in *Daily Report,* June 20, 1978, p. M-1.

14. See *Daily Report,* March 22, 1979, p. M-1; and *Excelsior*, June 13, 1978, p. 1-A.

15. Nathan Warman, subsecretary of patrimony, quoted in *Excelsior*, November 3, 1977, p. 5-A; and *Excelsior*, August 30, 1979, p. 17-A.

16. *Washington Post,* December 23, 1979, p. A-18.

17. Ibid.

18. *Proceso*, February 12, 1979, p. 9.

19. *Washington Post,* July 1, 1979, p. A-30.

20. *Bank of London & South America Review,* 11 (January 1977):3.

21. *Proceso*, January 1, 1979, p. 14.

22. *Excelsior*, May 25, 1977.

23. *Daily Report,* April 24, 1978, p. M-1.

24. Ibid., April 24, 1979, p. M-2.

25. Ibid., November 30, 1979, p. M-1, and December 5, 1979, p. M-1.

26. Ibid., January 15, 1979, p. M-3.

27. *Excelsior*, December 3, 1977, p. 1-A.

28. *New York Times,* November 30, 1975, p. 24.

29. Ibid., January 28, 1976, p. 2.

30. Ibid., December 13, 1975, p. 12.

31. Ibid., December 6, 1975, p. 10, and December 13, 1975, p. 12.

32. Samuel I. del Villar in *Excelsior,* quoted by the *New York Times,* December 30, 1975, p. 2; cited in the *Washington Post,* December 19, 1975, p. A-16.

33. *New York Times,* January 1, 1976, p. 2.

34. Ibid., January 28, 1976, p. 2.

35. Ibid., March 3, 1976, p. 3.

36. Ibid., February 16, 1977, p. 14.

37. *Facts on File,* July 9, 1977, p. 528.

38. *Daily Report,* November 28, 1977, pp. M-1, M-2.

39. *Military Balance 1977–1978* (London: International Institute for Strategic Studies, 1977), p. 101.

40. *Proceso*, March 10, 1980, pp. 23–24.

41. For an analysis of commercial ties, see "Mexican-Israeli Trade Relations," *Comercio exterior de México,* 25 (January 1979):28–33.

42. *Petroleum Economist,* 47 (January 1980):19.

43. *Events,* September 8, 1978, p. 37.

44. Ibid.

45. Ibid.

46. *Excelsior,* July 28, 1978, p. 1-A.

47. *Daily Report,* January 15, 1979, p. M-2.

48. *Excelsior,* March 16, 1979, pp. 1-A, 17-A.

49. *Proceso*, October 1, 1979, p. 28, and October 8, 1979, pp. 12-14; the text of the speecn appears in *Comercio exterior de México*, 25 (December 1979):445-49.

50. *Proceso*, October 15, 1979, p. 42.

51. Ibid.

52. Williams, "Mexico, Oil and OPEC," p. 6.

53. Carlos Andrés Pérez restated this argument in his last news conference as president of Venezuela; see *Excelsior*, March 11, 1979, p. 1-A.

54. *Daily Report,* September 6, 1978, p. M-6.

55. Ibid.

Chapter 7: Oil and U.S.-Mexican Relations

1. Olga Pellicer de Brody, *México y la revolución cubana* (Mexico: El Colegio de México, 1972), p. 124.

2. Karl M. Schmitt, *Mexico and the United States, 1828-1973: Conflict and Coexistence* (New York: John Wiley, 1974), p. x.

3. Exceptions to this policy have been Franco's Spain, from which thousands of refugees fled to Mexico, and Pinochet's Chile. After a May 20, 1979, meeting with President Rodrigo Carazo of Costa Rica, López Portillo severed diplomatic relations with the Somoza regime in Nicaragua.

4. George W. Grayson, "Mexico's Foreign Policy," *Current History,* 72 (March 1977):98.

5. Mario Ojeda, "Mexico ante los Estados Unidos en la coyuntura actual," *Foro Internacional,* 18 (1977):33-34.

6. *Daily Report,* September 9, 1977, p. M-16.

7. *Excelsior*, October 26, 1977, p. 1-A.

8. David Ronfeldt and Caesar Sereseres, "Immigration Issues Affecting U.S.-Mexican Relations," Paper presented to the Brookings Institution-El Colegio de México Symposium on Structural Factors Contributing to Current Patterns of Migration in Mexico and the Caribbean Basin, Washington, D.C., June 28-30, 1978, p. 9.

9. Ibid.

10. Ibid., p.18.

11. *Facts on File,* February 26, 1977, p. 136.

12. Involved in this working group on the U.S. side were the departments of State, Justice, and Health, Education and Welfare as well as the Environmental Protection Agency, the International Communication Agency, and the Southwest Boundary Development Commission; on the Mexican side were the Office of the Attorney General and the ministries of Foreign Relations, Labor, and Health and Environment.

13. *New York Times,* November 7, 1978, p. A-22.

14. Ibid., February 16, 1979, p. A-16.

15. *Excelsior*, March 15, 1979, p. 1-A.

16. *New York Times,* February 19, 1979, p. A-5.

17. Excerpts from the presidents' joint communiqué appeared in the *New York Times,* February 17, 1979, p. A-5.

18. March 24, 1979, p. 2.

19. *Washington Post,* February 14, 1979, p. D-8.

20. *Excelsior,* November 4, 1977, p. 17-A.

21. U.S. Congress, *Mexico's Oil and Gas Policy,* p. vii.

22. Ibid., p. x.

23. *Washington Post,* January 1, 1979, p. A-3.

24. Ibid., January 11, 1979, p. 3.

25. U.S. Congress, *Mexico's Oil and Gas Policy,* p. vii.

26. The activities of the congressional interchange are described in U.S. Congress, Senate, *The Mexico–United States Interparliamentary Group: A Sixteen-Year History,* Report prepared by Mike Mansfield, Majority Leader, U.S. Senate, for the Committee on Foreign Relations, 94th Cong., 2nd sess. (Washington, D.C.: Government Printing Office, 1976).

27. *Proceso,* March 19, 1979, p. 11.

28. Ibid.

29. *Washington Post,* February 7, 1979, p. A-17.

30. For articles on an "Energy Common Market," see *Fortune,* September 10, 1979, pp. 118–21; and *Business Week,* August 20, 1979, pp. 42–43.

31. *Excelsior,* May 20, 1979, pp. A-1, A-8.

32. *Facts on File,* June 11, 1977, p. 438, and December 17, 1977, p. 967.

33. *Business Week,* May 29, 1978, p. 50.

34. Ibid.

35. October 19, 1978, p. 22.

36. Wolf Grabendorff, "Review Essay: Mexico's Foreign Policy—Indeed a Foreign Policy?" *Journal of Inter-American Studies and World Affairs,* 20 (February 1978):88.

37. Pellicer de Brody, *México y la revolución cubana,* p. 124.

38. *Latin America Political Report,* September 9, 1977, p. 273.

39. *Daily Report,* October 5, 1979, pp. M-1, M-2.

40. *Excelsior,* August 30, 1979, p. 17-A.

41. *Daily Report,* November 19, 1979, p. M-3.

42. *Latin America Economic Report,* November 10, 1978, p. 346.

43. *Daily Report,* August 14, 1979, p. M-1.

44. *Excelsior,* November 25, 1979, pp. 5-A, 28-A.

45. *Daily Report,* February 3, 1978, p. M-1.

46. *Petroleum Intelligence Weekly,* December 10, 1979, p. 10.

47. Ibid., March 1, 1979, p. M-1.

48. *Latin America Weekly Review,* May 16, 1980, p. 6.

49. *Washington Post,* January 14, 1979, p. K-2.

50. *Excelsior,* May 29, 1977, p. 34-A.

51. *Daily Report,* November 16, 1979, p. D-1.

52. *Petroleum Intelligence Weekly,* August 27, 1979, p. 11.

53. *Excelsior,* May 21, 1979, pp. 15-A, 21-A; and *Comercio exterior de México,* 26 (January 1980):10.

54. *Daily Report,* January 25, 1978, p. M-1.

55. In 1975 Venezuela established a $500-million aid program whereby less-developed Central American and Caribbean nations received a credit of $6 on each barrel purchased. This credit, which remained in the purchaser's central bank, was repayable over a six-year period. The funds could be reborrowed as twenty-five-year development loans. Venezuela supplied about 156,000 bpd under this arrangement in 1979 to Panama (18,000), Costa Rica (8,000), Nicaragua (15,000), Honduras (8,000), El Salvador (18,000), Guatemala (17,000), Dominican Republic (30,000), Jamaica (28,000), and smaller islands (14,000); see *Petroleum Intelligence Weekly,* December 31, 1979, p. 9.

56. *Daily Report,* April 5, 1979, p. M-1.

57. *Excelsior,* January 25, 1980, pp. A-1, A-10.

58. *Proceso,* February 11, 1980, p. 26; and *Petroleum Intelligence Weekly,* February 18, 1980, p. 11.

Chapter 8: Mexico and the United States: The Natural Gas Controversy

1. For analyses of the legislation, see *Oil & Gas Journal,* August 28, 1978, pp. 31–34; the *Washington Post,* July 30, 1978, p. A-14; and the *Wall Street Journal,* October 16, 1978, pp. 2, 33. Portions of this chapter appeared originally in my "Mexico and the United States: The Natural Gas Controversy," *Inter-American Economic Affairs,* 32 (Winter 1978):3–27.

2. The administration began assigning highest priority to its gas bill after the demise of the proposed Crude Oil Equalization Tax became obvious.

3. Lloyd N. Unsell, executive vice-president, Independent Petroleum Association of America, in a letter to the editor, the *Washington Post,* August 19, 1978, p. A-10.

4. Kathleen F. O'Reilly, head of the energy task force of the Consumer Federation of America, ibid., August 17, 1978, p. A-23.

5. Estimate offered by Díaz Serrano and published in the *Daily Report,* October 12, 1977, p. V-5.

6. *Petroleum Intelligence Weekly,* February 4, 1980, p. 11.

7. *Oil & Gas Journal* (December 31, 1979, p. 70) estimated Mexico's gas holdings to be fifty-nine trillion cubic feet as of January 1, 1980.

8. United States, Department of the Interior, "Mexican Oil Potential," Draft paper summarizing the department's understanding of Mexico's hydrocarbon reserves, Washington, D.C., 1980 (mimeo), p. 7. The API designation refers to an index of gravity developed by the American Petroleum Institute.

9. *Daily Report,* August 16, 1978, p. M-1.

10. *Oil & Gas Journal,* September 19, 1977, p. 84.

11. The other 40 percent embraces such compounds as ethane (a key feedstock for petrochemicals), carbon dioxide, hydrogen sulfide, and water; see PEMEX, *Linea troncal de gas natural,* p. 21.

12. Ibid., pp. 22, 23.

13. *Oil & Gas Journal,* September 19, 1977, p. 85.

14. James Schlesinger, "Gas Prospects and Policy," Speech delivered in New York City, January 9, 1979, to the National Association of Petroleum Investment Analysts and the Oil Analysts Group of New York, and distributed by the Department of Energy, Department of Public Affairs; cited in Judith Gentleman, "Nationalism and Dependency: An Analysis of Their Impact in the U.S.-Mexican Gas Negotiations," Paper presented at the annual meeting of the Latin American Studies Association, Pittsburgh, April 5–7, 1979 (mimeo), p. 19.

15. Telephone interview with James A. West, energy consultant (Daytona Beach, Florida, and McLean, Virginia), August 31, 1978.

16. *Latin America Economic Report,* October 21, 1977, p. 180.

17. PEMEX, *Linea troncal de gas natural,* p. 26.

18. Ibid., pp. 31, 27.

19. Ibid., p. 17.

20. *Latin America Economic Report,* October 21, 1977, p. 180. Jesús Chavarria-García, PEMEX's production manager, reported that up to six hundred thousand Mcf per day had been flared between October 1, 1977, and October 1, 1978. He also stated that flaring would be reduced sharply in early 1979; see *Oil & Gas Journal,* October 9, 1978, p. 36.

21. Richard R. Fagen and Henry R. Nau, "Mexican Gas: The Northern Connection," Paper presented at Conference on the United States, U.S. Foreign Policy, and Latin American and Caribbean Regimes, Washington, D.C., March 27–31, 1978 (mimeo), pp. 22–23.

22. *Petroleum Economist,* 45 (February 1978):69, 70.

23. PEMEX, *Linea troncal de gas natural*, p. 28.

24. The bulk of this energy source is shipped under long-term contracts. The rest, sold on the "spot market," is subject to greater price variations than crude because the relatively small volume of sales often takes place under difficult or emergency conditions such as protracted cold spells.

25. Letter from Jack H. Ray, president of the Tennessee Gas Transmission Company, to Sen. Adlai E. Stevenson, III, October 28, 1977, as quoted in Fagen and Nau, "Mexican Gas," p. 33.

26. *Excelsior*, October 28, 1977, p. 12-A.

27. Interview with Jack H. Ray, Houston, May 20, 1980.

28. PEMEX, *Linea troncal de gas natural*, p. 19.

29. Ibid.

30. Fagen and Nau, "Mexican Gas," p. 24.

31. Ibid.

32. In mid-1978 the Government Finance Committee of the Chamber of Deputies placed this figure at $23.8 billion; see *Daily Report*, July 20, 1978, p. M-1.

33. *Excelsior*, September 29, 1977, p. 4-A.

34. Export-Import Bank of the United States, *Eximbank News*, December 16, 1977, n.p.; and *Excelsior*, September 15, 1977, p. A-15.

35. Edward J. Williams, *The Rebirth of the Mexican Petroleum Industry* (Lexington, Mass.: D. C. Heath, 1979), p. 106.

36. *Daily Report*, July 25, 1977, p. M-1.

37. Ibid., December 12, 1977, p. A-1.

38. Williams, *Rebirth of Mexican Petroleum Industry*, p. 106.

39. *Excelsior*, December 8, 1977, p. 1-A.

40. *Proceso*, January 2, 1978, p. 6ff.

41. *Daily Report*, March 21, 1978, p. M-1.

42. *Latin America Economic Report*, October 21, 1977, p. 180.

43. *Excelsior*, September 17, 1977, p. 4-A.

44. Ibid., October 26, 1977, p. 11-A.

45. Williams, *Rebirth of Mexican Petroleum Industry*, pp. 104–05.

46. On May 1, 1979, the Canadian National Energy Board replaced the Saudi charge with a composite price of all imported crudes. Prices have increased from $2.30 (May 1, 1979) to $2.80 (August 11, 1979) to $3.45 (November 4, 1979) to $4.47 (early 1980); see *Petroleum Economist*, 46 (November 1979):480.

47. The bank must formally report to Congress its intention to grant loans of $60 million or more. If the House and Senate have not "dictated otherwise" within twenty-five working days, the pending credit may be granted. To date Congress has never employed this procedure to block a transaction.

48. U.S. Congress, *Congressional Record* (Senate), October 19, 1977, pp. S17370–S17371, S19937.

49. While Senator Stevenson would not have used his legislative clout to block renewal of the bank's charter, he might have sought unwanted amendments. For example, the bank must take into consideration the possible harmful effects on the American economy of projects that it finances. More restrictive language could stipulate that there be no adverse impact. The bank was not pleased with a Stevenson-endorsed 1977 amendment requiring it to consider, in consultation with the secretary of state, the human-rights impact of exports supported by Eximbank loans and financial guarantees.

50. Disbursement of the $250-million credit has begun; the bank's governing board has also given final approval to the $400-million loan, but payouts will not start "until binding contracts

have been executed for the purchase of gas by U.S. natural gas transmission companies." See *Eximbank News,* December 16, 1977, n.p.

51. *Oil & Gas Journal,* September 4, 1978, p. 36.

52. For a chronology of U.S.-Mexican gas discussions, see U.S. Senate, *Mexico: Promise and Problems,* pp. 157–62.

53. Ibid., pp. 157, 158.

54. Telephone interview with Walter J. McDonald, former special assistant secretary of international affairs, Department of Energy, August 13, 1979.

55. *Latin America Economic Report,* January 6, 1977, p. 1.

56. *Washington Post,* February 13, 1979, p. D-7.

57. *Petroleum Economist,* 46 (August 1979):315.

58. *Washington Post,* February 12, 1979, p. A-8.

59. On February 4, 1977, the Federal Power Commission authorized up to forty million cubic feet a day of natural gas imports for sixty days at an approximate cost of $2.25 per Mcf.

60. *Latin America Economic Report,* January 6, 1977, p. 1.

61. *Daily Report,* January 12, 1979, p. M-3.

62. *Excelsior,* January 7, 1978, p. 10-A. It should be noted that Canadian producers had so much excess gas in later 1978 that they sought permission for increased exports to the United States; see *Oil & Gas Journal,* July 10, 1978, p. 34.

63. Interview with Jack H. Ray, May 20, 1980.

64. *Washington Post,* February 13, 1979, p. D-7.

65. Ibid., February 14, 1979, p. A-17.

66. Ibid., February 13, 1979, p. A-17.

67. Gentleman, "Nationalism and Dependency," p. 20.

68. Petróleos Mexicanos, *Informe del director general de Petróleos Mexicanos, 18 de marzo de 1978* (Mexico City: Petróleos Mexicanos, 1978), p. 21.

69. Steven Bosworth, deputy assistant secretary of state for international resources and food policy; cited in U.S. Senate, *Mexico: Promise and Problems,* p. 41.

70. Ibid., p. 36.

71. Gentleman, "Nationalism and Dependency," p. 20.

72. Petróleos Mexicanos, *Report of the Director General, Petróleos Mexicanos, March 18, 1979* (Mexico City: Petróleos Mexicanos, 1979), p. 11; *Oil & Gas Journal,* May 12, 1980, p. 5.

73. Gentleman, "Nationalism and Dependency," p. 20.

74. *Washington Post,* December 15, 1979, p. 1.

75. "Listen Yank! Mexico is a Nation, Not an Oil Well," in ibid., February 11, 1979, p. L-1.

76. Joseph Kraft, "A Reporter at Large: The Mexican Oil Puzzle," *New Yorker,* October 15, 1979, p. 174.

77. Cited in ibid., p. 175.

78. Ibid., p. 176.

79. Ibid.

80. U.S. Congress, *Mexico's Oil and Gas Policy,* p. 49; and *Washington Post,* August 6, 1979, p. A-20.

81. Kraft, "Mexican Oil Puzzle," pp. 176–77.

82. Ibid., pp. 177–78.

83. *Petroleum Intelligence Weekly,* October 1, 1979, p. 6.

84. The price for a given quarter will be the product of the base price and the crude price average for that quarter divided by the average for the previous quarter; see *Petroleum Economist,* 47 (January 1980):19–20.

85. *Petroleum Intelligence Weekly,* October 1, 1979, p. 6.

86. *Latin America Economic Report,* August 17, 1979, p. 252.

87. Ibid., September 28, 1979, p. 297.

88. Interview with Jack H. Ray, May 20, 1980.

89. "Take-or-pay" obligates the buyer to purchase the gas even if the domestic market is saturated; "take-and-pay" entails no such commitment.

Chapter 9: The Ixtoc 1 Blowout

1. *Oil & Gas Journal,* November 5, 1979 ("Newsletter"), and November 19, 1979, p. 208.

2. *Petroleum Economist,* 46 (August 1979):334.

3. The following description of the blowout draws heavily on an account presented by Steven Mahood, executive vice president, SEDCO, in U.S. Congress, Senate, *Campeche Oil Spill,* Report prepared for the Committee on Commerce, Science, and Transportation and the Committee on Energy and Natural Resources, 96th Cong., 1st sess. (Washington, D.C.: Government Printing Office, 1980), pp. 67–76.

4. Ibid., p. 70.

5. Ibid.

6. *Excelsior,* June 17, 1979, p. 1.

7. Ibid., March 19, 1980, p. 12-A. GORs in the Ixtoc field are believed to run between 750 and 1,250-to-one. The estimate that three billion cubic feet of gas were lost is based on a GOR of 1,000-to-one. This figure may be conservative in view of the estimate by Professor Jerome Milgram, Massachusetts Institute of Technology, that Ixtoc 1 had a GOR of 2,000-to-one; see U.S. Senate, *Campeche Oil Spill,* p. 89.

8. *Proceso,* September 3, 1979, p. 29.

9. *Oil & Gas Journal,* June 25, 1979, p. 46.

10. *Excelsior,* August 9, 1979, p. 1.

11. Ibid. In mid-August an American diver, Allen Anderson, died while working around the wellhead; see *Daily Report,* August 17, 1979, p. v-1.

12. *New York Times,* October 5, 1979, p. 18.

13. According to unconfirmed reports, Adair originally proposed filling the well with cement. PEMEX is believed to have rejected this strategy because while the action might have blocked its vertical flow, the highly pressurized oil and gas might simply have migrated through the deposit, seeking escape through any and all available apertures. Such releases would have been virtually impossible to control.

14. *Excelsior,* June 13, 1979, p. 15-A.

15. *Offshore,* 39 (August 1979):48; and *Ocean Industry,* 15 (April 1980):12.

16. *Excelsior,* June 12, 1979, p. 10, and July 14, 1979, p. 19.

17. *New York Times,* October 5, 1979, p. 1; *Excelsior,* March 25, 1980, p. 18.

18. U.S. Senate, *Campeche Oil Spill,* p. 4. For a summary of optimistic statements by PEMEX personnel, see *Proceso,* February 11, 1980, pp. 18–20.

19. Ibid., pp. 125–28.

20. One industry publication reported that "as with many exploratory holes—Ixtoc 1 downhole surveys provided only angles of deflection without direction; hence its coordinates could not be precisely located other than as being within a circle generated by angular deflection"; see *Ocean Industry,* 15 (April 1980):13.

21. Ibid.; and *Offshore,* 39 (August 1979):48.

22. *Excelsior,* July 11, 1979, p. 1.

23. This plan was developed in response to the mandate of section 311 of the Federal Water Pollution Control Act.

24. U.S. Senate, *Campeche Oil Spill,* p. 14.

25. *Excelsior,* August 2, 1979, p. 10.

26. *New York Times,* October 5, 1979, p. 18.

27. Ibid., October 5, 1979, p. 18; and *Excelsior,* October 13, 1979, p. 19-A.

28. *Oil & Gas Journal,* September 3, 1979, p. 33.

29. Ibid., December 17, 1979, p. 37.

30. *New York Times,* October 5, 1979, p. 18.

31. U.S. Senate, *Campeche Oil Spill,* p. 90.

32. Ibid., p. 89.

33. *Excelsior,* September 17, 1979, pp. 1, 20.

34. *Ocean Industry,* 15 (April 1980):13.

35. *Excelsior,* June 11, 1979, p. 10.

36. Ibid., June 16, 1979, p. 4. On March 18, 1980, Díaz Serrano stated that 50 percent of the oil and gas was burned off, 17 percent had evaporated on the surface of the sea, 4.5 percent was collected, and 28.5 percent was dispersed by chemicals; see *Excelsior,* March 25, 1980.

37. Ibid., June 17, 1979, p. 11.

38. Ibid., June 27, 1979, p. 14.

39. *Proceso,* July 30, 1979, p. 12.

40. *Excelsior,* August 6, 1979, p. 14.

41. Ibid., August 10, 1979, p. 11.

42. *Daily Report,* July 23, 1979, p. M-1.

43. *Proceso,* October 15, 1979, p. 34.

44. Ibid.

45. *Excelsior,* August 8, 1979, p. 1.

46. *New York Times,* July 22, 1979, p. 19.

47. For a day-by-day account of these activities, see U.S. Fish and Wildlife Service, "Ixtoc 1 Oil Spill," Summary report of response activities, June 3, 1979 to November 30, 1979, prepared by Charlie Sanchez, Jr., regional pollution response coordinator, Albuquerque, New Mexico, 1980 (mimeo).

48. *New York Times,* August 5, 1979, p. 26.

49. Ibid., August 17, 1979, p. A-10.

50. U.S. Senate, *Campeche Oil Spill,* pp. 20-21; and *New York Times,* August 8, 1979, p. A-10.

51. Ibid., August 21, 1979, p. A-17.

52. Ibid., August 17, 1979, p. A-10.

53. Documents related to this case (C.A. No. H-79-1880) and on which this section relies are filed in the office of the clerk for the United States District Court for the Southern District of Texas, Houston Division, Federal Office Building, Houston, Texas.

54. *New York Times,* September 23, 1979, p. 31.

55. Ibid.

56. Cited in U.S. Congress, Senate and House, "On Liability for the Mexican Oil Spill," Report prepared by the Congressional Research Service, Library of Congress, published in *Inter-American Economic Affairs,* 33 (Autumn 1979):94.

57. Ibid.

58. Cited in James Barros and Douglas M. Johnston, *The International Law of Pollution* (New York: Free Press, 1974), p. 76.

59. Ibid., p. 301.

60. *Proceso*, September 3, 1979, p. 9.

61. G. Pope Atkins, *Latin America in the International System* (New York: Free Press, 1977), p. 212.

62. *Proceso*, September 3, 1979, p. 12.

63. Ibid., pp. 12–13.

64. *Excelsior*, August 11, 1979, p. A-20.

65. *New York Times*, August 26, 1979, p. 13.

66. *Washington Star*, September 30, 1979, p. A-3.

67. Ibid.

68. *Excelsior*, October 1, 1979, p. 13.

69. Ibid., July 26, 1979, pp. A-1, A-15, A-18.

70. *Offshore*, 39 (May 1979):135–36.

71. *Proceso*, September 24, 1979, p. 10. On March 18, 1980, Díaz Serrano observed that the costs of the Ixtoc spill could be recouped within ten days from earnings derived from Campeche's production; see *Excelsior*, March 19, 1980, p. 12-A.

72. *Excelsior*, September 21, 1979, p. A-6.

Chapter 10: The Prospects for Mexican Oil

1. George W. Grayson, "Oil Rich Mexico: The New Saudi Arabia?" *Washington Post*, October 27, 1978, p. A-17.

2. Central Intelligence Agency, "The World Oil Market in the Years Ahead," Research paper prepared by the CIA's National Foreign Assessment Center, ER79-10327U, August 1979, p. 45.

3. Ibid., p. 46.

4. Robert Stombaugh and Daniel Yergin, eds., *Energy Future* (New York: Random House, 1979), p. 31.

5. Ibid., p. 37.

6. Organization for Economic Cooperation and Development, *World Energy Outlook* (Paris: OECD, 1977), p. 37; and John R. Brodman and Richard E. Hamilton, *A Comparison of Energy Projections to 1985*, International Energy Agency Monograph (Paris: OECD, 1979), p. 7.

7. Stombaugh and Yergin, *Energy Future*, p. 50.

8. Telephone interview with Ruth Blevins, public affairs officer, Defense Fuel Supply Center, Defense Logistics Agency, Department of Defense, Alexandria, Virginia, January 8, 1980.

9. U.S. Congress, *Mexico's Oil and Gas Policy*, p. 25.

10. September 26, 1979; and cited in *Facts on File*, October 19, 1979, p. 781.

11. Stombaugh and Yergin, *Energy Future*, p. 5.

12. This point is developed in Mancke, *Mexican Oil*, p. 128ff.

13. Christopher Buckley, "Mexico's Oil Boom," p. 47.

14. Mancke, *Mexican Oil*, p. 138.

15. Peter R. Huessy and Gerda Bikales, "Trading Oil for Immigrants," *Washington Post*, January 30, 1979, p. A-17.

16. Stombaugh and Yergin, *Energy Future*, pp. 11–12.

17. George W. Grayson, "The Promise of Mexican Oil," *Newsday*, September 11, 1978.

18. Wayne A. Cornelius, Jr., "Urbanization as an Agent in Latin American Political Instability: The Case of Mexico," *American Political Science Review*, 63 (September 1969):852.

19. "Mexico: Crisis of Poverty," *Los Angeles Times* supp., July 15, 1979, p. 12.

Bibliography

Books, Chapters in Books, and Monographs

Adams, Samuel Hopkins. *Incredible Era: The Life and Times of Warren Gamaliel Harding*. Boston: Houghton Mifflin, 1939.

Alemán Valdés, Miguel. *La verdad del petróleo en México*. Mexico City: Editorial Grijalbo, 1977.

Alonso González, Francisco. *Historia y petróleo, México: El problema del petróleo*. Madrid: Editorial Ayuso, 1972.

Ashby, Joe C. *Organized Labor and the Mexican Revolution Under Lázaro Cárdenas*. Chapel Hill: University of North Carolina Press, 1967.

Atkins, G. Pope. *Latin America in the International System*. New York, Free Press, 1977.

Avery, Peter. *Modern Iran*. New York: Praeger, 1965.

Bailey, Thomas A. *A Diplomatic History of the American People*. New York: Appleton-Century-Crofts, 1964.

Baker, Ron. *A Primer of Oil-Well Drilling*. 4th ed. Austin: Petroleum Extension Service of the University of Texas, 1979.

Ball, George W. *Diplomacy for a Crowded World*. Boston: Little, Brown, 1976.

Barros, James, and Johnston, Douglas M. *The International Law of Pollution*. New York: Free Press, 1974.

Beals, Carleton. *Porfirio Díaz: Dictator of Mexico*. Westport, Conn.: Greenwood Press, 1971.

Bermúdez, Antonio J. *The Mexican National Petroleum Industry: A Case Study in Nationalization*. Palo Alto, Calif.: Institute of Hispanic American and Luso-Brazilian Studies, Stanford University, 1963.

——. *La política petrolera mexicana*. Mexico City: Cuadernos de Joaquín Mortiz, 1976.

Bond, Robert D., ed. *Contemporary Venezuela and Its Role in International Affairs*. New York: New York University Press, 1977.

Brandenburg, Frank. *The Making of Modern Mexico*. Englewood Cliffs, N.J.: Prentice-Hall, 1967.

Briseño, Jaime Aguilar. *"La Quina."* Privately printed, n.d.

Brodman, John R., and Hamilton, Richard E. *A Comparison of Energy Projections to 1985*. International Energy Agency Monograph. Paris: OECD, 1979.

Cline, Howard F. *The United States and Mexico*. New York: Atheneum, 1965.

Cosio Villegas, Daniel. *Historical moderna de México*. 7 vols. Mexico City: Editorial Hermes, 1965.

Cumberland, Charles C. *Mexican Revolution: The Constitutionalist Years*. Austin: University of Texas Press, 1972.

Daniels, Josephus. *Shirt-Sleeve Diplomat*. Chapel Hill: University of North Carolina Press, 1947.

Grindle, Merilee Serrill. *Bureaucrats, Politicians, and Peasants in Mexico: A Case Study in Public Policy*. Berkeley and Los Angeles: University of California Press, 1977.

Gruenig, Ernest. *Mexico and Its Heritage*. New York: Appleton-Century, 1934.

Hansen, Roger D. *The Politics of Mexican Development*. Baltimore and London: Johns Hopkins University Press, 1971.

Hargrove, John Lawrence, ed. *Who Protects the Ocean?* St. Paul, Minn.: West, 1975.

Hellman, Judith Adler. *Mexico in Crisis*. New York: Holmes & Meier, 1978.

Hobson, G. D., and Pohl, W., eds. *Modern Petroleum Technology*. New York and Toronto: John Wiley, 1973.

Kaufman, Edy, et al. *Israeli-Latin American Relations*. New Brunswick, N.J.: Transaction Books, 1979.

Lavín, José Domingo. *Petróleo: Pasado, presente y futuro, de una industria mexicana*. Mexico: Fondo de Cultura Económica, 1976.

Lerner, Ira T. *Mexican Jewry in the Land of the Aztecs*. Mexico City: B. Costa-AMIC, 1973.

López Portillo y Weber, José. *El petróleo de México: Su importancia, sus problemas*. Mexico City: Fondo de Cultura Económica, 1975.

Mancke, Richard B. *Mexican Oil and Natural Gas: Political, Strategic, and Economic Implications*. New York: Praeger, 1979.

Meyer, Lorenzo. *Mexico and the United States in the Oil Controversy, 1917–1942*. Austin: University of Texas Press, 1977.

Mikdashi, Zuhayr. *The Community of Oil Exporting Countries*. Ithaca: Cornell University Press, 1972.

Noggle, Burt. *Teapot Dome: Oil and Politics in the 1920's*. Baton Rouge, La.: Louisiana State University Press, 1962.

O'Keane, Josephine. *Thomas J. Walsh: A Senator from Montana*. Francistown, N.H.: Marshall Jones, 1955.

Organization for Economic Cooperation and Development. *World Energy Outlook*. Paris: OECD, 1977.

Owen, Edgar Wesley. *Trek of the Oil Finders: A History of Exploration for Petroleum*. Tulsa: American Association of Petroleum Geologists, 1975.

Padgett, L. Vincent. *The Mexican Political System*. Boston: Houghton Mifflin, 1976.

Parkes, Henry Bamford. *A History of Mexico*. Boston: Houghton Mifflin, 1960.

Pazos, Luis. *Mitos y realidades del petróleo mexicano*. Mexico: Editorial Diana, 1979.

Pellicer de Brody, Olga. *México y la revolución cubana*. Mexico: El Colegio de México, 1972.

———. "Mexico in the 1970s and Its Relations with the United States." In *Latin America and the United States: The Changing Political Realities*, ed. by Julio Cotler and Richard R. Fagen. Stanford: Stanford University Press, 1974.

Petroleum Economist. Latin America and Caribbean Oil Report. London: Woodcote Press, 1979.

Potter, Jeffrey. *Disaster by Oil*. New York: Macmillan, 1973.

Powell, J. Richard. *The Mexican Petroleum Industry: 1938–1950*. New York: Russell & Russell, 1972.

Ramirez Heredia, Rafael. *La otra cara del petróleo*. Mexico City: Editorial Diana, 1979.

Rippy, J. Fred. *British Investment in Latin America, 1822–1949*. Minneapolis: University of Minnesota Press, 1959.

Rippy, Merrill. *Oil and the Mexican Revolution*. Leiden: Brill, 1972.

Rodríguez, Antonio. *El rescate del petróleo*. Mexico City: Ediciones El Caballito, 1975.

Ross, Stanley R., ed. *Views Across the Border: The United States and Mexico*. Albuquerque: University of New Mexico Press, 1978.

Rouhani, Fuad. *A History of O.P.E.C.* New York: Praeger, 1971.

Rustow, Dankwart A., and Mugno, John F. *OPEC: Success and Prospects*. New York: Council on Foreign Relations, New York University Press, 1976.

Sampson, Anthony. *The Seven Sisters.* New York: Viking Press, 1975.

Schmitt, Karl M. *Mexico and the United States, 1828–1973: Conflict and Coexistence.* New York: John Wiley, 1974.

Scott, Robert E. *Mexican Government in Transition.* Urbana, Ill.: University of Illinois Press, 1964.

Silvia Herzog, Jesús. *Historia de la expropriación de las empresas petroleras.* Mexico: Instituto Mexicano de Investigaciones Económicas, 1964.

Sindicato de Trabajadores Petroleros de la República Mexicana. *Acta constitutiva y estatudos generales.* 1977.

Stombaugh, Robert, and Yergin, Daniel, eds. *Energy Future.* New York: Random House, 1979.

Tiratsoo, Eric N. *Oilfields of the World.* Houston: Gulf, 1976.

Townsend, William Cameron. *Lázaro Cárdenas, Mexican Democrat.* Ann Arbor, Mich.: George Wahr, 1952.

Tugwell, Franklin. *The Politics of Oil in Venezuela.* Stanford: Stanford University Press, 1975.

Vargas MacDonald, Antonio. *Hacia una nueva política petrolera.* Mexico: Editorial Promoción, 1959.

Vernon, Raymond. *The Dilemma of Mexican Development.* Cambridge, Mass.: Harvard University Press, 1963.

Weyl, Nathaniel, and Weyl, Sylvia. *The Reconquest of Mexico: The Years of Lázaro Cárdenas.* New York: Oxford University Press, 1939.

Wilber, Donald N. *Contemporary Iran.* New York: Praeger, 1963.

Wilkins, Mira. *The Emergence of Multinational Enterprise: American Business Abroad from the Colonial Era to 1914.* Cambridge, Mass.: Harvard University Press, 1970.

Williams, Edward J. *The Rebirth of the Mexican Petroleum Industry.* Lexington, Mass.: D. C. Heath, 1979.

Williams, Howard R., and Meyers, Charles J. *Oil and Gas Terms.* 4th ed. New York: Bender, 1976.

Wise, David, and Ross, Thomas B. *The Invisible Government.* New York: Random House, 1964.

Government Documents

Mexico. Banco de México, S.A. *Informe anual, 1977.* Mexico City, 1978.

———. Consejo Nacional de Población. *La revolución demográfica.* Mexico, 1975.

———. Secretaria del Patrimonio Nacional. *El petróleo de México.* Mexico, 1963.

———. Secretaria de Programación y Presupuesto. *Trabajo y salarios industriales, 1975.* Mexico City: Dirección General de Estadística, 1977.

Petróleos Mexicanos. *Anuario estadístico.* Mexico City: Petróleos Mexicanos, 1975–1978.

———. *Contrato colectivo de trabajo.* Celebrado entre Petróleos Mexicanos y el Sindicato de Trabajadores Petroleros de la República Mexicana. Mexico, 1977.

———. *Informe del director general* [Report of the Director General]. Mexico City: Petróleos Mexicanos, 1942–1980.

———. *Linea troncal nacional de distribución de gas natural. El director general de Petróleos Mexicanos ante la H. Cámara de Diputados.* Mexico City: Petróleos Mexicanos, 1977.

———. *Memoria de labores.* Mexico City: Petróleos Mexicanos, 1976–1979.

———. *Rescisión de los contratos con Sharmex, Isthmus y Pauley Noreste.* Mexico City: Petróleos Mexicanos, 1970.

U.S. Arms Control and Disarmament Agency. *World Military Expenditures and Arms Transfers 1967–76.* Pub. 98 (1978).

U.S. Central Intelligence Agency. *The International Energy Situation: Outlook to 1985.* ER 77-10240 U, April 1977.

———. "The World Oil Market in the Years Ahead." Research paper prepared by the CIA's National Foreign Assessment Center, ER79-10327U, August 1979.

U.S. Congress. Senate. *Campeche Oil Spill.* Report prepared for Committee on Commerce, Science, and Transportation and Committee on Energy and Natural Resources. 96th Cong., 1st sess., 1979, Serial No. 96-66.

———. Senate. *Mexico: The Promise and Problems of Petroleum.* Report prepared for Committee on Energy and Natural Resources. 96th Cong., 1st. sess., 1979, Pub. No. 96-2.

———. Senate. *The Mexico-United States Interparliamentary Group: A Sixteen Year History.* Report prepared by Mike Mansfield, Majority Leader, U.S. Senate, for the Committee on Foreign Relations. 94th Cong., 2nd sess. Washington, D.C.: Government Printing Office, 1976.

———. Senate and House. *Mexico's Oil and Gas Policy: An Analysis.* Report prepared by the Congressional Research Service, Library of Congress, for the Committee on Foreign Relations and the Joint Economic Committee. 95th Cong., 2nd sess. Washington, D.C.: Government Printing Office, 1978.

———. "On Liability for the Mexican Oil Spill." Report prepared by the Congressional Research Service, Library of Congress, published in *Inter-American Economic Affairs,* 33 (Autumn 1979):93–96.

U.S. Export-Import Bank. *Eximbank News.* December 16, 1977.

U.S. Fish and Wildlife Service. "Ixtoc 1 Oil Spill." Summary report of response activities, June 3, 1979, to November 30, 1979. Prepared by Charlie Sanchez, Jr., regional pollution response coordinator. Albuquerque, N.M. 1980 (mimeo).

U.S. Foreign Broadcast Information Service. *Daily Report* (Latin America). 1976–April 1980.

Magazine, Newspaper, and Journal Articles

Bank of London & South America Review. 1970–1979.

Berryman, Phillip. "Ecuador's Oil: No Bonanza for the People." *Christian Century,* June 11, 1975, pp. 604–05.

Buckley, Christopher. "Mexico's Oil Boom and What's In It for Us." *Esquire,* Dec. 19, 1978, pp. 44–52.

Business Week. 1978.

Christiano, Michael. " 'A Tidal Wave' of Illegal Aliens." Letter to the editor. *Washington Post,* December 28, 1978, p. A-22.

Comercio exterior de México. 1976–1979.

Cornelius, Wayne A., Jr. "Urbanization as an Agent in Latin American Political Instability: The Case of Mexico." *American Political Science Review,* 63 (September 1969):833–57.

Daily Press (Newport News, Va.). 1977–1980.

Díaz Serrano, Jorge. "¿En que consiste una reserva petrolera?" *Ciencia y desarrollo,* No. 13 (March–April 1977):5–9.

Dodge, Stephen C. "Venezuela's Bright Future." *Current History,* 70 (February 1976):65.

Economist (London). 1976–1979.

Events (London). September 8, 1978, p. 37.

Excelsior. 1970–1979.

Facts on File. 1947–1979.

Felix, David. "Income Inequality in Mexico." *Current History,* 72 (March 1977):112.

Financial Times (London). April 10, 1979, p. 4.

Gallegos, Alfredo, et al. "Recent Trends in Contraceptive Use in Mexico." *Studies in Family Planning*, 8 (August 1977).

Gollás, Manuel. "El desempleo en México: Soluciones posibles." *Ciencia y desarrollo*, No. 14 (May–June 1978).

Gordon, David. "Mexico: A Survey." *Economist*, April 22, 1978.

Grabendorff, Wolf. "Review Essay: Mexico's Foreign Policy—Indeed a Foreign Policy?" *Journal of Inter-American Studies and World Affairs*, 20 (February 1978):85–92.

Grayson, George W. "Mexico's Foreign Policy." *Current History* 72 (March 1977):97–101, 134–35.

———. "Will Mexico's Oil Light the Lamps of Her Poor?" *Washington Post*, August 13, 1977, p. A-15.

———. "Mexico's Opportunity: The Oil Boom." *Foreign Policy*, No. 29 (Winter 1977–1978):65–89.

———. "The Deft Hand of López Portillo." *Washington Post*, July 12, 1978, p. A-19.

———. "In Mexico, Black Gold for a Green Revolution." *Washington Post*, August 19, 1978, p. A-11.

———. "The Promise of Mexican Oil." *Newsday*, Sept. 11, 1978.

———. "Oil-Rich Mexico: The New Saudi Arabia?" *Washington Post*, October 27, 1978, p. A-17.

———. "Mexico and the United States: The Natural Gas Controversy." *Inter-American Economic Affairs*, 32 (Winter 1978):3–27.

———. "Mexico's Oil—No Reason to Tolerate Illegal Aliens." *Christian Science Monitor*, January 3, 1979, p. 22.

———. "PEMEX: Threat to Mexico's Environment." *Washington Post*, June 25, 1979, p. A-19.

———. "Mexico: Overestimating the Oil Cornucopia." *Christian Science Monitor*, August 8, 1979, p. 22.

———. "Oil and U.S.-Mexican Relations." *Journal of Interamerican Studies and World Affairs*, 21 (November 1979):427–56.

———. "Oil and Politics in Mexico." *Current History*, 78 (February 1980):53–56, 83.

Hickey, John. "Pemex: A Study in Public Policy." *Inter-American Economic Affairs*, 14 (Autumn 1960):73.

Huessy, Peter R., and Bikales, Gerda. "Trading Oil for Immigrants." *Washington Post*, Jan. 30, 1979, p. A-17.

Kraft, Joseph. "A Reporter at Large: The Mexican Oil Puzzle." *New Yorker*, October 15, 1979, pp. 150–81.

Latin America Economic Report. 1977–1979.

Latin America Political Report. 1977–1979.

Latin America Regional Reports: Andean Group. February 29, 1980.

Latin America Regional Reports: Mexico & Central America. January 11, 1980, and March 21, 1980.

Latin America Weekly Report. 1979–April 1980.

Los Angeles Times. 1977–1979.

MacMahon, Arthur W., and Dittmar, W. R. "The Mexican Oil Industry Since Expropriation." Part I. *Political Science Quarterly*, 57 (March 1942):28–50.

Metz, William D. "Mexico: The Premier Oil Discovery in the Western Hemisphere." *Science*, December 22, 1978, pp. 1261–65.

El Nacional (Caracas). March 26, 1980.

Near East Report. 1978–1979.

New York Times. 1940–1979.

News (Mexico City). January 12–31, 1959.

Newsweek. 1977–1979.

Ocean Industry. May 1978, July 1979, November 1979, and April 1980.

Offshore. January 1978, June 20, 1979, and August 1979.

Oil & Gas Journal. 1970–April 1980.

"Oil of Olé!" *New Republic*. August 19, 1978, pp. 5, 6, and 8.

Ojeda, Mario. "Mexico ante los Estados Unidos en la coyuntura actual." *Foro internacional*, 18 (1977).

Petroleum Economist. 1977–March 1980.

Petroleum Intelligence Weekly. 1979–April 1980.

Proceso. 1976–May 1980.

Purcell, Susan Kaufman, and Purcell, John F. H. "State and Society in Mexico: Must a Stable Polity Be Institutionalized?" *World Politics*, 32 (January 1980):194–227.

Quarterly Economic Report (Mexico City). March 1979.

Ranis, Gustav. " ¿Se está tornado amargo el milagro mexicano?" *Demográfia y Economía*, 8 (1974).

Reyes, Saul Trejo. "El desempleo en México: Características generales." *El Trimestre económico*, 42.

St. Petersburg Times. July 27, 1978, p. 14-A.

"Survey of Venezuela." *Economist*. December 27, 1975.

Time. June 16, 1947.

Times-Herald (Newport News, Va.). 1977–1980.

Ultimas noticias. May 27, 1977.

U.S. News & World Report. April 25, 1977, p. 35.

Virginian-Pilot (Norfolk). November 24, 1979, p. A-21.

Wall Street Journal. 1976–1979.

Washington Post. 1975–1979.

Washington Star. September 30, 1979, p. A-3.

Williams, Edward J. "Mexico, Oil and OPEC." *Latin American Digest*, 11 and 12 (Winter 1977–78):4–6.

World Oil. 1977–March 1980.

Encyclopedias

Americana Annual 1979. Danbury, Conn.: Grolier, 1979.

Encyclopedia Yearbook 1980. Danbury, Conn.: Grolier, 1980.

Unpublished Papers and Reports

Adleson, S. Lief. "Coyuntura y conciencia: Factores convergentes en la fundación de los Sindicatos Petroleros de Tampico durante la década de 1920." Paper presented at the Fifth Mexican–North American Historians' Conference. Pátzcuaro, Michoacán, October 1977.

American Federation of Labor and Congress of Industrial Organizations. "Statement of Rudy Oswald, Director, Department of Research, American Federation of Labor and Congress of Industrial Organizations, Before the Senate Committee on Judiciary on S. 2252, 'Alien Adjustment and Employment Act of 1977.' " Washington, D.C., May 17, 1978.

Bailey, John J. "Presidency, Bureaucracy, and Administrative Reform in Mexico: The Secretariat of Programming and Budget." Paper presented at the annual meeting of the American Society for Public Administration. San Francisco, April 14–17, 1980.

Bentsen, Sen. Lloyd. "Survey of President Carter's Illegal Alien Problem." 1978.

Cornelius, Wayne A. "Mexican Migration to the United States: Causes, Consequences, and U.S. Responses." Report Prepared for the Migration and Development Study Group, Massachusetts Institute of Technology. Cambridge, Mass.: Center for International Studies, 1978.

Duff, Ernest. "Mexico's Reforma Política." Paper presented at the first annual meeting of the Middle Atlantic Council on Latin American Studies. University of Delaware, Newark, Delaware, April 17–19, 1980.

Evans, John S., and James, Dilmus D. "Increasing Productive Employment in Mexico." Paper presented at the Brookings Institution–El Colegio de México Symposium on Structural Factors in Migration in Mexico and the Caribbean Basin. Washington, D.C., June 28–30, 1978.

Fagen, Richard R., and Nau, Henry R. "Mexican Gas: The Northern Connection." Paper presented at Conference on the United States, U.S. Foreign Policy, and Latin American and Caribbean Regimes. Washington, D.C., March 27–31, 1978.

Gentleman, Judith. "Nationalism and Dependency: An Analysis of Their Impact in the U.S.-Mexican Gas Negotiations." Paper presented at the annual meeting of the Latin American Studies Association. Pittsburgh, April 5–7, 1979.

Hart, Jeffrey A. "Industrialization and the Fulfillment of Basic Human Needs in Venezuela." Paper presented to an Institute for World Order Colloquium. October 1979 (revised).

National Association for the Advancement of Colored People. "NAACP 68th Annual Convention Resolutions." Adopted at convention held in St. Louis, Missouri, June 27–July 1, 1977.

North, David S. "The Migration Issue in U.S.-Mexican Relations." Paper presented at the second annual symposium on Mexico–U.S. Economic Relations. Xochimilco University, Mexico City, May 23–25, 1979.

Ronfeldt, David, and Sereseres, Caesar. "Immigration Issues Affecting U.S.-Mexican Relations." Paper presented at the Brookings Institution–El Colegio de México Symposium on Structural Factors Contributing to Current Patterns of Migration in Mexico and the Caribbean Basin. Washington, D.C., June 28–30, 1978.

Solís, Leopoldo. "The External Sector of the Mexican Economy." Paper presented at the Center for International Affairs, Harvard University. Cambridge, Mass., March 1977.

Street, James H. "The Mexican Paradox, Other Models and the Technological Challenge." Paper presented at the North American Economic Studies Association. Mexico City, December 27, 1978.

Street, James H. "Prospects for Mexico's Industrial Development Plan in the 1980s." Paper presented at the first annual meeting of the Middle Atlantic Council on Latin American Studies. University of Delaware, Newark, Delaware, April 17–19, 1980.

"Structural Factors in Mexican and Caribbean Basin Migration." Proceedings of a Brookings Institution–El Colegio de México Symposium. Washington, D.C., June 28–30, 1978.

Woodside, Philip R. "Comments on Oil and Gas Reserves in Mexico." Paper presented at the Workshop on Oil and Gas Supply from Foreign Sources. Reston, Virginia, June 12–13, 1979.

Interviews

This manuscript profited from confidential personal and telephone interviews with individuals on congressional staffs, in the Export-Import Bank, in Petróleos Mexicanos, in the departments of Defense, Energy, State, and Treasury, and elsewhere.

Adleson, S. Lief. Historian. Mexico City. Interview, May 24, 1978.

Alonso González, Francisco. Author and petroleum engineer. Mexico City. Interviews, June 1, 1979, August 17, 1980.

Blevins, Ruth. Public Affairs Officer, Defense Fuel Supply Center, Defense Logistical Agency, Department of Defense, Alexandria, Virginia. Telephone interview, January 8, 1980.

Brown, Edmund G., Jr. Governor of California. Interview by Tom Snyder on *Prime-Time Sunday*, National Broadcasting Company, July 15, 1979.

Fentress, A. Lee. Attorney. Washington, D.C. Interview, August 16, 1978.

Friedeberg Merzbach, Walter. Manager for production, PEMEX, 1970–1976. Mexico City. Interview, March 27, 1980.

Gutiérrez Roldán, Pascual. Director general of PEMEX, 1958–1964. Mexico City. Interviews, June 1, 1977, and May 24, 1978.

Hernández Galicia, Joaquín. Secretary general, STPRM, 1962-1964. Ciudad Madero. Interview, May 31, 1978.

López Juárez, Alfonso. Coordinator, CONAPO sex-education programs. Mexico City. Interview, June 1, 1977.

Lucey, Patrick J. U.S. ambassador. Mexico City. Interview, June 7, 1979.

McDonald, Walter J. Special assistant secretary for international affairs, DOE, 1977–1978. Telephone interview, August 13, 1979.

Mier y Terán de Muñoz, Lucia. Social anthropologist, CONAPO. Mexico City. Interview, June 1, 1977.

Rase, Glen. Petroleum attaché, U.S. Embassy. Mexico City. Interview, May 23, 1979.

Ray, Jack H. President, Tennessee Gas Transmission Company. Houston. Interview, May 20, 1980.

Roldán Vargas, Gustavo. Secretary of education and social security, STPRM. Mexico City. Interview, May 29, 1978.

Terrazas Zozaya, Samuel. Secretary general, STPRM, 1968-1970. Mexico City. Interview, June 6, 1978.

Vargas MacDonald, Antonio. Author. Mexico City. Interview, June 2, 1979.

Vázquez Gutiérrez, Hebraicaz. President, the National Petroleum Movement. Mexico City. Interview, May 31, 1978.

Verdugo, Ernesto. President of the Mexican Society of Chemical Engineers, 1977. Mexico City. Interview, May 23, 1978.

West, James A. Energy consultant (Daytona Beach, Florida, and McLean, Virginia). Telephone interview, August 31, 1978.

Index

Oteyza, José Andrés, 132, 134, 137, 138, 139, 180, 193

Palestine Liberation Organization, 153, 154
Panama, 176
Panama Canal, 190
Panhandle Eastern Pipeline Corporation, 202
Pauley Noreste, 31, 44
Peace Corps, 160
Pearson, Weetman Dickinson, 5, 7–9, 10, 12
Peláez, Manuel, 12
PEMEX, 16, 19, 20, 23, 56, 57, 98, 132, 136, 146, 154, 155; business techniques applied to, 61; and conservation, 48, 51, 52, 55–56, 185–86; corruption within, 22, 32–33, 36, 39, 59, 78, 95, 97, 99–100, 101; criticisms of, 32–33, 42, 97, 101; employee growth in, 26, 34, 49; explorations and discoveries by, 22, 26–27, 29, 31, 36, 40-41, 46, 47–48, 56, 64, 67–68, 69, 70; as exporter, 48, 176, 177–81, 183, 187, 188, 189, 191, 194–96, 198–99, 202, 228–29, 231; facilities constructed by, 36, 39, 48, 49, 51, 63, 76-77, 143–44, 179–80, 196, 197, 203, 223; financial problems of, 29–30, 34, 35, 93; fleet of, 43; and foreign companies and suppliers, 29–30, 31, 39, 40, 44, 48–49, 52, 174, 187–88, 189; and Indians, 75; jurisdictions of, 25–26; labor/management relations within, 24–26, 34, 36, 49–51, 77, 87, 88, 93, 95, 98, 99–100, 101; loans to, foreign, 35, 52, 63, 73, 77, 131, 169, 173, 185–86, 189, 192; and local residents, 203, 215; and Mexican Petroleum Institute, 42; and natural gas, 27, 29, 38, 41, 62, 63, 69, 70, 71, 73, 75–76, 183, 184, 185, 186, 191, 194, 197–98; and OPEC, 144, 145–46, 147, 148, 149; personnel changes within, 59, 61; and petrochemicals, 38–39, 40, 44, 47, 48–49, 62–63, 184–85; and pollution, 51–52, 77–78, 79–80, 185, 214; popular support for, 22–23; postexpropriation problems of, 21–22, 24, 39, 61, 76–77; presidential involvement in, 32, 33–34, 35, 47; and prices, 35–36, 41, 48, 52, 67, 75, 125, 146, 172, 186, 187–88, 199–200, 201, 202; production increases of, 30, 31, 32, 62, 63–64, 70-71, 101-02, 227–28, 229;

profits of, government use of, 136, 139; reforms within, 33, 34, 36-37, 44, 46; related industries and, 30–31, 32, 36, 73; self-sufficiency sought by, 40, 41, 48–49, 62, 143, 179, 197, 198; and six-year plan, 62–63, 75, 76; social projects of, 26, 49, 86; stabilization of, 26–27, 29, 31, 32; technical experts of, 126, 144; and U.S., 163, 169, 170, 172–73, 174–75. *See also* Drilling equipment and techniques; Ixtoc 1 spill; Oilfields and reserves; Pipeline construction
Pérez, Carlos Andrés, 118, 120, 121
Pérez Alfonso, Juan, 141, 142
Pérez Jiménez, Marcos, 118, 119
PERMARGO, 73, 203, 204, 211, 212, 217, 218–19, 224
Petrochemicals, 38–39, 40, 41, 44, 47, 48–49, 62–63, 184–85
PETROBRAS, 223
Petróleos Mexicanos. *See* PEMEX
Petroleum. *See* Oil industry; PEMEX
Petroleum Law of 1901, 9
PIDER program, 233
Pierce, Henry Clay, 5–6
Pipeline construction, 36, 187, 188–91, 192, 193, 196, 197–98, 202
Poli-Rey, S.A., 39
Political system (Mexico). *See* Governmental system (Mexico).
Popular Socialist party (PPS), 190, 221
Population growth, 54, 104, 108–09, 110–11, 112, 138, 225
Portes Gil, Emilio, 83
Poverty, 53, 108, 109
Poza Rica massacre, 96
Poza Rica oilfields, 13, 67, 68, 77–78, 88, 95, 96, 97, 205
PRI, 40, 46, 190, 224; and absentee votes, 53–54; criticism of, 53, 54; decline in power of, 53–54, 75; earlier names of, 248; economic program of, 135–36; and foreign companies, 44; opposition to, 74; and population growth, 109, 110–11; and poverty, 53, 109; presidential elections of, 53; reform of, 74; stability of, 100
Private enterprise, 43, 129, 130, 131, 132–33, 154, 227
Proceso, 73, 78, 172

PITT LATIN AMERICAN SERIES

Cole Blasier, Editor

PLAS